Catalysis in the Refining of Fischer–Tropsch Syncrude

RSC Catalysis Series

Titles in the Series:
1: Carbons and Carbon Supported Catalysts in Hydroprocessing
2: Chiral Sulfur Ligands: Asymmetric Catalysis
3: Recent Developments in Asymmetric Organocatalysis
4: Catalysis in the Refining of Fischer–Tropsch Syncrude

How to obtain future titles on publication:
A standing order plan is available for this series. A standing order will bring delivery of each new volume immediately on publication.

For further information please contact:
Book Sales Department, Royal Society of Chemistry, Thomas Graham House, Science Park, Milton Road, Cambridge, CB4 0WF, UK
Telephone: +44 (0)1223 420066, Fax: +44 (0)1223 420247, Email: books@rsc.org
Visit our website at http://www.rsc.org/Shop/Books/

Catalysis in the Refining of Fischer–Tropsch Syncrude

Arno de Klerk
Department of Chemical and Materials Engineering, University of Alberta, Edmonton, Alberta, Canada

Edward Furimsky
IMAF Group, 184 Marlborugh Avenue, Ottawa, Ontario, Canada

RSC Publishing

RSC Catalysis Series No. 4

ISBN: 978-1-84973-080-8
ISSN: 1757-6725

A catalogue record for this book is available from the British Library

Published by The Royal Society of Chemistry,
Thomas Graham House, Science Park, Milton Road,
Cambridge CB4 0WF, UK

Registered Charity Number 207890

For further information see our web site at www.rsc.org

Preface

Fischer–Tropsch synthesis (FTS) has been used on a commercial scale for more than 80 years. Three countries stand out in the history of FTS, namely Germany, the United States of America and South Africa. FTS was developed and commercialised in Germany for strategic reasons. It provided a source of transportation fuels that was independent from crude oil. The strategic advantage of such technology was realised in the USA, but commercial production was short lived. Crude oil was too readily available and too cheap. Nevertheless, initial developments in the field of high-temperature FTS took place in the USA. For much the same reason as Germany, South Africa invested in FTS. It provided a secure source of transportation fuels when its political dispensation resulted in an economic embargo limiting its access to crude oil. Initially the technology for FTS employed in South Africa was of German and US origin, but over the course of more than half a century, considerable experience was gained in the operation of Fischer–Tropsch-based facilities. This ultimately led to improvements in FTS and the development of some new technologies for FTS.

Today, interest in FTS is more global. Many of the oil majors invested in Fischer–Tropsch research. Some of these programmes resulted in demonstration- and even commercial-scale facilities. However, FTS is by no means a mainstream technology yet. Several technologies have been commercialised, which can be broadly classified as iron-based high-temperature Fischer–Tropsch (Fe-HTFT), iron-based low-temperature Fischer–Tropsch (Fe-LTFT) and cobalt-based low-temperature Fischer–Tropsch (Co-LTFT) synthesis.

The product distribution obtained during LTFT synthesis differs markedly from that obtained from HTFT synthesis. The synthetic crude from LTFT is dominated by *n*-alkanes with a wide carbon number distribution and a sizeable fraction of waxes. The lighter product fraction also contains some alkenes and oxygenates. The synthetic crude from HTFT has a narrower carbon number distribution and is rich in alkenes, the remainder being alkanes, aromatics and oxygenates. Neither of the synthetic crudes contains sulfur- or nitrogen-containing compounds. The composition of Fischer–Tropsch synthetic crude

RSC Catalysis Series No. 4
Catalysis in the Refining of Fischer–Tropsch Syncrude
By Arno de Klerk and Edward Furimsky

Published by the Royal Society of Chemistry, www.rsc.org

(syncrude) is consequently different from that of conventional crude oil in a number of respects.

Since the primary hydrocarbons from FT processes contain little sulfur and nitrogen, but are rich in acyclic hydrocarbons, they may be suitable blending components with petroleum-derived fuels. In this way, the overall costs of refining conventional crude oil fractions may be decreased. The integration of FTS with conventional crude oil refining may be an attractive option for improving the efficiency of fuels production from both. FTS also holds promise as an enabling technology for biomass upgrading. Small-scale biomass-to-liquids facilities may overcome the logistic problems associated with the transportation of low energy density biomass. These and other economic and environmental drivers may stimulate interest in FTS and this book is partly justified by our belief that there is indeed a growing interest in FTS.

The main justification for this work is the lack of a general overview of the catalysis that will be needed to convert Fischer–Tropsch syncrude into useful products. Much of the research in the field of Fischer–Tropsch technology has been devoted to FTS. However, the real value addition is not in converting alternative carbon sources into a syncrude, but in delivering final products to the market. Converting syncrude into final products requires catalysts that can convert oxygenates, exploit the reactivity of alkenes and benefit from the low coking propensity of *n*-alkanes. Clearly, the catalysis of Fischer–Tropsch syncrude refining is not the same as that of crude oil refining. Although Fischer–Tropsch syncrude can also be employed for the production of various chemicals, the primary focus of this book is on the catalysis needed for the upgrading of syncrude to transportation fuels.

Alkenes dominate the lighter fractions of Fischer–Tropsch syncrude. The conversion of light alkenes to liquid fuels via oligomerisation is an important part of FT refining. Isomerisation, hydroisomerisation and hydrocracking are equally important reactions for converting *n*-alkanes and *n*-alkenes into fuels and lubricants. Hydrotreating is likewise necessary to ensure that final product specifications are met. The catalysis of these conversion processes will therefore be covered in detail. In this respect, specific attention is given to the conversion of oxygenates and waxes. Other types of catalysis relevant to the refining of Fischer–Tropsch syncrude are also covered, but in less detail. Thus, only a cursory account is provided of FTS and Fischer–Tropsch technology in general, with focus on the aspects that determine the composition of primary products relevant to refining catalysis. Theoretical, engineering and commercial aspects related to FTS have been extensively covered in other books and authoritative reviews and will not be duplicated.

A review of the catalysis in the refining of Fischer-Tropsch syncrude is the main objective of this book. This is the first time that such an extensive study dealing with the upgrading of Fischer–Tropsch syncrude to commercial fuels, lubricants and other products has been undertaken. We hope that this book is a useful, if not overdue, addition to the literature on Fischer–Tropsch technology.

Contents

RSC Catalysis Series No. 4
Catalysis in the Refining of Fischer–Tropsch Syncrude
By Arno de Klerk and Edward Furimsky

Published by the Royal Society of Chemistry, www.rsc.org

Abbreviations and Symbols

ASA	Amorphous silica–alumina
ASF	Anderson–Schulz–Flory
ASTM	American Society for Testing and Materials
CFPP	Cold filter plugging point
CN	Cetane number
DO	Decanted oil
EPA	Environmental Protection Agency
FBP	Final boiling point
FCC	Fluid catalytic cracking
FT	Fischer–Tropsch
FTS	Fischer–Tropsch synthesis
GTL	Gas-to-liquids
HDAr	Hydrodearomatisation
HCR	Hydrocracking
HDM	Hydrodemetallisation
HDN	Hydrodenitrogenation
HDO	Hydrodeoxygenation
HDS	Hydrodesulfurisation
HFRR	High frequency reciprocating rig (ASTM D6079 test method)
HIS	Hydroisomerisation
HTFT	High-temperature Fischer–Tropsch
HVGO	Heavy vacuum gas oil
HYD	Hydrogenation
IBP	Initial boiling point
IFP	Institut Français du Pétrole
IS	Isomerisation
LHSV	Liquid hourly space velocity
LPA	Liquid phosphoric acid
LSR	Light straight run
LTFT	Low-temperature Fischer–Tropsch
LVGO	Light vacuum gas oil
MAPO	Magnesium aluminophosphate
MEK	Methyl ethyl ketone (2-butanone)
MOGD	Mobil olefins to gasoline and distillates
MON	Motor octane number

MOR	Mordenite
MSA	Mesoporous silica–alumina
MTBE	Methyl *tert*-butyl ether (2-methoxy-2-methylpropane)
OLI	Oligomerisation
P	Pressure
PAO	Polyalphaolefin
PCP	Protonated cyclopropane
PM	Particulate matter
RFCC	Residue fluid catalytic cracking
RON	Research octane number
SAPO	Silico-aluminophosphate
SLO	Stabilised light oil
SPA	Solid phosphoric acid
SZ	Sulfated zirconia
T	Temperature
TAME	*tert*-Amyl methyl ether (2-methoxy-2-methylbutane)
TPA	Tungstophosphoric acid
TZ	Tungstated zirconia
ULO	Unstabilised light oil
UOP	Universal Oil Products
VGO	Vacuum gas oil
WGS	Water gas shift
WHSV	Weight hourly space velocity

CHAPTER 1

Introduction

1.1 Overview of Fischer–Tropsch-based Facilities

It has been more than 80 years since the Fischer–Tropsch synthesis (FTS) was first described in the literature.[1] Advances in the development of this technology have been documented in numerous books and review papers dealing with FTS.[2–20]

During FTS, synthesis gas (H_2 and CO) is converted into a mixture of hydrocarbons, oxygenates, water and carbon dioxide. The hydrocarbon and oxygenate fraction is commonly referred to as a synthetic crude oil or syncrude for short. This syncrude, just like conventional crude oil, has to be refined in order to produce useful products, such as transportation fuels and chemicals. A simplified flow diagram of an FTS facility is shown in Figure 1.1.

In principle any carbon-containing raw material may be employed as feed for synthesis gas production. The nature of the raw material will determine the nature of the feed-to-syngas conversion technology and appropriate feed preparation. When solid feed, such as coal or biomass, is used as raw material, the synthesis gas is produced by gasification. There are various gasification technologies to choose from,[21,22] and the choice depends on the nature of the feed and also the Fischer–Tropsch technology that has been selected. During gasification, some liquid pyrolysis products may be produced that can be refined with the syncrude, as indicated by the dashed line in Figure 1.1. When natural gas is used as raw material, synthesis gas is typically produced by gas reforming. Impurities in the raw synthesis gas are removed before FTS and synthesis gas conditioning may include processes such as water gas shift (WGS) conversion and CO_2 removal. After FTS, the product is cooled stepwise and separated into different syncrude fractions. Some of the light gases may be recycled and the synthesis gas conditioning steps (gas cleaning and H_2:CO ratio adjustment), FTS and product cooling are together called the gas loop. The syncrude from FTS forms the feed to the Fischer–Tropsch refinery, where the syncrude is upgraded to intermediate or final products.

RSC Catalysis Series No. 4
Catalysis in the Refining of Fischer–Tropsch Syncrude
By Arno de Klerk and Edward Furimsky

Published by the Royal Society of Chemistry, www.rsc.org

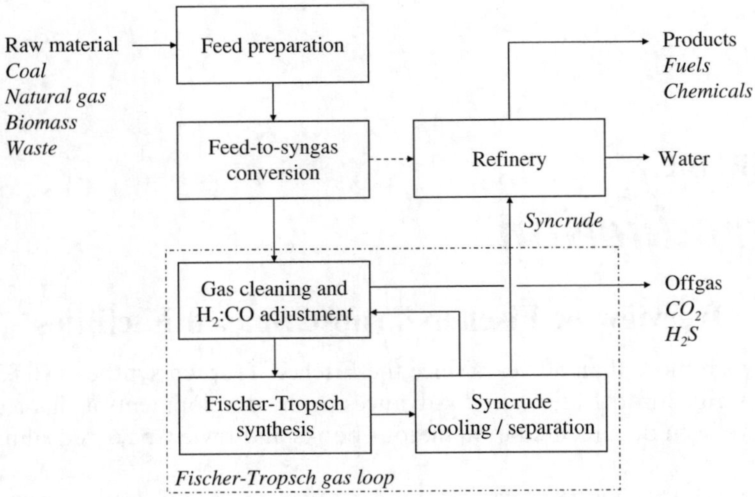

Figure 1.1 Simplified flow diagram of a Fischer–Tropsch-based facility.

The composition and carbon number distribution of syncrude depend on the type of FTS employed (Table 1.1).[23] The gas-phase products from FTS consist only of hydrocarbons, with very little oxygenates. The oil phase contains hydrocarbons and oxygenates. In the oil, the hydrocarbons are dominated by *n*-alkanes and *n*-alkenes. The combined aromatics, cycloalkane and cycloalkene content in the oil varies from 0 to 15%, depending on the type of process. The oxygenate content varies over the same range and the main oxygenate classes are alcohols, carbonyls and carboxylic acids. The concentration of a compound class in a specific fraction may, of course, fall outside the indicated ranges. Low-temperature Fischer–Tropsch (LTFT) synthesis also produces a wax fraction that is rich in *n*-alkanes, and is a solid under ambient conditions. The aqueous fraction obtained from FTS contains mainly short carbon chain oxygenates and very little hydrocarbons. Usually, the primary products from FTS contain practically no sulfur- and nitrogen-containing compounds. Gas cleaning ensures that the synthesis gas contains very little sulfur (parts per billion) and nitrogen; the Fischer–Tropsch catalyst itself is also an excellent sulfur trap. The heteroatom content of Fischer–Tropsch syncrude is consequently limited to oxygen.

1.2 Refining of Fischer–Tropsch Syncrude

Historically, FTS has been used mostly for the production of transportation fuels. Despite some of the positive attributes of syncrude, such as being sulfur free, the primary liquids from FTS cannot be used directly as transportation fuels. Various quality issues must be addressed. For example, syncrude has poor cold flow properties and relatively low thermal and storage stability. Also,

Table 1.1 Syncrude compositions representative of cobalt-based low-temperature Fischer-Tropsch (Co-LTFT), iron-based low-temperature Fischer-Tropsch (Fe-LTFT) and iron-based high-temperature Fischer-Tropsch (Fe-HTFT) synthesis.[a]

Product fraction	*Carbon range*	*Compound class*	*Syncrude composition (mass%)*[b,c]		
			Co-LTFT	*Fe-LTFT*	*Fe-HTFT*
Gas phase					
Tail gas	C_1	Alkane	5.6	4.3	12.7
	C_2	Alkene	0.1	1.0	5.6
		Alkane	1.0	1.0	4.5
LPG	C_3–C_4	Alkene	3.4	6.0	21.2
		Alkane	1.8	1.8	3.0
Oil and wax phases					
Naphtha	C_5–C_{10}	Alkene	7.8	7.7	25.8
		Alkane	12.0	3.3	4.3
		Aromatic	0	0	1.7
		Oxygenate	0.2	1.3	1.6
Distillate	C_{11}–C_{22}	Alkene	1.1	5.7	4.8
		Alkane	20.8	13.5	0.9
		Aromatic	0	0	0.8
		Oxygenate	0	0.3	0.5
Residue	C_{22+}	Alkene	0	0.7	1.6
		Alkane	44.6	49.2	0.4
		Aromatic	0	0	0.7
		Oxygenate	0	0	0.2
Aqueous phase					
Reaction water	C_1–C_5	Alcohol	1.4	3.9	4.5
		Carbonyl	0	0	3.9
		Carboxylic acid	0.2	0.3	1.3

[a]Syncrude composition is affected by factors such as the deactivation state of the Fischer–Tropsch catalyst, operating conditions and reactor technology.
[b]The syncrude composition is based on the total product from FTS, excluding inert gases and water gas shift products (H_2O, CO, CO_2 and H_2).
[c]Zero values indicate a low concentration and not necessarily a total absence of such compounds.

key performance parameters such as the octane number for motor gasoline need some adjustments. It is therefore necessary to refine the syncrude in order to meet the specification requirements of commercial transportation fuels.

One way of approaching this is to integrate FTS with crude oil refining. This integration can alleviate some problems associated with the use of refinery residues, such as petroleum coke from coking and asphalt from deasphalting. In some specific cases, it may be beneficial to produce sufficient quantities of vacuum residue to be used as the feed for gasification to produce synthesis gas. On the refinery site, the upgraded FTS liquids can be blended with the liquids of petroleum origin. By doing so, one can exploit the blending synergies available to mixtures of Fischer–Tropsch liquids, coal liquids and petroleum liquids.[24] For example, due to the low aromatics content of syncrude, blending FTS liquids with similar petroleum-derived fractions can decrease the costs associated with deep hydrodearomatisation (HDAr) of distillates. This offers some flexibility in response to ever-changing environmental regulations.

The industrial approach followed thus far is to construct stand-alone FTS facilities. This implies on-site refining or off-site blending in order to produce marketable transportation fuels. With the continuous developments in catalysis and conversion processes, Fischer–Tropsch refining presents an ever-changing landscape. One can learn a lot by studying older Fischer–Tropsch refinery designs and technologies,[25] despite the fact that fuel specifications and engine technology have changed considerably since the first industrial applications of FTS in Germany.

Fischer–Tropsch syncrude can be used, with appropriate pretreatment, in conjunction with any catalytic process that is employed for the conversion of conventional crude oil. Yet Fischer–Tropsch syncrude is in many respects different from crude oil.[26] Efficient refining of Fischer–Tropsch syncrude requires a different combination of refining technologies.[27] These technologies exploit the unique properties of syncrude (Table 1.1). Fischer–Tropsch syncrude can also be refined to a variety of chemicals.[28–33]

1.3 Catalysis in Fischer–Tropsch Refining

Although industrial-scale FTS has been practised in conjunction with syncrude refining since its inception, the literature on Fischer–Tropsch refining catalysis is less abundant than that dealing with the catalysis of FTS. The purpose of this book is to address this deficiency and provide an overview of the catalysis relevant to the refining of Fischer–Tropsch syncrude. The focus will be mainly on refining catalysis for the production of transportation fuels, although the catalytic conversion of syncrude to other products will also be touched upon. The main interest is in Fischer–Tropsch-derived materials, but other relevant studies are also included in the discussion. For example, studies using *n*-alkanes and *n*-alkenes, and also branched hydrocarbons, as model compounds have a direct bearing on the catalysis of Fischer–Tropsch-derived feeds.

Three of the most important catalytic conversions in Fischer–Tropsch refining catalysis are (a) oligomerisation (OLI) for the conversion of light alkenes into liquid products, (b) hydrocracking (HCR) for the conversion of heavy alkanes into lighter liquid products and (c) hydroisomerisation (HIS) to introduce some branching into the linear hydrocarbons for applications such as lubricating oil and jet fuel production. The catalysis of these conversions will be discussed in detail. Moreover, the information in the literature on OLI, HCR and HIS is so extensive that a separate book could be written on each topic. It is hoped that the studies that were selected for discussion here will give a good indication of the type of research that is relevant to the upgrading of the Fischer–Tropsch syncrude. Specific attention is paid to the influence of oxygenates, since this is one of the main differentiating features of syncrude compared with crude oil. Other types of catalysis relevant to syncrude conversion are also covered, albeit in less detail.

References

1. F. Fischer and H. Tropsch, *Brennst.-Chem.*, 1923, **3**, 276.
2. V. I. Komarewsky, C. H. Riesz and F. L. Estes, *The Fischer–Tropsch Process. An Annotated Bibliography*, Institute of Gas Technology, Chicago, 1945.
3. B. H. Weil and J. C. Lane, *The Technology of the Fischer-Tropsch Process*, Constable, London, 1949.
4. H. H. Storch, N. Golumbic and R. B. Anderson, *The Fischer–Tropsch and Related Syntheses*, Wiley, New York, 1951.
5. R. B. Anderson, in *Catalysis. Volume IV. Hydrocarbon Synthesis, Hydrogenation and Cyclization*, ed. P. H. Emmett, Reinhold, New York, 1956, p. 1.
6. F. Asinger, *Paraffins Chemistry and Technology*, Pergamon Press, Oxford, 1968.
7. I. Wender, *Catal. Rev. Sci. Eng.*, 1976, **14**, 97.
8. H. Kölbel and M. Rálek, *Catal. Rev. Sci. Eng.*, 1980, **21**, 225.
9. A. T. Bell, *Catal. Rev. Sci. Eng.*, 1981, **23**, 203.
10. M. E. Dry and J. C. Hoogendoorn, *Catal. Rev. Sci. Eng.*, 1981, **23**, 265.
11. P. Biloen and W. M. M. Sachtler, *Adv. Catal.*, 1981, **30**, 165.
12. M. E. Dry, in *Catalysis Science and Technology*, Vol. **1**, ed. J. R. Anderson and M. Boudart, Springer, Berlin, 1981, p. 159.
13. V. Ponec, *Catalysis*, 1982, **5**, 48.
14. M. E. Dry, in *Applied Industrial Catalysis*, Vol. **2**, ed. B. E. Leach, Academic Press, New York, 1983, p. 167.
15. R. B. Anderson, *The Fischer–Tropsch Synthesis*, Academic Press, Orlando, FL, 1984.
16. J. C. W. Kuo, in *The Science and Technology of Coal and Coal Utilization*, ed. B. R. Cooper and W. A. Ellingson, Plenum Press, New York, 1984, p. 163.
17. A. P. Steynberg and M. E. Dry (eds), *Fischer–Tropsch Technology*, Studies in Surface Science and Catalysis, Vol. **152**, Elsevier, Amsterdam, 2004.
18. B. H. Davis and M. L. Occelli (eds), *Fischer–Tropsch Synthesis, Catalysts and Catalysis*, Studies in Surface Science and Catalysis, Vol. **163**, Elsevier, Amsterdam, 2007.
19. P. M. Maitlis and V. Zanotti, *Chem. Commun.*, 2009, 1619.
20. B. H. Davis and M. L. Occelli, (eds), *Advances in Fischer–Tropsch Synthesis, Catalysts and Catalysis*, Taylor and Francis, Boca Raton, FL, 2009.
21. J. Rezaiyan and N. P. Cheremisinoff, *Gasification Technologies. A Primer for Engineers and Scientists*, Taylor and Francis, Boca Raton, FL, 2005.
22. C. Higman and M. van der Burgt, *Gasification*, 2nd edn, Gulf Professional Publishing, Oxford, 2008.
23. A. de Klerk, *Energy Fuels*, 2009, **23**, 4593.
24. D. Lamprecht and P. N. J. Roets, *Prepr. Pap. Am. Chem. Soc. Div. Pet. Chem.*, 2004, **49** (4), 426.
25. A. de Klerk, *Prepr. Pap. Am. Chem. Soc. Div. Pet. Chem.*, 2008, **53** (2), 105.

26. A. de Klerk, *Green Chem.*, 2007, **9**, 560.
27. A. de Klerk, *Green Chem.*, 2008, **10**, 1249.
28. M. E. Dry, *ACS Symp. Ser.*, 1987, **328**, 18.
29. J. H. Gregor, *Catal. Lett.*, 1990, **7**, 317.
30. J. Collings, *Mind Over Matter. The Sasol Story: a Half Century of Technological Innovation*, Sasol, Johannesburg, 2002.
31. A. P. Steynberg, W. U. Nel and M. A. Desmet, *Stud. Surf. Sci. Catal.*, 2004, **147**, 37.
32. A. Redman, in *Proceedings of the 18th World Petroleum Congress*, Johannesburg, 2005, cd179.
33. A. de Klerk, L. P. Dancuart and D. O. Leckel, in *Proceedings of the 18th World Petroleum Congress*, Johannesburg, 2005, cd185.

CHAPTER 2

Production of Synthesis Gas

All indirect liquefaction technologies make use of synthesis gas (a mixture of H_2 and CO) as intermediate product. Ideally, synthesis gas, or syngas for short, should make Fischer–Tropsch synthesis (FTS) and other syngas-to-syncrude technologies independent of the raw feed material. This is a commonly held perception, but not entirely true. It is not possible to view FTS independently from the gas loop (Figure 1.1). In the gas loop, the raw synthesis gas has to be purified to remove compounds that may poison the catalyst used for FTS. The synthesis gas composition is also adjusted in the gas loop in order to provide FTS with a synthesis gas feed that has the desired H_2:CO ratio. The optimal H_2:CO ratio depends on the Fischer–Tropsch technology, and although a usage ratio of 2:1 is implied by the generic expression of FTS [Equation (2.1)], the real usage ratio depends on the real product selectivity (Table 1.1). The H_2:CO ratio of synthesis gas is adjusted by making use of the water gas shift (WGS) reaction:

$$2H_2 + CO \rightarrow -(CH_2)- + H_2O \quad (2.1)$$

The production of synthesis gas will be considered in the context of the gas loop, with its component parts being discussed separately.

2.1 Synthesis Gas from Gaseous Feed

The steam reforming of natural gas and/or refinery gases has been the most common source of synthesis gas. Although steam reforming is mainly used to produce a hydrogen-rich synthesis gas as a source of refinery hydrogen, it is also useful for applications such as ammonia synthesis and syngas-to-methanol conversion. Theoretically, synthesis gas having a H_2:CO ratio of 3:1 can be produced from steam reforming of methane:

$$CH_4 + H_2O \rightarrow CO + 3H_2 \quad (2.2)$$

Synthesis gas production from methane is endothermic and a portion of feed material has to be combusted to supply the heat necessary for the reforming

RSC Catalysis Series No. 4
Catalysis in the Refining of Fischer–Tropsch Syncrude
By Arno de Klerk and Edward Furimsky

Published by the Royal Society of Chemistry, www.rsc.org

reactions. Neither steam reforming nor the WGS reaction that is needed to adjust the H_2:CO ratio proceed to completion.

The view has been expressed that steam reforming by itself is not the preferred technology for synthesis gas production in large-scale gas-to-liquids (GTL) based on FTS.[1] This view is supported by the poor economy of scale compared with partial oxidation processes and the hydrogen-rich synthesis gas that is well above the usage ratio required by FTS. In partial oxidation processes, such as autothermal reforming (ATR), the energy to drive the reforming reaction is provided by partial combustion of the feed in the reformer. The synthesis gas thus produced typically has an H_2:CO ratio in the range 1.6–1.9, which is closer to the usage ratio required by FTS.

It was pointed out that the conversion of natural gas to syncrude, starting with steam reforming, through WGS, CO_2 scrubbing and ending with FTS, may not be accomplished without a negative overall energy balance.[2] On a global scale, the direct utilisation, in either energy applications or transportation, may be the most efficient use for a high-value fuel such as natural gas. Natural gas inherently has a high H:C ratio, which is degraded when it is employed for syncrude production.

2.2 Synthesis Gas from Liquid and Solid Feed

Synthesis gas may be produced from a variety of solid carbon sources by gasification. Higman and van der Burgt listed various raw materials that have been investigated for gasification.[3] These include coal, bitumen–water emulsions, oil sand residues, biomass, heavy petroleum fractions and wastes. Of these, only coal is at present used industrially in conjunction with FTS. Instances where coal can be obtained by low-cost surface mining are of particular importance. Coal gasification is capital intensive and a low raw material cost is necessary to make the construction and operation of such facilities economically viable. Irrespective, gasification of the solid and/or semi-solid feeds to produce synthesis gas, which is followed by WGS and FTS, can be employed to convert a low-value feed material into higher value products.[4]

The composition of the synthesis gas obtained by gasification depends on the feed material. The approximate concentrations of gasification products obtained from a lignite, vacuum residue, asphalt from deasphalting and fluid coke (petcoke) are given in Table 2.1.[4] The lignite and coke were fed as ~50:50 water slurries, whereas vacuum residue and asphalt were in a liquid form. It is evident that with respect to the H_2:CO ratio, vacuum residue and asphalt are more suitable feeds for gasification with FTS in mind. Thus, in order to obtain an H_2:CO ratio of around 2:1 from a synthesis gas with a ratio of around 0.4:1, such as the gaseous mixture obtained from lignite in a British Gas Lurgi (BGL) gasifier, the synthesis gas has to be subjected to substantial WGS:

$$2.5CO + H_2 + 1.4H_2O \rightarrow 1.1CO + 2.4H_2 + 1.4CO_2 \tag{2.3}$$

Table 2.1 Composition of clean and dry synthesis gas produced by gasification in British Gas Lurgi (BGL) and Texaco gasifiers employing different liquid and solid feed materials.

Composition	*Lignite coal*		*Vacuum residue*	*Asphalt*	*Fluid coke (Petcoke)*
	BGL	*Texaco*[a]	*Texaco*[b]	*Texaco*[b]	*Texaco*[a]
H_2:CO ratio	0.4	0.8	1.0	1.0	0.5
H_2 (%)	26	35	47	47	28
CO (%)	63	45	47	47	54
CO_2 (%)	3	18	4	4	15
CH_4 (%)	5	Trace	1	1	Trace
N_2 + Ar (%)	3	2	1	1	1

[a]Fed as a water slurry.
[b]Fed in a liquid form.

Much less extensive WGS is required for gaseous mixtures obtained from vacuum residue and asphalt:

$$H_2 + CO + 0.35H_2O \rightarrow 1.35H_2 + 0.65CO + 0.35CO_2 \tag{2.4}$$

2.3 Water Gas Shift Conversion

The composition of the synthesis gas can be adjusted by employing the water gas shift reaction [Equation (2.5)]. The WGS reaction is reversible. Lower temperatures favour CO_2 and H_2, whereas higher temperatures favour CO and H_2O.

$$CO + H_2O \rightleftharpoons CO_2 + H_2 (\Delta H = -41.1\,kJ \cdot mol^{-1}) \tag{2.5}$$

At very high temperatures (> 900 °C), WGS does not require a catalyst, but for most industrial applications it is conducted over a catalyst. Low-temperature catalytic WGS conversion (200–270 °C) employs alumina-supported copper–zinc oxide (Cu–ZnO–Al_2O_3) catalysts. These catalysts are sensitive to sulfur poisoning and the synthesis gas must first be purified (see Section 2.4) to remove acid gases. The sulfur content in the feed should preferably be less than 0.1 $\mu g\, g^{-1}$ for low-temperature WGS catalysts.[3] High-temperature catalytic WGS conversion (300–500 °C) employs combined iron oxide and chromium oxide (Fe_2O_3–Cr_2O_3) catalysts, which may include stabilisers and promoters, such as copper oxide.[5] It is not necessary to remove all the acid gases before high-temperature WGS and catalysts are tolerant of sulfur levels up to 100 $\mu g\, g^{-1}$.[3] High-temperature WGS reactors may therefore be operated either as 'sweet' shift or as 'sour' shift processes. For true 'sour' shift, it is best to employ a sulfided CoMo-based catalyst that requires the sulfur to remain in its sulfided state.[3] These catalysts can be considered medium-temperature WGS catalysts and typically operate in the range 250–350 °C.[5] In an FTS gas loop, any sulfur in the synthesis gas must be removed to avoid poisoning of the

Fischer–Tropsch catalyst and there is no need to employ a 'sour' shift. It is also possible to make use of noble metal-based catalysts for WGS and numerous examples of noble metal-based WGS catalysts were described in a review paper by Ratnasamy and Wagner.[5]

2.4 Synthesis Gas Purification

An integral part of synthesis gas production is gas purification. Gas purification is mainly required to remove sulfur-containing compounds that are catalyst poisons for Ni-based reforming catalysts, WGS catalysts and Fe- or Co-based Fischer–Tropsch catalysts.

When natural gas is used as a feed material, the natural gas can be desulfurised by hydrotreating, followed by absorption on ZnO.[1] When coal is gasified, the raw synthesis gas from gasification contains, amongst other compounds, sulfur and nitrogen species. The raw synthesis gas can be purified by a cold methanol wash, such as employed in the Rectisol technology,[6] which has the added benefit of removing the CO_2. Other gas cleaning technologies may also be considered depending on the feed type and synthesis gas purity requirements.[7]

The production of synthesis gas may be accompanied by the co-production of pyrolysis products. Although it does not have a direct impact on FTS or the gas loop configuration, it will affect the design of the gas purification section. The condensable products may be recovered during gas purification and used as feed for chemical extraction, fuel or further refining.

References

1. K. Aasberg-Petersen, T. S. Christensen, I. Dybkjær, J. Sehested, M. Østberg, R. M. Coertzen, M. J. Keyser and A. P. Steynberg, *Stud. Surf. Sci. Catal.*, 2004, **152**, 258.
2. E. Furimsky, *Energy Sources A*, 2008, **30**, 119.
3. C. Higman and M. van der Burgt, *Gasification*, Gulf Professional Publishing, Oxford, 2008.
4. E. Furimsky, *Oil Gas Sci. Technol. Rev. IFP*, 1999, **54**, 597.
5. C. Ratnasamy and J. P. Wagner, *Catal. Rev. Sci. Eng.*, 2009, **51**(3), 325.
6. H. Weiss, *Gas Sep. Purif.*, 1988, **2**, 171.
7. M. J. Richardson and J. P. O'Connell, *Ind. Eng. Chem. Process Des. Dev.*, 1975, **14**, 467.

CHAPTER 3

Fischer–Tropsch Synthesis

Up-to-date information on Fischer–Tropsch synthesis (FTS) can be found in recent textbooks.[1–3] The purpose of this chapter is not to duplicate this literature, but rather to provide a brief overview and to highlight aspects that affect the syncrude composition. The syncrude composition directly influences the catalysis of Fischer–Tropsch syncrude refining and is pertinent to the topic of this book.

3.1 Chemistry of Fischer–Tropsch Synthesis

When synthesis gas is converted over a Fischer–Tropsch catalyst, the following stoichiometric reactions yield hydrocarbons and oxygenates as primary products:

$$(2n+1)H_2 + nCO \rightarrow C_nH_{2n+2} + nH_2O \tag{3.1}$$

$$2nH_2 + nCO \rightarrow C_nH_{2n} + nH_2O \tag{3.2}$$

$$2nH_2 + nCO \rightarrow C_nH_{2n+2}O + (n-1)H_2O \tag{3.3}$$

$$(2n-1)H_2 + nCO \rightarrow C_nH_{2n}O + (n-1)H_2O \tag{3.4}$$

$$(2n-2)H_2 + nCO \rightarrow C_nH_{2n}O_2 + (n-2)H_2O \tag{3.5}$$

In these reactions, the first two represent the formation of alkanes [Equation (3.1)] and alkenes [Equation (3.2)]. The last three reactions represent the formation of various oxygenates, namely alcohols and ethers [Equation (3.3)], aldehydes and ketones [Equation (3.4)] and carboxylic acids and esters [Equation (3.5)]. Of these, the compounds with functional groups on the terminal carbon are generally considered primary products from FTS.

All Fischer–Tropsch reactions are highly exothermic; an average value for the heat of reaction is around 10 kJ g^{-1} of hydrocarbon product.

RSC Catalysis Series No. 4
Catalysis in the Refining of Fischer–Tropsch Syncrude
By Arno de Klerk and Edward Furimsky

Published by the Royal Society of Chemistry, www.rsc.org

3.2 Factors Influencing Fischer–Tropsch Syncrude Composition

The syncrude composition that is obtained from FTS is influenced by many variables. The values in Table 1.1[4] and Table 3.1[5] are consequently only indicative of the syncrude compositions obtained from the main classes of FTS that are practised industrially. Factors that significantly affect syncrude composition are the Fischer–Tropsch catalyst type, the reactor technology employed for FTS, Fischer–Tropsch catalyst deactivation and the operating conditions of FTS.

3.2.1 Fischer–Tropsch Catalyst Type

The main products produced over different Fischer–Tropsch-active metals (Table 3.2) show the effect of catalyst type on product composition.[6,7] Apart from the main FTS-active metal, Fischer–Tropsch catalysts include various promoters and may be combined with a support. In fact, for the same active metals, the support can have a pronounced effect on conversion and selectivity of the catalyst.[8]

There have been many reports dealing with the two most frequently used Fischer–Tropsch-active metals, namely iron and cobalt. The comparison by Schulz (Table 3.3)[9] illustrates the significant difference between iron-based

Table 3.1 Selectivity changes during industrial Fe-HTFT synthesis with increasing time on stream, illustrating how catalyst deactivation affects the composition of syncrude. The selectivity values do not reflect water gas shift products (H_2, H_2O, CO and CO_2) that are also affected by deactivation.

Compound or fraction	*Selectivity (%)*		
	Start of run	*Average*	*End of run*
Methane	7	10	13
Ethene	4	4	3
Ethane	3	6	9
Propene	10	12	13
Propane	1	2	3
Butenes	7	8	9
Butanes	1	1	2
C_5 and heavier condensate	6	8	9
Light oil	40	35	30
Decanted oil	14	7	2
Aqueous product	7	7	7

Table 3.2 Effect of Fischer–Tropsch active metals and operating range on the nature of the products.

Metal	*Temperature (°C)*	*Pressure (MPa)*	*Nature of products*
Fe	200–250	1.0–3.0	Alkanes, alkenes, oxygenates
	320–340	1.0–3.0	Alkanes, alkenes, aromatics, oxygenates
Co	170–220	0.5–3.0	Alkanes, some alkenes and oxygenates
Ru	150–250	10–100	Paraffin wax
ThO_2	300–450	10–100	Isoalkanes
Ni	170–205	0.1[a]	Alkanes, some alkenes

[a]At higher pressures, loss of Ni through $Ni(CO)_4$ formation becomes too high.

Table 3.3 Comparison of low-temperature Fischer–Tropsch synthesis over potassium-promoted iron-based and cobalt-based catalysts.

Catalysis property	*Fe-LTFT*	*Co-LTFT*
Extensive methanation	No	At increasing temperature and decreasing CO partial pressure
Alkali promoters	Essential	No
Monomers	CH_2	CH_2 (CO, C_2H_4)
Water gas shift activity	Yes	No
Branching reaction	Static, increases with time	Dynamic, decreases with time
Alkene hydrogenation	No (little)	Extensive
Alkene isomerisation	No (little)	Extensive

low-temperature Fischer–Tropsch (Fe-LTFT) and cobalt-based low-temperature Fischer–Tropsch (Co-LTFT) synthesis. In addition to differences in catalysis listed in Table 3.3, differences in product distributions are also evident (*e.g.* Tables 1.1 and 3.1). It has further been noted that the Co-LTFT catalysts give a higher conversion rate (depending on synthesis gas conditions) and reportedly have a longer catalyst life. Co-LTFT catalysts are also more active for hydrogenation (HYD) and consequently produce less unsaturated hydrocarbons and oxygenates than Fe-based catalysts. On the other hand, Fe-LTFT catalysts are more easily prepared, cheaper, more robust and more tolerant to poisoning by sulfur.

Details of selectivity control during FTS in relation to catalyst design can be found in the literature, for example the review published by Iglesia *et al.*[10] Valuable insights into the Fischer–Tropsch mechanism in relation to the nature and structure of the catalyst can be found in, among others, publications by Fahey,[11] Davis[12] and Maitlis and Zanotti.[13]

3.2.2 Fischer–Tropsch Reactor Technology

There are four main types of reactor technology that have been employed industrially for FTS (Figure 3.1). The high heat release during FTS is a crucial consideration in the design of commercial reactors for FTS. Provision of cooling through steam generation is evident in all of the reactor types. The operating temperature of FTS determines the steam pressure and in this respect a higher operating temperature is beneficial.

Iron-based high-temperature Fischer–Tropsch (Fe-HTFT) processes make use of fluidised bed reactor technology and FTS takes place entirely in the gas phase. The product distribution from FTS does not seem to be significantly affected by the reactor technology *per se*, with similarly operated circulating fluidised bed and fixed fluidised bed reactors yielding similar product distributions.

The same is not true of low-temperature Fischer–Tropsch processes. The product distributions from fixed bed and slurry bubble column FTS are different. This is to be expected, since a fixed bed reactor approximates plug flow behaviour, whereas a slurry bubble column reactor approximates continuous stirred tank behaviour.

Satterfield *et al.* directly compared Fe-LTFT in fixed bed and slurry bubble column reactors.[14] Little difference in methane selectivity and carbon number distribution was observed, but the alkene to alkane ratio from the fixed bed reactor was much lower than that from the slurry bubble column reactor. Jager and Espinoza,[15] who compared data from industrial operation of Fe-LTFT in these two reactor types, corroborated these findings. Fixed bed Fe-LTFT was more hydrogenating and produced a syncrude with a lower alkene to alkane ratio. Operation with a fixed bed reactor was also found to be 1.5–2 times less sensitive to sulfur poisoning than operation with a slurry bubble column

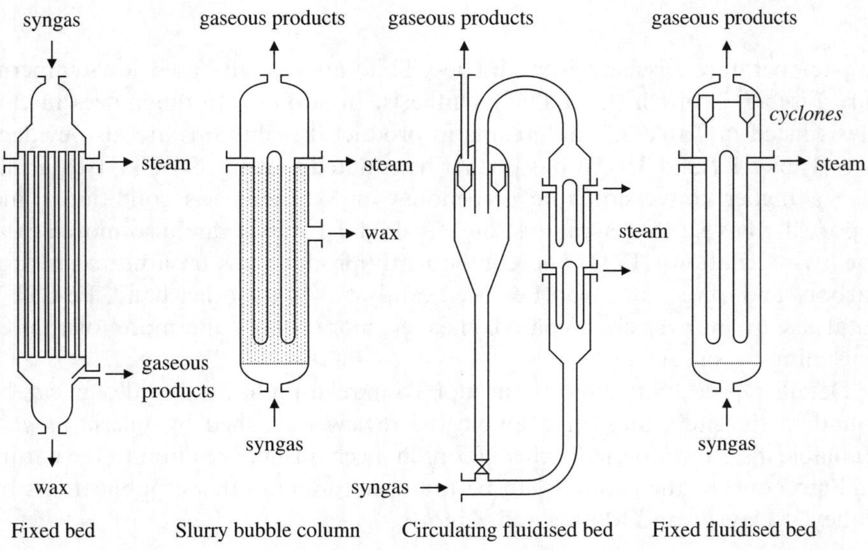

Figure 3.1 Industrially applied Fischer–Tropsch reactor technologies.

reactor. Moderate sulfur poisoning of Fe-LTFT catalysts mainly affects activity and not product selectivity. Slurry bubble column operation led to more productive use of the catalyst. In terms of product produced per unit mass of catalyst, the slurry bubble column reactor could achieve the same productivity with 30% or less catalyst mass than required for a fixed bed reactor.

The reactor technology places different demands on the mechanical strength of the Fischer–Tropsch catalyst. Slurry bubble column operation leads to higher levels of catalyst attrition and care should be taken during Fischer–Tropsch catalyst development to ensure that the working catalyst has sufficient attrition resistance.[16] Catalyst attrition affects the syncrude composition by increasing the level of solids present in the syncrude. It may also contribute to increased levels of dissolved metals in the syncrude.

3.2.3 Fischer–Tropsch Catalyst Deactivation

Syncrude composition is dependent on the age and deactivation history of the Fischer–Tropsch catalyst. As a consequence, the products from FTS may vary with time. These variations can be reduced when fluidised bed and slurry bubble column reactor technologies are employed, since these reactor technologies allow continuous catalyst addition and removal. This is not possible with fixed bed reactor technology, although the impact of such time-dependent changes may be reduced by the parallel operation of multiple fixed bed reactors with different age profiles.

The impact of deactivation on the composition of syncrude is different for the three main classes of Fischer–Tropsch catalysts:

1. An Fe-LTFT catalyst may deactivate until it reaches a stable 'equilibrium' catalyst that shows little further deactivation. During the initial period of deactivation, the carbon number distribution becomes lighter with time-on-stream and then stabilises (Figure 3.2).[17] Deactivation is accompanied by a slight increase in alkene and oxygenate (alcohol and carboxylic acid) selectivity. Methane increases and then stabilises at around 3.5% (Figure 3.2) and much of the increase in lighter products is in the C_2–C_4 carbon number range. It was pointed out that Fe-LTFT deactivation is actually beneficial for product refining.[18]
2. Co-LTFT catalyst deactivation takes place by various mechanisms.[19] The most prominent of these are poisoning, notably by sulphur compounds, sintering and coalescence of Co crystallites, carbon formation and fouling. Other deactivation mechanisms that may be active include re-oxidation, carbidisation, metal-support reactions, surface reconstruction, leaching of Co and catalyst attrition. It has been found that Co-LTFT catalysts are very sensitive to part per million levels of impurities, even during preparation, which can markedly affect regenerability and deactivation rate.[16,20] Deactivation with time-on-stream leads to a shift in the carbon number distribution. The relationship between increased methane selectivity and decreased liquid product yield seems to be independent of Co-LTFT catalyst type,[21] and has a detrimental impact on product refining.[22]

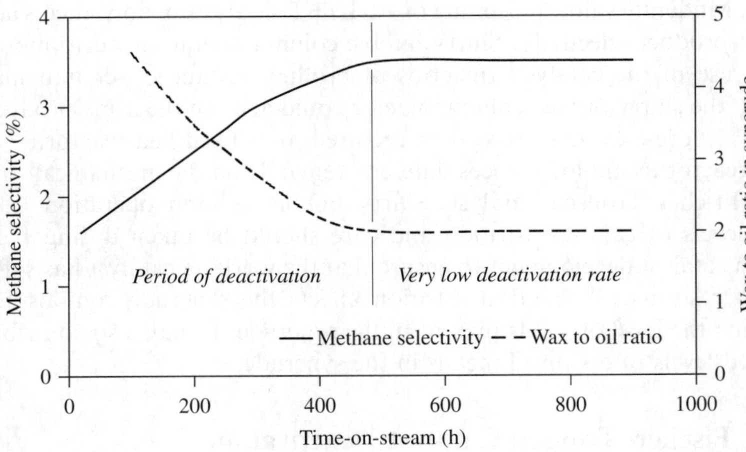

Figure 3.2 Influence of deactivation on the product distribution from an iron-based low-temperature Fischer–Tropsch (Fe-LTFT) catalyst.

3. Fe-HTFT catalysts deactivate mainly through loss of alkali metal silicate promoter, poisoning by sulfur present in the synthesis gas feed and coke deposits forming on the more active alkali/Fe sites.[23] Of these, perhaps the loss of the small loose alkali metal silicate promoter is the most important industrial deactivation mechanism, which causes the syncrude product distribution to become lighter and more saturated with increasing catalyst deactivation.

3.2.4 Fischer–Tropsch Operating Conditions

The classification of Fischer–Tropsch technologies based on their operating temperature into LTFT and HTFT indicates that operating temperature has a significant influence on product selectivity. Increasing the operating temperature is always accompanied by a shift in the carbon number distribution to lighter products. The response of syncrude composition is not as straightforward. Under typical LTFT operating conditions (<250 °C), an increase in temperature may initially decrease the alkene to alkane ratio, but ultimately hydrogenation has to compete with endothermic processes such as desorption and dehydrogenation, leading to an increase in alkene to alkane ratio. Under typical HTFT operating conditions (>320 °C), the syncrude has a high alkene to alkane ratio and the syncrude also contains aromatics. Side-reactions generally increase with increasing temperature. HTFT syncrude therefore contains ketones, branched hydrocarbons and internal alkenes in much higher concentration than found in LTFT syncrude.

The influence of pressure and the synthesis gas composition on the syncrude composition depends on the catalyst and operating regime.[24] On cobalt-based Fischer–Tropsch catalysts, a decrease in the H_2:CO ratio and an increase in total pressure of the synthesis gas result in a shift in the carbon number distribution to heavier products. On iron-based Fischer–Tropsch catalysts, the relationship is more complex, because iron-based catalysts can catalyse the water gas shift (WGS) reaction and this markedly affects their behaviour. The WGS reaction causes a change in the partial pressures of H_2 and CO beyond the change caused by FTS itself. Under LTFT conditions (liquid and gas phase), the carbon number distribution is influenced mainly by the H_2:CO ratio, and not by the total or partial pressure of the synthesis gas components. Under HTFT conditions (gas phase only), the H_2:CO ratio and pressure influence selectivity. An increase in pressure results in a heavier product and less methane.

3.3 Carbon Number Distribution of Fischer–Tropsch Syncrude

The information on composition of the primary FT gases, liquids, heavy oil and wax is necessary for designing and optimising product upgrading. As in crude oil refining, the boiling point or carbon number distribution from FTS determines the relative amounts of straight run product fractions and the size of different refinery units.

Attempts to predict the composition from FTS are based on the condensation–polymerisation hypothesis of Flory,[25,26] which requires only a single parameter, namely the probability of chain growth, or α-value. The probability of chain growth (α) is defined in terms of the rate of polymerisation (r_p) and the rate of termination (r_t) of the growing chains:

$$\alpha = r_p/(r_p + r_t) \tag{3.6}$$

The product distribution can then be represented in terms of x_n, the mole fraction of all products having carbon number n:

$$x_n = (1 - \alpha)\alpha^{n-1} \tag{3.7a}$$

$$\log x_n = \log[(1 - \alpha)/\alpha] + n\log\alpha \tag{3.7b}$$

Similar representations were used in the study of Dictor and Bell.[27] Further modifications resulted in the development of the Anderson–Schulz–Flory (ASF) description of the carbon number distribution (Figure 3.3).[25,28,29]

Both negative and positive deviations of the experimental data from those predicted by the theory have been reported and were ascribed to various parameters, such as pressure, temperature, type of catalyst, product analysis, time-on-stream, hydrocarbon chain cracking and secondary reactions. Among others, this led to the development of the two-α-model to explain the deviation in the carbon number distribution around C_8–C_{12} often reported for LTFT products. In this model, it is assumed that two different sites or growth mechanisms occur in parallel, with different chain growth probabilities (α_i) and

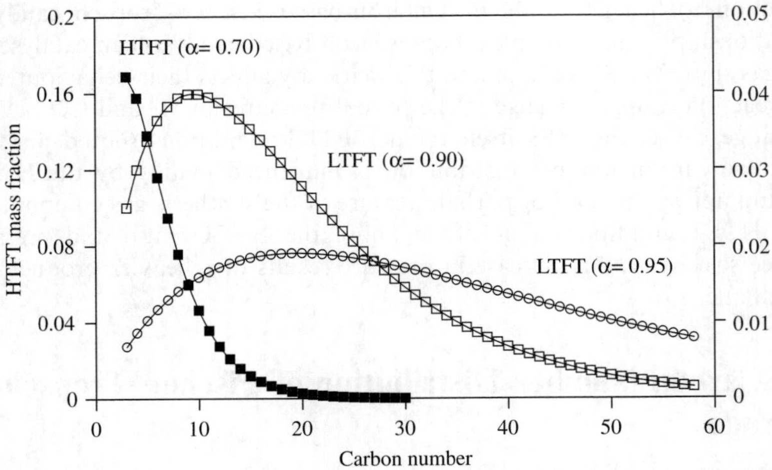

Figure 3.3 Calculated Anderson–Schulz–Flory (ASF) carbon number distribution of C_3 and heavier products showing typical values for the chain growth probability (α value) during high-temperature Fischer–Tropsch (HTFT) and low-temperature Fischer–Tropsch (LTFT) processes.

contributions (κ_i) to the overall product formation:[30]

$$x_n = \kappa_1 \alpha_1^{n-1} + \kappa_2 \alpha_2^{n-1} \tag{3.8}$$

The prediction of the product distribution, and deviations from it, and also the probabilistic calculation of product distribution based on mechanistic assumptions, have been the focus of a number of studies.[31–42] The work of Botes is noteworthy, as he was able to propose a model for Fe-LTFT synthesis that accounts for the alkane to alkene ratio and that describes the deviation of C_1 and C_2 compounds from the ASF distribution.[41]

3.4 Industrially Applied Fischer–Tropsch Processes

Over the years, a number of different Fischer–Tropsch technologies have been applied industrially (Table 3.4). Of these, there are six Fischer–Tropsch technologies that are being operated industrially at present. These processes differ mainly in terms of their operating conditions, reactor type and the base metal selected for the Fischer–Tropsch catalyst.

Various new technologies for FTS are in different stages of development, with much of the focus on a decrease in capital and operating costs. Dancuart and Steynberg assessed their potential in relation to the currently used technologies for FTS.[43] In many instances, the developments in FTS are paralleled by developments in hydrocracking, but little attention is devoted to other refining technologies.

Table 3.4 Industrially applied Fischer–Tropsch technologies, including the first year of industrial production and their present status.

Type	*FT catalyst*	*Reactor-type*	*Technology*	*Year*	*Status*
LTFT	Precipitated Co	Fixed bed	German normal pressure	1936	Ruhr, Germany (no longer used)[a]
LTFT	Precipitated Co	Fixed bed	German medium pressure	1937	Ruhr, Germany (no longer used)
HTFT	Fused Fe	Fixed fluidised bed	Hydrocol	1951	Brownsville, TX, USA (no longer used)
LTFT	Precipitated Fe	Fixed bed	Arge[b]	1955	Sasolburg, South Africa
HTFT	Fused Fe	Circulating fluidised bed	Kellogg Synthol	1955	South Africa (no longer used)
HTFT	Fused Fe	Circulating fluidised bed	Sasol Synthol	1980	Secunda, South Africa (no longer used); Mossel Bay, South Africa
LTFT	Supported Co	Fixed bed	Shell Middle Distillate Synthesis	1993	Bintulu, Malaysia; Ras Laffan, Qatar, under construction
LTFT	Precipitated Fe	Slurry bubble column	Sasol Slurry Bed Process	1993	Sasolburg, South Africa
HTFT	Fused Fe	Fixed fluidised bed	Sasol Advanced Synthol	1995	Secunda, South Africa
LTFT	Supported Co	Slurry bubble column	Sasol Slurry Bed Process	2007	Ras Laffan, Qatar; Escravos, Nigeria, under construction

[a]History is not clear on whether Rheinpreussen in the Niederrhein area or Wintershall in the Ruhr area was the first to start production.
[b]Arbeitsgemeinschaft Ruhrchemie-Lurgi.

3.4.1 Industrial Fe-LTFT Synthesis

Sasol has been operating Fischer–Tropsch plants on a commercial scale since 1955. Two different Fe-LTFT processes are operated by Sasol at Sasolburg in South Africa, producing predominantly high molecular mass linear alkanes and waxes (Table 1.1). The α-values for the LTFT technologies are typically higher than 0.90. The Arbeitsgemeinschaft Ruhrchemie-Lurgi (Arge) fixed bed process is the longest operating industrial process for FTS and has been in operation since the commissioning of the original Sasol 1 facility.[44] This provides testimony to the stability and operability of fixed bed technology for FTS. In 1993, a process based on slurry bubble column reactor technology was commissioned and this process has been operating well ever since. Despite the success of the Fe-LTFT technologies, Fe-LTFT is industrially applied only at the Sasol 1 facility.

Until mid-2004, coal gasification using Lurgi dry-ash gasifiers was the primary source of synthesis gas for the Fe-LTFT processes. Coal has since been replaced as the feed for the Sasol 1 facility by natural gas, which is imported via pipeline from Mozambique.

3.4.2 Industrial Fe-HTFT Synthesis

The Sasol Synfuels plants in Secunda, South Africa, employ coal as feed material and make use of Fe-HTFT technology for the production of transportation fuels and chemicals. The α-value for HTFT synthesis is around 0.65–0.70. The syncrude from Fe-HTFT synthesis therefore has a lower molecular weight distribution and it contains more alkenes and oxygenates than the syncrudes from LTFT synthesis (Table 1.1 and Table 3.1). The original Fe-HTFT reactors at Secunda were circulating fluidised bed reactors that were modified from the Kellogg Synthol reactor design.[45] These reactors have since been replaced by fixed fluidised bed reactors.[46]

The PetroSA facility in Mossel Bay, South Africa, is a gas-to-liquids facility that employs Fe-HTFT technology. FTS takes place in circulating fluidised bed reactors. The refinery has been designed to produce transportation fuels, with only limited chemical co-production.

3.4.3 Industrial Co-LTFT Synthesis

Shell developed a cobalt-based LTFT fixed bed process that was used for the gas-to-liquids (GTL) plant in Bintulu, Malaysia.[47,48] The syncrude resembles that of German Co-LTFT, but it is heavier and more saturated. In many respects, the syncrude resembles that from Fe-LTFT, but it is somewhat lighter and contains less alkenes and oxygenates (Table 1.1). The refinery design is uncomplicated and the only conversion units are a hydrotreater and a hydrocracker. This allows the production of waxes and *n*-alkanes (paraffins) in addition to distillate, naphtha and liquefied petroleum gas (LPG).

A scaled-up, but similar fixed bed Co-LTFT facility, called Pearl GTL, is under construction at Ras Laffan in Qatar.[49] The product slate of the Peal GTL facility also includes lubricating base oils.

The Oryx GTL facility at Ras Laffan in Qatar uses a cobalt-based LTFT catalyst in a slurry bed reactor. The reactor technology is similar to that employed for Fe-LTFT. However, unlike operation with the iron-based catalyst, the cobalt-based catalyst resulted in operating problems and catalyst attrition has been an issue since start-up of the facility.[50] The Co-LTFT syncrude is similar to that of the Shell process. The associated refinery consists of a single conversion unit, namely a hydrocracker. The syncrude from FTS is hydrocracked to distillate, naphtha and LPG. Superficially, the Oryx GTL refinery design has much in common with the Shell GTL design, but there are important differences. There is no separate hydrotreater, which limits the production of chemicals, such as waxes. The hydrocracker in the Oryx GTL uses a sulfided base metal catalyst that was designed for conventional petroleum feeds and it does not employ a noble metal catalyst designed for Fischer–Tropsch waxes as is the case in the Shell process.[51]

A similar slurry bubble column-based Co-LTFT facility is under construction at Escravos in Nigeria.[52] The plant is essentially a copy of the Oryx GTL facility. However, it is expected that the modifications necessary to deal with Co-LTFT catalyst attrition will be implemented in the basic design.

References

1. A. P. Steynberg and M. E. Dry (eds), *Fischer–Tropsch Technology*, Studies in Surface Science and Catalysis, Vol. **152**, Elsevier, Amsterdam, 2004.
2. B. H. Davis and M. L. Occelli (eds), *Fischer–Tropsch Synthesis, Catalysts and Catalysis*, Studies in Surface Science and Catalysis, Vol. **163**, Elsevier, Amsterdam, 2007.
3. B. H. Davis and M. L. Occelli (eds), *Advances in Fischer–Tropsch Synthesis, Catalysts and Catalysis*, Taylor and Francis, Boca Raton, FL, 2009.
4. A. de Klerk, *Energy Fuels*, 2009, **23**, 4593.
5. J. C. Hoogendoorn, *Clean Fuels Coal Symp.*, Chicago, Sept 1973, 353–365.
6. F. Asinger, *Paraffins Chemistry and Technology*, Pergamon Press, Oxford, 1968.
7. J. C. W. Kuo, in *The Science and Technology of Coal and Coal Utilization*, ed. B. R. Cooper and W. A. Ellingson, Plenum Press, New York, 1984, p. 163.
8. S. Bessel, *Appl. Catal.*, 1993, **96**, 253.
9. H. Schulz, *Stud. Surf. Sci. Catal.*, 2007, **163**, 177.
10. E. Iglesia, S. C. Reyes, R. J. Madon and S. L. Soled, *Adv. Catal.*, 1993, **39**, 221.
11. D. R. Fahey, *J. Am. Chem. Soc.*, 1981, **103**, 136.
12. B. H. Davis, *Fuel Process. Technol.*, 2001, **71**, 157.
13. P. M. Maitlis and V. Zanotti, *Chem. Commun.*, 2009, 1619.

14. C. N. Satterfield, G. A. Huff Jr., H. G. Stenger, J. L. Carter and R. J. Madon, *Ind. Eng. Chem. Fundam.*, 1985, **24**, 450.
15. B. Jager and R. Espinoza, *Catal. Today*, 1995, **23**, 17.
16. E. Rytter, D. Schanke, S. Eri, H. Wigum, T. H. Skagseth and E. Bergene, *Stud. Surf. Sci. Catal.*, 2007, **163**, 327.
17. M. J. Janse van Vuuren, J. Huyser, G. Kupi and T. Grobler, *Prepr. Pap. Am. Chem. Soc. Div. Pet. Chem.*, 2008, **53** (2), 129.
18. A. de Klerk, *Prepr. Pap. Am. Chem. Soc. Div. Pet. Chem.*, 2010, **55** (1), 86.
19. N. E. Tsakoumis, M. Ronning, Ø. Borg, E. Rytter and A. Holmen, *Catal. Today*, 2010, **154**, 162.
20. J. J. H. M. Font Friede, J. P. Collins, B. Nay and C. Sharp, *Stud. Surf. Sci. Catal.*, 2007, **163**, 37.
21. E. Rytter, T. H. Skagseth, S. Eri and A. O. Sjåstad, *Ind. Eng. Chem. Res.*, 2010, **49**, 4140.
22. A. de Klerk, *Prepr. Pap.-Am. Chem. Soc., Div. Petrol. Chem.*, 2010, **55**(1), 86.
23. M. E. Dry, *Stud. Surf. Sci. Catal.*, 2004, **152**, 533.
24. M. E. Dry, *Stud. Surf. Sci. Catal.*, 2004, **152**, 196.
25. P. J. Flory, *Principles of Polymer Chemistry*, Cornel University Press, Ithaca, NY, 1953.
26. H. G. Stenger Jr. and C. F. Askonas, *Ind. Eng. Chem. Fundam.*, 1986, **25**, 410.
27. R. A. Dictor and A. T. Bell, *Ind. Eng. Chem. Fundam.*, 1983, **22**, 678.
28. R. B. Anderson, R. A. Friedel and H. H. Storch, *J. Phys. Chem.*, 1951, **19**, 313.
29. H. Schulz, K. Bek and E. Erich, *Stud. Surf. Sci. Catal.*, 1988, **36**, 457.
30. M. Claeys and E. van Steen, *Stud. Surf. Sci. Catal.*, 2004, **152**, 601.
31. G. Henrici-Olive and S. Olive, *Angew. Chem. Int. Ed. Engl.*, 1976, **15**, 138.
32. C. S. Kellner and A. T. Bell, *J. Catal.*, 1981, **70**, 418.
33. C. N. Satterfield and G. A. Huff Jr., *J. Catal.*, 1982, **73**, 187.
34. G. A. Huff Jr. and C. N. Satterfield, *J. Catal.*, 1984, **85**, 370.
35. S. Novak and R. J. Madon, *Ind. Eng. Chem. Fundam.*, 1984, **23**, 274.
36. G. H. Stenger Jr., *J. Catal.*, 1985, **92**, 426.
37. B. W. Wojciechowski, *Can. J. Chem. Eng.*, 1986, **64**, 149.
38. N. M. Rice and B. W. Wojciechowski, *Can. J. Chem. Eng.*, 1987, **65**, 102.
39. L. Basini, *Ind. Eng. Chem. Fundam.*, 1989, **28**, 659.
40. Y. Liu, J. Patzlaff and J. Gaube, *Prepr. Pap. Am. Chem. Soc. Div. Pet. Chem.*, 2004, **49** (2), 165.
41. F. G. Botes, *Energy Fuels*, 2007, **21**, 1379.
42. J. Huyser, M. J. Janse van Vuuren and G. Kupi, *Prepr. Pap. Am. Chem. Soc. Div. Pet. Chem.*, 2008, **53** (2), 90.
43. L. P. Dancuart and A. P. Steynberg, *Stud. Surf. Sci. Catal.*, 2007, **163**, 379.
44. J. Meintjes, *Sasol 1950–1975*, Tafelberg, Cape Town, 1975.
45. W. C. A. Holtkamp, F. T. Kelly and T. Shingles, *ChemSA*, 1977, March, 44.

46. A. P. Steynberg, R. L. Espinoza, B. Jager and A. C. Vosloo, *Appl. Catal. A*, 1999, **186**, 41.
47. P. J. A. Tijm, *Prepr. Pap. Am. Chem. Soc. Div. Fuel Chem.*, 1994, **39** (4), 1146.
48. J. Ansorge, *Prepr. Pap. Am. Chem. Soc. Div. Fuel Chem.*, 1997, **42** (2), 654.
49. N. Fabricius, in *Fundamentals of Gas to Liquids,* 2nd edn., ed. E. Soutar, Petroleum Economist, London, 2005, p. 12.
50. Anon., *Pet. Econ.*, 2008, **75** (6), 36.
51. A. de Klerk, *Prepr. Pap. Am. Chem. Soc. Div. Pet. Chem.*, 2008, **53** (2), 105.
52. K. Fraser, in *Fundamentals of Gas to Liquids*, 2nd edn., ed. E. Soutar, Petroleum Economist, London, 2005, p. 15.

CHAPTER 4

Fischer–Tropsch Syncrude

Information on the composition of primary products from FTS is necessary in order to determine the extent and means of upgrading required to achieve the product quality needed for marketable commercial products. These commercial products may be transportation fuels or chemicals.

Refining is an important step for both fuel and non-fuel applications. The product from FTS is a synthetic crude oil (syncrude). Although it can in principle be marketed as such, FTS produces significant gaseous, aqueous and solid product fractions. In order to convert these fractions into an oil phase liquid product, some upgrading is required. Consequently, all industrial Fischer–Tropsch facilities have at least some refining units to upgrade the syncrude, even though some facilities produce mainly intermediate commodities and not final products.

The subsequent discussion will focus predominantly on the products from FTS for fuel applications. There are also various options for exploiting FTS products in petrochemical applications,[1–7] and where appropriate some of these will be mentioned.

4.1 Pretreatment of Fischer–Tropsch Primary Products

After exiting the FTS reactor, primary products may contain suspended fine particles from catalyst abrasion and attrition. Such contamination is more evident in the case of fluidised bed and slurry bubble column reactors than in fixed bed reactors. Design of reactors downstream of FTS has to take this fact into consideration, unless these solids are removed from the syncrude prior to refining.

Industrially, catalyst fines are removed by cyclones from gaseous products and by filtration from liquid products. These are only two of the technologies that can be considered.

Sarkar *et al.*[8] described a continuous process for the separation of ultrafine (3–5 nm) Fe-based catalyst particles from a simulated FT wax/catalyst mixture.

RSC Catalysis Series No. 4
Catalysis in the Refining of Fischer–Tropsch Syncrude
By Arno de Klerk and Edward Furimsky

Published by the Royal Society of Chemistry, www.rsc.org

A prototype stainless-steel cross-flow filtration module that has a nominal pore opening of 0.1 μm, was used. Using this process, an iron concentration in the wax of less than 35 $\mu g\,g^{-1}$ was attained compared with almost 1600 $\mu g\,g^{-1}$ in the slurry. A related approach, also based on magnetic separation, was suggested by Oder.[9] A magnetic field is employed to retard the flow of catalyst containing wax in a separating vessel to allow an overflow of wax with a 95–98% reduction in iron. These are examples to illustrate of some of the non-traditional approaches to separate catalyst from the wax produced during slurry bubble column FTS. The separation of catalyst-derived material from the FTS primary product is critical to the operation of slurry phase FTS and attracted considerable attention in the patent literature (see Section 10.1).

During the stepwise cooling of FTS syncrude, it is advantageous to strip the carbon oxides from the products. A good separation between the aqueous phase and organic phase is also beneficial, especially if complete removal of the corrosive short-chain carboxylic acids from the organic phase can be achieved. The short-chain carboxylic acids preferentially dissolve in the aqueous phase, but quantitative removal of these acids from the organic phase requires proper design and operation. The longer chain carboxylic acids preferentially dissolve in the organic phase and are less corrosive. In fact, the carboxylic acids in the distillate range provide boundary layer lubricity and are beneficial fuel components.

Although it is no longer practised industrially, the removal of oxygenates from primary products in some cases improved the efficiency of hydrocarbon upgrading.[10] The oxygenates, on account of their higher polarity, have a stronger interaction with most refining catalysts. This may lead to catalyst inhibition by the oxygenates. Furthermore, oxygenates are reactive molecules and oxygenates may also result in unwanted side-reactions. However, judicious selection of refining catalysts may turn the oxygenates present in the syncrude into an advantage, as noted by Leckel and others.[11,12]

Removal of oxygenates may be accomplished via selective extraction, for example using a sodium hydroxide–methanol–water solution.[13] The removal of oxygenates is industrially practised during 1-alkene (linear α-olefin) extraction from FTS products,[6,14–16] but it is generally too costly to be considered for fuels refining.

4.2 Composition of Fischer–Tropsch Syncrude

The Anderson–Schulz–Flory (ASF) description of the carbon number distribution from FTS is often used to characterise FTS in terms of a single parameter, the α-value (see Section 3.3). It is the large difference in α-value that gives rise to the clear distinction between the syncrude from high-temperature Fischer–Tropsch (HTFT) and low-temperature Fischer–Tropsch (LTFT) processes. This is shown in Figure 3.3.

The syncrude from HTFT synthesis consists mainly of lighter hydrocarbons and oxygenates (Table 4.1).[7] Many of these molecules are commodity

Table 4.1 Rank order of the 10 most abundant chemicals in HTFT syncrude, excluding methane and water gas shift products.

Rank	*Compound*	*Yield (mass%)*[a]
1	Propene	13.1
2	1-Butene	9.2
3	1-Pentene	6.8
4	Ethene	6.3
5	1-Hexene	5.3
6	Ethane	5.1
7	1-Heptene	3.5
8	Ethanol	3.4
9	Acetone	2.5
10	1-Octene	2.4

[a]Calculated on a total syncrude basis, including methane (12.7% of total syncrude), but excluding water gas shift products.

chemicals. Although HTFT synthesis was originally developed for the production of motor gasoline, it is clearly an excellent platform for petrochemical production.

LTFT synthesis employing Fe based catalysts produces long-chain hydrocarbons, up to C_{120} and possibly longer. The heavier hydrocarbons (waxes) consist mainly of *n*-alkanes, but the lower hydrocarbon fractions contain significant amounts of alkenes. For example, for Fe-LTFT, the naphtha range contains >30% alkenes and the distillate range >20% alkenes, the exact amounts depending on the reactor technology and operating temperature. Even for Co-LTFT, which is more hydrogenating, the distillate range product contains 10–20% alkenes and the remainder consists mostly of *n*-alkanes and some branched alkanes. The formation of gaseous hydrocarbons (C_1–C_4) is also evident, although the yield is much lower than that from HTFT synthesis.

The analysis of Fischer–Tropsch syncrude is not trivial and an accurate mass balance of the different phases and also proper analysis of each product phase are required.[17] In order to overcome some of these problems, Dictor and Bell used on-line gas chromatography with mass spectrometry (GC–MS) for a detailed characterisation of the total syncrude product.[18] Straight-chains alkanes and alkenes accounted for most of the hydrocarbon groups up to C_{30}. With advances in analytical instrumentation, it is possible to obtain increasingly detailed analyses.[19] However, only limited isomer identification is possible with mass spectrometry. Detailed oxygenate analyses of Fischer–Tropsch products have been attempted by gas chromatography,[20] but are especially difficult due to the widely differing flame ionisation detector response factors for different oxygenates.[21–25]

4.2.1 Primary Separation of Fischer–Tropsch Syncrude

After FTS, the products are condensed in different fractions by stepwise cooling of the primary Fischer–Tropsch products. The design of the product

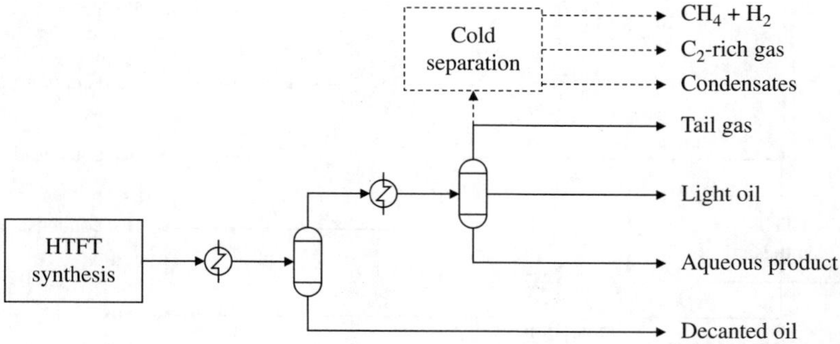

Figure 4.1 Primary separation of high-temperature Fischer–Tropsch (HTFT) syncrude typically employed to produce different product fractions for refining.

cooling steps influences the feed fractions to the refinery. This can be seen as a convenient pre-fractionation and one of the potential advantages that syncrude has over conventional crude oil.

In HTFT synthesis, the complete product leaving FTS is in the gas phase, which is then cooled to condense the different product fractions (Figure 4.1). The heaviest product fraction, which is condensed first, is called the decanted oil (DO), and this fraction contains most of the catalyst particles that were not removed by the cyclones after FTS. Most of the oil products are condensed in a second step and the product is called unstabilised light oil (ULO). After stabilisation, that is, the removal of dissolved light gases, it is called stabilised light oil (SLO). The water produced during FTS condenses with the light oil and is phase separated to produce the aqueous product. There is consequently a natural partitioning of compounds between the organic and aqueous phases, with the more polar light oxygenates preferentially dissolving in the aqueous phase. The uncondensed products in the tail gas may be further separated by cold separation.

In LTFT synthesis, part of the product is liquid under the synthesis conditions. Depending on the reactor technology, the product leaving FTS is a two-phase mixture from fixed bed synthesis or a three-phase mixture from slurry bubble column synthesis. In the latter case, the catalyst must be separated from the liquid product. Primary gas–liquid phase separation takes place in the Fischer–Tropsch reactor (Figure 4.2). The hot gaseous product is typically condensed in two stages. The first stage condenses the heavy organic products that is not liquid under the synthesis conditions, such as the wax fraction. This is called the hot condensate. (The terminology used industrially is somewhat confusing, since HTFT condensates are the products from cold separation, whereas LTFT condensates are the products from standard cooling.) In the second stage, the water produced during FTS condenses with lighter organic products. The three-phase mixture is separated into an aqueous product, cold condensate (organic liquid phase) and tail gas. As in the case of HTFT, the tail gas may be further separated in a cold separation section, which is not shown.

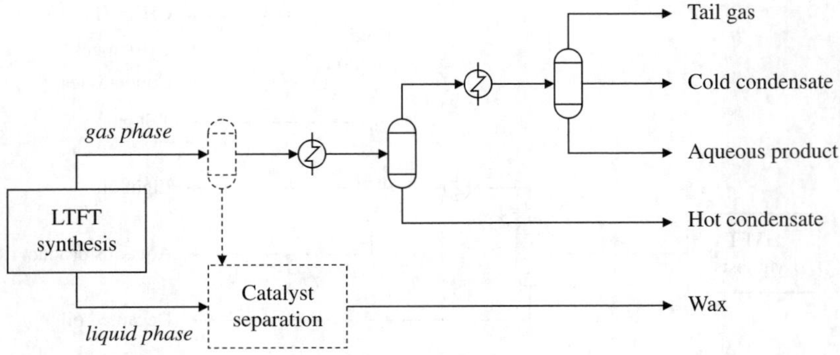

Figure 4.2 Primary separation of low-temperature Fischer–Tropsch (LTFT) syncrude typically employed to produce different product fractions for refining.

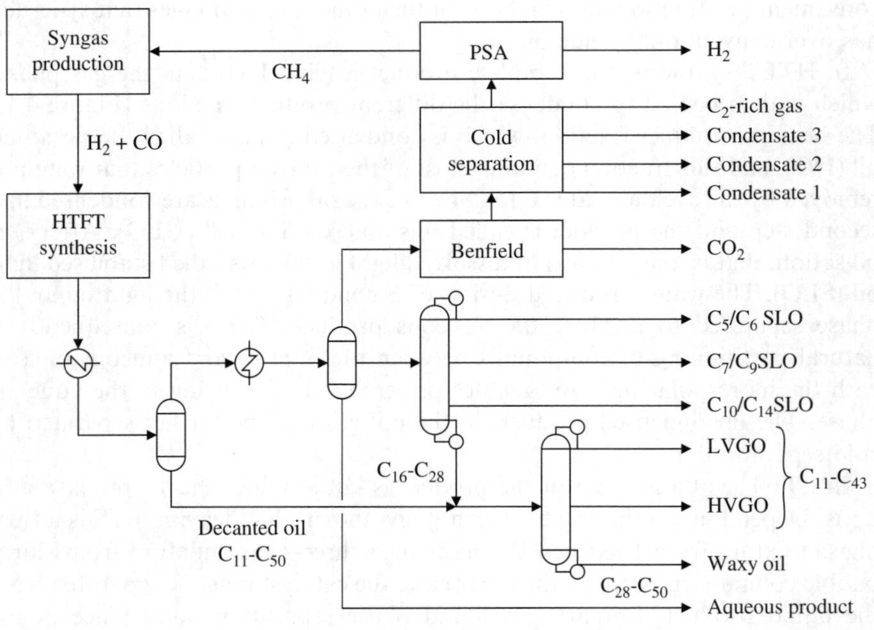

Figure 4.3 Separation of HTFT syncrude as applied in the Sasol Synfuels facility. The stabilised light oil (SLO) fractions are retained as separate feeds to the refinery, but the light vacuum gas oil (LVGO) and heavy vacuum gas oil (HVGO) are recombined after distillation.

4.2.2 Gaseous and Liquid Hydrocarbons

The flow diagram in Figure 4.3 shows the origin of various liquid streams in the Sasol Synfuels plant in Secunda, South Africa, that employs HTFT synthesis.[26,27] Although this separation strategy is fairly inefficient and only one of many possible design approaches, the published studies dealing with HTFT refining catalysis often made use of these feed fractions. The fractions of this specific separation strategy are identified, because they are referred to in subsequent chapters.

Atmospheric distillation of the HTFT light oil can be used to separate the light distillate from the heavier material, but the reboiler temperature is typically limited to 300–320 °C in order to avoid thermal cracking of the oxygenate-rich material. This leads to poor separation and significant carbon number overlap. Even with the bottom product from atmospheric distillation being well within the distillate range, there is not much material in the residue fraction of HTFT syncrude to serve as feed for vacuum distillation.[28]

The composition of some light fractions obtained according to Figure 4.3 are shown in Table 4.2.[26,29,30] Hydrogen can be recovered by pressure swing absorption (PSA), with the rest of the methane-rich gas being used for reforming. The C_2-rich gas (not listed in Table 4.2) is a mixture of ethene and ethane, from which polymer-grade ethene is recovered by distillation. Propene is recovered from the Condensate 3 and 2 streams and the combined product listed in Table 4.2 is after propene recovery. A more detailed analysis of the C_4 fraction is given in Table 4.3.[31] The composition of this fraction is consequently dependent on the amount of propene recovered. The Condensate 1 stream and C_5/C_6 SLO (light naphtha) have an overlapping carbon number distribution, but the SLO light naphtha contains much more oxygenates. The C_7/C_9 SLO (heavy naphtha) contains even more oxygenates, and also some aromatics.

Table 4.2 Gaseous and naphtha streams from HTFT synthesis at the Sasol Synfuels facility.

Compound	*Condensate 2 + 3*	*Condensate 1*	*Stabilised light oil*	
	C_3–C_5[a]	C_5–C_6	C_5–C_6 *SLO*	C_7–C_9 *SLO*
Propene	26	–	–	–
Propane	18	–	–	–
Butenes	36	1	–	–
Butanes	13	2	–	–
Pentenes	7	52	27	–
Pentanes	–	3	5	–
Hexenes	–	29	48	3
Hexanes	–	5	9	<1
Heptenes	–	6.5	8	31
Heptanes	–	1	–	6
Octenes	–	–	–	35
Octanes	–	–	–	7
Nonenes	–	–	–	4
Nonanes	–	–	–	<1
Aromatics	–	–	–	5
Oxygenates	–	0.5	3[b]	9

[a]Condensate 2 and 3 after some propene recovery by distillation. The propene-to-propane ratio in HTFT syncrude is typically around 7:1.

[b]The most abundant oxygenate in this fraction is 2-butanone (methyl ethyl ketone).

Table 4.3 Composition of an HTFT C_4 fraction obtained from the Sasol Synfuels facility.

Compound	*HTFT C_4 cut (mass%)*[a]
C_3 and lighter material	<0.1
Methylpropane (isobutane)	4.5
n-Butane	22.1
trans-2-Butene	2.0
1-Butene	54.8
Methylpropene (isobutene)	5.7
cis-2-Butene	3.4
2-Methylbutane (isopentane)	0.4
n-Pentane	0.2
3-Methyl-1-butene	1.7
2-Methyl-2-butene	3.7
2-Methyl-1-butene	0.2
Other C_5 and heavier material[b]	1.3

[a]Also contains 0.03% 1,3-butadiene and 0.02% oxygenates (mainly 2-butanone).
[b]Mainly 1-pentene.

The HTFT distillate (C_{11}–C_{22}), comprising C_{10}/C_{14} SLO, part of the C_{15} and heavier SLO fraction and part of the decanted oil, accounts for about 10% of the FT product. The residue (material boiling above 360 °C) represents only about 3% of the HTFT product. Although the heavy material is fractionated into light vacuum gas oil (LVGO) and heavy vacuum gas oil (HVGO), these fractions are recombined after distillation. These streams contain distillate and residue material. The HTFT distillate and residue cuts are different in composition from the predominantly linear alkanes and waxes that are found in equivalent LTFT cuts. The combined LVGO and HVGO fraction from HTFT synthesis (Table 4.4) contains a significant amount of aromatics (but little polynuclear aromatics), oxygenates and alkenes, but it is almost sulfur and nitrogen free.[32]

The formation of C_1–C_4 hydrocarbons always accompanies the formation of liquid and solid hydrocarbons. Even with the much heavier syncrude from LTFT synthesis, this cannot be avoided. However, cryogenic cooling is not currently applied industrially in conjunction with LTFT and the tail gas is typically used as fuel gas. The LTFT condensate fractions are rich in linear alkanes and linear alkenes, with some alcohols and carboxylic acids. The syncrude from LTFT synthesis contains a much higher fraction of linear material than that from HTFT synthesis (Table 4.5).[33]

4.2.3 Waxes

It has already been pointed out that HTFT does not produce waxes, but an aromatic residue product (Table 4.4). Fischer–Tropsch waxes are produced exclusively by LTFT synthesis. Depending on the process conditions and α-value of the catalyst, the upper range of hydrocarbons in wax is generally above

Table 4.4 Composition of the vacuum gas oil from HTFT synthesis at the Sasol Synfuels facility, which contains both distillate and residue material on account of poor separation.

Property	*Vacuum gas oil*[a]
Alkene content (g Br per 100 g)	63
Aromatics content (mass%)	27
monoaromatics (mass%)	26.3
binuclear aromatics (mass%)	0.6
polycyclic aromatics (mass%)	0.1
Oxygen content (mass% O)	3.3
acid content (mg KOH g^{-1})	12.8
Nitrogen content ($\mu g \cdot g^{-1}$)	6
Sulfur content ($\mu g \cdot g^{-1}$)	<1

[a]Boiling range: 139–496 °C (T10 = 175 °C, T50 = 251 °C, T90 = 390 °C).

C_{60}, but may reach or exceed C_{120}. The wax fraction undergoes vacuum distillation to produce medium wax and hard wax. The high concentration of straight-chain alkanes is the main reason for wax being in a solid form under ambient conditions. Industrially produced, unrefined medium wax is usually white, whereas the hard wax has a yellow to brown colour.[34] This may be the result of partial cracking during vacuum distillation or due to trace levels of impurities. On a carbon atom basis, the selectivity for hard wax boiling above 500 °C at atmospheric pressure is about 27% from Fe-LTFT synthesis.

Wax products are normally characterised by different congealing points, for example, 55–60 °C for medium wax and 94–99 °C for hard wax. However, it is possible to produce various waxes, wax grades and waxes with intermediate congealing points from LTFT, as illustrated by the industrial product ranges of Sasol and Shell.[35,36]

4.2.4 Organic Phase Oxygenates

The oxygenates are usually concentrated in the carbon number fractions below C_{20} and in the study of Dictor and Bell oxygenates higher than 1–undecanol and 1–dodecanal were not evident.[18] The organic phase oxygenate composition of Fe-HTFT and Fe-LTFT synthesis is compared in Table 4.6 to illustrate the differences in selectivity.[37,38] The dominant oxygenate class is alcohols. Based on oxygenates only, the alcohol selectivity in LTFT syncrude is around 90%, but in HTFT syncrude the alcohol selectivity is only 40–60%, with carboxylic acids and carbonyl compounds being more significant contributors to the overall oxygenate composition.

An interesting observation by Janse van Vuuren *et al.* is that the carboxylic acid selectivity over Fe-LTFT is inversely proportional to the double bond isomerisation selectivity.[39] This implies that high 1–alkene selectivity goes hand in hand with high carboxylic acid selectivity.

Table 4.5 Hydrogenated C_4–C_8 products from fixed bed low-temperature Fischer–Tropsch synthesis over a Co–ThO_2–kieselguhr catalyst at 190 °C and 100 kPa (Co-LTFT), fixed bed low-temperature Fischer–Tropsch synthesis over a commercial Sasol precipitated iron catalyst (Fe-LTFT) and circulating fluidised bed high-temperature Fischer–Tropsch synthesis over a commercial Sasol fused iron catalyst (Fe-HTFT).

Carbon number	*Compound*	*Hydrogenated products (mass% per C_n)*		
		Co-LTFT	*Fe-LTFT*	*Fe-HTFT*
C_4	*n*-Butane	95.4	95.9	91.6
	2-Methylpropane	4.6	4.1	8.4
C_5	*n*-Pentane	87.8	93.1	80.5
	2-Methylbutane	12.2	6.7	18.8
	Cyclopentane	–	0.14	0.7
C_6	*n*-Hexane	80.6	90.5	70.7
	2-Methylpentane	12.5	4.6	14.6
	3-Methylpentane	6.8	4	10.1
	2,3-Dimethylbutane	0.1	0.3	0.8
	C_6 cyclic compounds[a]	–	0.6	3.6
C_7	*n*-Heptane	73.6	90.6	58.7
	2-Methylhexane	11.3	3.3	11.1
	3-Methylhexane	14.3	3.9	16.7
	3-Ethylpentane	0.4	0.4	0.8
	Dimethylpentanes	0.4	0.7	2.2
	C_7 cyclic compounds	–	1	7
	Toluene	–	0.14	3.5
C_8	*n*-Octane	67.9	90.1	53.6
	2-Methylheptane	10.1	2.7	10.4
	3-Methylheptane	12.3	3.3	12.3
	4-Methylheptane	6.8	1.2	5.2
	3-Ethylhexane	1.7	0.6	1.5
	Dimethylhexanes	1.2	0.8	3.5
	Other C_8 branched aliphatics	–	0.07	0.4
	C_8 cyclic compounds	–	0.9	7.9
	C_8 aromatics	–	0.3	5.2

[a]Contains benzene.

Due to the more hydrogenating nature of cobalt, there is generally less oxygenates in Co-LTFT syncrude than Fe-LTFT operated under similar conditions. However, Co-LTFT catalysts are capable of producing significant quantities of oxygenates when operated at lower temperatures.[40] As in the case of Fe-LTFT, the dominant oxygenate class in Co-LTFT syncrude is the alcohols.

4.2.5 Aqueous Phase Oxygenates

Water is invariably co-produced during FTS [Equations (3.1)–(3.5)] and it is often referred to as reaction water. Since its is only the short-chain oxygenates

Table 4.6 Product composition of straight run (unrefined) C_5–C_{11} naphtha and C_{12}–C_{18} distillate cuts from fixed bed Fe-LTFT and circulating fluidized bed Fe-HTFT synthesis.

Compound class	*C_5–C_{11} naphtha*		*C_{12}–C_{18} distillate*	
	Fe-LTFT	*Fe-HTFT*	*Fe-LTFT*[a]	*Fe-HTFT*
Alkenes	32	57	25	73
n-Alkanes	57	8	61	6
Branched alkanes	3	6	4	4
Cycloalkanes	0	8	–	–
Aromatics	0	7	0	10
Alcohols	7	6	6	4
Carbonyls	0.6	6	0.3	2
Carboxylic acids	0.4	2	0.05	1

[a]Data from primary reference do not add up to 100%.

that preferentially dissolve in the water, the amount of oxygenates in the aqueous phase product from HTFT synthesis is more than that from LTFT synthesis. The amounts of oxygenates contained in the aqueous products from Fe-HTFT, Fe-LTFT and Co-LTFT as percentages of the total syncrudes are 10, 4 and 2%, respectively (Table 1.1).[41] The distribution of oxygenate classes follows the same trend as in the organic phase, with alcohols being the most abundant (Table 4.7).[37,38,42,43]

The separation of the aqueous product from the rest of the syncrude has been shown in Figures 4.1 and 4.2. The aqueous phase product can be thought of as a dilute aqueous solution of oxygenates. The aqueous product from iron-based FTS contains 5–10% oxygenates and that from cobalt based FTS <5% oxygenates.

Many short-chain oxygenates have value as chemicals, and in the case of Fe-HTFT synthesis, these compounds constitute a significant fraction of the overall syncrude (Table 4.8).[7] The chemicals with boiling temperatures below 100 °C can be recovered by distillation, but it is too energy intensive to recover remainder by distillation. Extraction has been employed on the pilot scale to recover carboxylic acids from the aqueous phase, but it was found to be very solvent intensive and the process was never scaled up.[44] Recovering carboxylic acids from the aqueous product is challenging and generally the acid water is treated as an industrial wastewater stream.

4.3 Comparison of Fischer–Tropsch Syncrude with Conventional Crude Oil

In order to appreciate the differences between Fischer–Tropsch syncrudes and conventional crude oil, it is instructive to compare them in general terms. There is, of course, no such thing as a single syncrude composition or a single crude oil composition, but some characteristics may be generalised. Such a comparison is presented in Table 4.9.[45]

Table 4.7 Composition of the aqueous phase oxygenates from different industrial iron-based Fischer–Tropsch processes.

Compound	*Normal boiling point (°C)*	*Composition (mass%)*		
		Fe-LTFT	*Fe-HTFT*	
		Fixed bed	*Circulating fluidised bed*	*Fixed fluidised bed*
Non-acid chemicals				
Methanol	65	24	1.2	0.5
Ethanol	79	45	46.4	28.8
1-Propanol	97	13	10.7	7.9
2-Propanol (isopropanol)	82	1	2.5	3.2
1-Butanol	117	5	3.5	2.9
2-Butanol	98	–	0.7	0.9
2-Methyl-1-propanol	108	–	3.5	1.0
Other alcohols		3.6	1.6	1.4
Ethanal (acetaldehyde)	21	0.5	2.5	3.9
Propanal	49	0.1	0.8	1.1
Other aldehydes		–	0.5	0.4
Propanone (acetone)	56	4	8.9	22.1
Butanone (methyl ethyl ketone)	80	0.3	2.5	9.0
Other ketones		–	0.8	3.4
Carboxylic acids				
Ethanoic acid (acetic)	117	3.5[a]	9.8	8.5
Propanoic acid	141		2.2	3.0
Butanoic acid	166		1.2	1.1
Other acids			0.7	0.9

[a]Acid content calculated by difference, no breakdown by species given.

Table 4.8 Rank order of the five most abundant chemicals in the aqueous product from Fe-HTFT synthesis.

Rank	*Compound*	*Yield (mass%)*[a]
1	Ethanol	3.4
2	Propanone (acetone)	2.5
3	Butanone (MEK)	1.2
4	1-Propanol	1.0
5	Ethanoic acid (acetic acid)	0.9

[a]Calculated on a total syncrude basis, including methane (12.7% of total syncrude), but excluding water gas shift products.

Some of these differences have a significant impact on the catalysis and conversion processes needed to refine syncrude. The following important differences can be noted:

1. Fischer–Tropsch syncrudes contain little metals and almost no sulfur- or nitrogen-containing compounds. It can therefore be expected that

Table 4.9 Generalised property comparison of Fischer–Tropsch syncrudes and conventional crude oil.

Property	*HTFT*	*LTFT*	*Crude oil*
Alkanes	> 10%	Major product	Major product
Cycloalkanes	< 1%	< 1%	Major product
Alkenes	Major product	> 10%	None
Aromatics	5–10%	< 1%	Major product
Oxygenates	5–15 %	5–15%	< 1% O (heavy)
Sulfur species	None	None	0.1–5% S
Nitrogen species	None	None	< 1% N
Organometallics	Carboxylates	Carboxylates	Phorphyrins
Water	Major by-product	Major by-product	0–2%

hydrodesulfurisation (HDS), hydrodenitrogenation (HDN) and hydrodemetallisation (HDM) reactions, which are crucial for upgrading crude oil, play no role during upgrading of FTS liquids. Most of the sulfur and nitrogen that are always present in petroleum feeds are in the form of very stable and refractory heterocyclic rings. For conventional fuels, current fuel specifications can only be met under severe hydroprocessing conditions, which is not necessary for syncrude.

2. In crude oil refining oxygenates play only a minor role, but in Fischer–Tropsch refining they play a key role. Oxygenates, which are usually present in Fischer–Tropsch liquids, are of an aliphatic nature, rather than in the form of furanic rings, phenols and aromatic ethers, as is the case with liquids of petroleum origin.[46] This suggests that much less severe conditions are needed to achieve a high level of hydrodeoxygenation (HDO). However, in the case of syncrude, the removal of a large amount of oxygen as water may affect hydrogen consumption. Also, water can modify the catalyst surface, causing competitive adsorption, hydration and deactivation, depending on the type of catalyst.
3. The primary product from HTFT synthesis contains some aromatics and cycloalkanes, but considerably less than most crude oils, whereas these compounds are almost absent from LTFT syncrudes. Cyclic compounds are important to provide energy density for all transportation fuels and this deficiency must be addressed during refining. For example, aromatics are high octane number compounds that are necessary for motor gasoline production, whereas cycloalkanes have balanced diesel fuel properties that are important in the production of on-specification diesel fuel from syncrude.[41]
4. Straight-run syncrude contains light alkenes, whereas alkenes are absent from unrefined crude oil. The catalysis of light alkene conversion, such as oligomerisation (OLI), is consequently of paramount importance to Fischer–Tropsch refining, but plays a less important role in crude oil refining, where such alkenes are only produced during some refining processes.

5. The organometallic compounds in syncrude are much more stable than the porphyrins in conventional crude. Among the former, metal acetates are the main species.
6. In FTS, water is a major by-product and contains up to 10% of the syncrude as a dilute aqueous solution. Refining of the Fischer–Tropsch aqueous product is important to improve the overall refinery yield, whereas the water in crude oil contains little dissolved oil and it can be treated as a wastewater without refining yield loss.
7. The acyclic hydrocarbons in syncrude have little branching. In this respect, the acyclic alkanes from FTS are similar to those present in crude oil. Hydroisomerisation (HIS) and hydrocracking (HCR) are consequently important to Fischer–Tropsch refining.

When the refining of Fischer–Tropsch syncrude (excluding FTS and syngas production) is compared with conventional crude oil refining, Fischer–Tropsch refining is on the balance more environmentally friendly than crude oil refining:[45]

1. Syncrude has inherently better properties than conventional crude oil for the production of most transportation fuels.
2. The carbon number distribution of HTFT syncrude is such that it is the easiest feed material to refine to on-specification transportation fuels.
3. Motor gasoline production from syncrude- and crude oil-derived naphtha requires similar refinery complexity, but crude oil refining requires technologies that are less environmentally friendly, for example, the use of halogenated compounds (chloroalkanes) in standard catalytic reforming technology and aliphatic alkylation with liquid acids (H_2SO_4 or HF).
4. Distillate refining from syncrude and crude oil is of comparable complexity and environmental impact.
5. The conversion of crude oil residue, on account of its volume and high heteroatom content, requires significantly more effort than the conversion of the FTS syncrude residue. The lower H:C ratio of crude oil and the absence of alkenes in crude oil (alkenes are required for motor gasoline production) necessitate the inclusion of at least one carbon rejection technology. Such a unit typically operates at high temperature ($\geq 440\ ^\circ C$) and is not needed for syncrude. This makes syncrude residue conversion less energy intensive.
6. The separation complexity of FTS syncrude is less than that of crude oil on account of the pre-fractionation that takes place during stepwise syncrude cooling.
7. Extraction of chemicals can simplify the design of a refinery to reduce its environmental footprint. In comparison with crude oil, FTS syncrude has more opportunities for extraction of chemicals to reduce its environmental footprint.

This comparison is only valid if the Fischer–Tropsch syncrude is refined using best syncrude refining practice. It has been pointed out that attempts to refine

Fischer–Tropsch syncrude using a crude oil refining approach lead to many inefficiencies and may ultimately lead to a refinery design that has a larger environmental impact than a conventional crude oil refinery.[28] This is also true when considering Fischer–Tropsch syncrude as feed material for the production of products for which it is not well suited, the production of EN590:2004 diesel fuel being a case in point.[41]

4.4 Fischer–Tropsch Refining Requirements

The refining needs for the conversion of Fischer–Tropsch syncrude into transportation fuels is somewhat dependent on the country where the fuels will be marketed. Although all fuels of the same type have similar properties in order to ensure compatibility with engine technology, the fuel specifications addressing emissions and environmental standards differ from country to country. This affects refinery design, but does not detract from the general refining requirements. For each fuel type, two key aspects must be considered in developing a refinery design, namely:

1. How can the carbon number distribution of the syncrude be manipulated in an efficient way to maximise the production of each transportation fuel type?
2. How can the molecular composition be manipulated in an efficient way to ensure that each fuel type meets the relevant fuel specifications?

An approach that has proven valuable in determining Fischer–Tropsch syncrude refining requirements is the combination of technology pre-selection with carbon number-based conversion. The aim of technology pre-selection is to identify the types of catalysis and conversion processes that are on a molecular level best suited to the upgrading of Fischer–Tropsch syncrude.[47] Although it restricts the list of possible technologies based on their compatibility with syncrude, it does not give an indication of their usefulness or need in a Fischer–Tropsch refinery. This is where carbon number-based conversion comes in useful. Carbon number-based conversion evaluates each carbon number in terms of its usefulness and quality for the different types of transportation fuel.[48]

The discussion of the role of catalysis in the upgrading of Fischer–Tropsch syncrude will focus on four important conversions: oligomerisation to increase the carbon number distribution, cracking/hydrocracking to decrease the carbon number distribution, isomerisation/hydroisomerisation to improve the fuel quality by increasing the degree of branching, and hydroprocessing to improve fuel stability. Each of these topics will be covered in detail (Chapter 5). There are also two primary product classes from FTS where the upgrading will be considered in more depth, namely the LTFT waxes (Chapter 6) and the oxygenates that are present in both the Fischer–Tropsch aqueous product and Fischer–Tropsch oil fractions (Chapter 7).

The aforementioned conversions are necessary to upgrade syncrude to blending stocks for mixing with crude oil-derived transportation fuels. This upgrading strategy (partial refining strategy) is employed in many of the recent industrial applications of FTS, such as the SMDS plant in Bintulu, Malaysia, the Oryx GTL and Pearl GTL facilities in Ras Laffan, Qatar, and Escravos GTL in Nigeria. However, one may also want to produce final products from FTS. The catalysis relevant for the refining of syncrude into final on-specification transportation fuels is covered separately (Chapter 8).

References

1. B. Juguin, B. Torck and G. Martino, *Stud. Surf. Sci. Catal.*, 1985, **20**, 253.
2. M. E. Dry, *ACS Symp. Ser.*, 1987, **328**, 18.
3. T. M. Leib, J. C. W. Kuo, W. E. Garwood, D. M. Nace, W. R. Derr and S. A. Tabak, presented at the AIChE Annual Meeting, Washington, DC, 27 November–2 December 1988, paper 61d.
4. J. H. Gregor, *Catal. Lett.*, 1990, **7**, 317.
5. A. P. Steynberg, W. U. Nel and M. A. Desmet, *Stud. Surf. Sci. Catal.*, 2004, **147**, 37.
6. A. Redman, in *Proceedings of the 18th World Petroleum Congress*, Johannesburg, 2005, cd179.
7. A. de Klerk, L. P. Dancuart and D. O. Leckel, in *Proceedings of the 18th World Petroleum Congress*, Johannesburg, 2005, cd185.
8. A. Sarkar, J. K. Neathery, R. L. Spicer and B. H. Davis, *Prepr. Pap. Am. Chem. Soc. Div. Pet. Chem.*, 2008, **53** (2), 101.
9. R. R. Oder, *Stud. Surf. Sci. Catal.*, 2007, **163**, 337.
10. A. de Klerk, *Prepr. Pap. Am. Chem. Soc. Div. Pet. Chem.*, 2008, **53** (2), 105.
11. D. O. Leckel, *Energy Fuels*, 2007, **21**, 662.
12. A. de Klerk and M. J. Strauss, *Prepr. Pap. Am. Chem. Soc. Div. Fuel. Chem.*, 2008, **53** (1), 313.
13. D. D. Link, J. P. Baltrus, P. H. Zandhuis and D. Hreha, *Prepr. Pap. Am. Chem. Soc. Div. Pet. Chem.*, 2004, **49** (4), 418.
14. T. Hahn, presented at the South African Chemical Engineering Congress, Sun City, 2003, paper cd013.
15. K. McGurk, presented at the South African Chemical Engineering Congress, Sun City, 2003, paper cd082.
16. D. Diamond, T. Hahn, H. Becker and G. Patterson, *Chem. Eng. Process.*, 2004, **43**, 483.
17. G. A. Huff Jr., C. N. Satterfield and M. H. Wolf, *Ind. Eng. Chem. Fundam.*, 1983, **22**, 258.
18. R. A. Dictor and A. T. Bell, *Ind. Eng. Chem. Fundam.*, 1984, **23**, 252.
19. R. van der Westhuizen, A. Crouch and P. Sandra, *J. Sep. Sci.*, 2008, **31**, 3423.
20. F. P. di Sanzo, J. L. Lane, P. M. Bergquist, S. A. Mooney and B. G. Wu, *J. Chromatogr.*, 1983, **280**, 101.

21. J. C. Sternberg, W. S. Gallaway and D. T. L. Jones, in *Gas Chromatography*, ed. N. Brenner, J. E. Callen and M. D. Weiss, Academic Press, New York, 1962, p. 231.
22. G. Perkins Jr., G. M. Rouayheb, L. D. Lively and W. C. Hamilton, in *Gas Chromatography*, ed. N. Brenner, J. E. Callen and M. D. Weiss, Academic Press, New York, 1962, p. 269.
23. R. G. Ackman, *J. Gas Chromatogr.*, 1964, **2**, 173.
24. W. A. Dietz, *J. Gas Chromatogr.*, 1967, **5**, 68.
25. J. T. Scanlon and D. E. Willis, *J. Chromatogr. Sci.*, 1985, **23**, 333.
26. A. de Klerk, *Energy Fuels*, 2006, **20**, 439.
27. D. O. Leckel, *Energy Fuels*, 2009, **23**, 2342.
28. A. de Klerk, in: *Advances in Fischer–Tropsch Synthesis, Catalysts and Catalysis*, ed. B. H. Davis and M. L. Occelli, Taylor and Francis, Boca Raton, FL, 2009, p. 331.
29. A. de Klerk, D. J. Engelbrecht and H. Boikanyo, *Ind. Eng. Chem. Res.*, 2004, **43**, 7449.
30. A. de Klerk, *Energy Fuels*, 2007, **21**, 3084.
31. A. de Klerk, D. O. Leckel and N. M. Prinsloo, *Ind. Eng. Chem. Res.*, 2006, **45**, 6127.
32. D. O. Leckel, *Energy Fuels*, 2009, **23**, 38.
33. R. B. Anderson, *The Fischer–Tropsch Synthesis*, Academic Press, Orlando, FL, 1984.
34. F. H. A. Bolder, *Energy Fuels*, 2007, **21**, 1396.
35. J. H. le Roux and S. Oranje, *Fischer–Tropsch Waxes*, Sasol, Sasolburg, 1984.
36. J. Ansorge, *Prepr. Pap. Am. Chem. Soc. Div. Fuel Chem.*, 1997, **42** (2), 654.
37. M. E. Dry, in *Applied Industrial Catalysis*, Vol. **2**, ed. B. E. Leach, Academic Press, New York, 1983, p. 167.
38. M. E. Dry, *Stud. Surf. Sci. Catal.*, 2004, **152**, 196.
39. M. J. Janse van Vuuren, G. N. S. Govender, R. Kotze, G. J. Masters and T. P. Pete, *Prepr. Pap. Am. Chem. Soc. Div. Pet. Chem.*, 2005, **50** (2), 200.
40. R. B. Anderson, in *Catalysis. Vol. IV. Hydrocarbon Synthesis, Hydrogenation and Cyclization*, ed. P. H. Emmett, Reinhold, New York, 1956, p. 29.
41. A. de Klerk, *Energy Fuels*, 2009, **23**, 4593.
42. J. G. Kronseder and M. J. P. Bogart, *Encycl. Chem. Process. Des.*, 1979, **9**, 299.
43. R. J. J. Nel and A. de Klerk, *Ind. Eng. Chem. Res.*, 2007, **46**, 3558.
44. J. Collings, *Mind Over Matter. The Sasol Story: a Half Century of Technological Innovation*, Sasol, Johannesburg, 2002.
45. A. de Klerk, *Green Chem.*, 2007, **9**, 560.
46. E. Furimsky, *Appl. Catal. A*, 2000, **199**, 147.
47. A. de Klerk, *Green Chem.*, 2008, **10**, 1249.
48. L. P. Dancuart, R. de Haan and A. de Klerk, *Stud. Surf. Sci. Catal.*, 2004, **152**, 482.

CHAPTER 5

Catalysis in the Upgrading of Fischer–Tropsch Syncrude

The conversion units that are employed in a Fischer–Tropsch refinery depend on the product slate that is being targeted. From an analysis of commercial Fischer–Tropsch refineries, it has been pointed out[1] that:

1. Syncrude is best refined to transportation fuels with co-production of chemicals, although it is possible to refine syncrude to only fuels or only chemicals.
2. Refining of high-temperature Fischer–Tropsch (HTFT) and low-temperature Fischer–Tropsch (LTFT) syncrudes requires different refinery designs.
3. Oxygenates present in syncrude have to be dealt with specifically.
4. Alkenes give syncrude synthetic capability and oligomerisation is a key technology.

The catalysis of conversion technologies that are found in most commercial Fischer–Tropsch upgrading and refining facilities will be discussed in detail. These are oligomerisation (OLI), isomerisation (IS), hydroisomerisation (HIS), cracking, hydrocracking (HCR) and hydrotreating. Additional detail on the upgrading of LTFT waxes, which includes some non-catalytic upgrading pathways, is provided separately (Chapter 6). Oxygenate processing will be highlighted throughout the discussion and additional detail is likewise provided separately (Chapter 7).

Limiting the discussion to only four types of catalytic conversion does not imply that other conversion technologies are not important. The current trend in the design of industrial Fischer–Tropsch facilities is to include only an upgrading section, not a full refinery. Such facilities produce mainly intermediate products by upgrading the syncrude, rather than final products by refining the syncrude. The distinction between upgrading and refining is blurred somewhat by the co-production of non-fuel products, such as waxes and

RSC Catalysis Series No. 4
Catalysis in the Refining of Fischer–Tropsch Syncrude
By Arno de Klerk and Edward Furimsky

Published by the Royal Society of Chemistry, www.rsc.org

lubricating oils. Refining syncrude to on-specification transportation fuels requires more than just the above-mentioned catalytic conversions. There are many other conversion processes, such as aromatisation (catalytic reforming) and aromatic alkylation, that play important roles. The catalysis of all the conversion processes that are necessary for Fischer–Tropsch refining have been reviewed recently,[2] and are discussed separately (Chapter 8).

5.1 Oligomerisation

Oligomerisation converts olefinic monomers predominantly to dimers, trimers and tetramers, whereas polymerisation produces high molecular weight plastics. The term 'oligomerisation' is therefore preferred to describe those conversions that in some instances are limited to dimerisation. In older literature, the term 'polymerisation' was frequently used to describe the same and is still colloquially used in conjunction with some processes, such as the 'Catalytic Polymerisation' (CatPoly) technology of Universal Oil Products (UOP).

At equilibrium, the shape and average molecular weight of the alkene distribution are dictated by thermodynamics. This is often a limited thermodynamic equilibrium that equilibrates only the carbon number distribution of the products and not the isomer distribution. Limited information on the properties of hydrocarbons hampers accurate prediction of the product distribution during OLI. With increasing molecular weight, the number of possible hydrocarbon isomers increases astronomically. Moreover, oligomers may undergo secondary reactions, such as cracking, isomerisation and aromatisation, particularly at high temperatures. Under conditions where secondary reactions occur, oligomers with carbon numbers that are integral multiples of the monomer may still be the dominant products, but products with intermediate carbon numbers are also present. Detailed discussions on the catalysis of OLI can be found in a number of reviews.[3–9]

Thermodynamic calculations performed by Quann *et al.* suggested that high temperatures favoured lower molecular weight alkenes, whereas low temperatures favoured high molecular weight products.[10] In a practical situation, the kinetics of OLI, disproportionation and cracking reactions determine the carbon number distribution of the product. It was reported by O'Connor that an increase in total pressure favours the formation of products with higher carbon numbers, although an equilibrium product distribution cannot be approached because of the kinetic constraints.[11] These observations can be rationalised by invoking Le Chatelier's principle, noting that OLI is very exothermic, with the heat release exceeding $60\,kJ\,mol^{-1}$ for each dimerisation step, and that the number of moles decreases with reaction.

The high heat release during OLI makes it difficult even for laboratory-scale reactors with bed dilution to maintain an isothermal temperature profile over the catalyst bed. This is not a problem at low conversion, but it is a problem during experimental investigations reporting on industrial levels of conversion with alkene-rich feed materials in fixed bed reactors. The reaction temperature reported is often a bed average temperature and not a true isothermal condition. Experimentally, one way in which this problem can be overcome is to operate the fixed bed reactor in an up-flow mode. Liquid has a better thermal conductivity than gas and operating in the up-flow mode, with a liquid-filled catalyst bed, results in better temperature control than with the down-flow mode. However, this would change the effective liquid holdup in the catalyst bed for a reaction that may be gas-phase mass transfer limited. The reactor hydrodynamics are important and may influence the reported conversion and selectivity values for OLI compared with industrial operation.

5.1.1 Mechanism and Reaction Network of Oligomerisation

The type of catalyst has a significant effect on the OLI mechanism (Figure 5.1). For many acidic catalysts, such as zeolites and acidic resins, the initial step involves the formation of carbocation by protonation of the alkene on a Brønsted acid site. The addition of a second alkene and the possible rearrangement of the adsorbed product are described by the classic Whitmore carbocation mechanism. On the other hand, for some catalysts, such as solid phosphoric acid (SPA), an intermediate phosphoric acid ester product is formed during the adsorption of an alkene and the stability of this intermediate and/or transition state governs rearrangements and the addition of a second alkene. These two mechanisms are discussed at length in the classic paper on the mechanism of OLI by Schmerling and Ipatieff.[12] Oligomerisation can also take place by 1,2-insertion and β-hydride elimination as is commonly found in organometallic catalysis, and also by non-catalytic means through free radical addition. The latter two mechanisms are often encountered in polymerisation.

A simplified reaction network involving OLI of alkenes and other related acid-catalysed reactions is shown in Figure 5.2.[13] Double bond isomerisation, skeletal isomerisation and cracking are all monomolecular acid-catalysed reactions, whereas OLI is a bimolecular reaction. A detailed account of the different reactions that are possible over solid acid catalysts can be found in a review by Corma.[14]

A reaction that is sometimes neglected during OLI is hydrogen transfer, because hydrogen transfer reactions are generally associated with high-temperature conversion. Aromatics formation becomes significant only at temperatures above 300 °C. Under typical OLI reaction conditions, hydrogen transfer can take place, however. In Figure 5.2, hydrogen transfer reactions were only indicated for C_7 and heavier compounds, since aromatic compounds are formed mainly by hydrogen transfer from C_7 and heavier alkenes. Hydrogen transfer from short-chain alkenes cannot be ruled out, but such reactions will produce dienes. This does not imply that hexenes cannot form aromatics by hydrogen transfer, but the aromatisation of hexene in this way requires the formation of a primary carbocation, making it an unfavourable reaction pathway.[15] Kanai and Kawata indeed found that aromatisation of 1-hexene occurred mainly by OLI to longer chain alkenes and subsequent

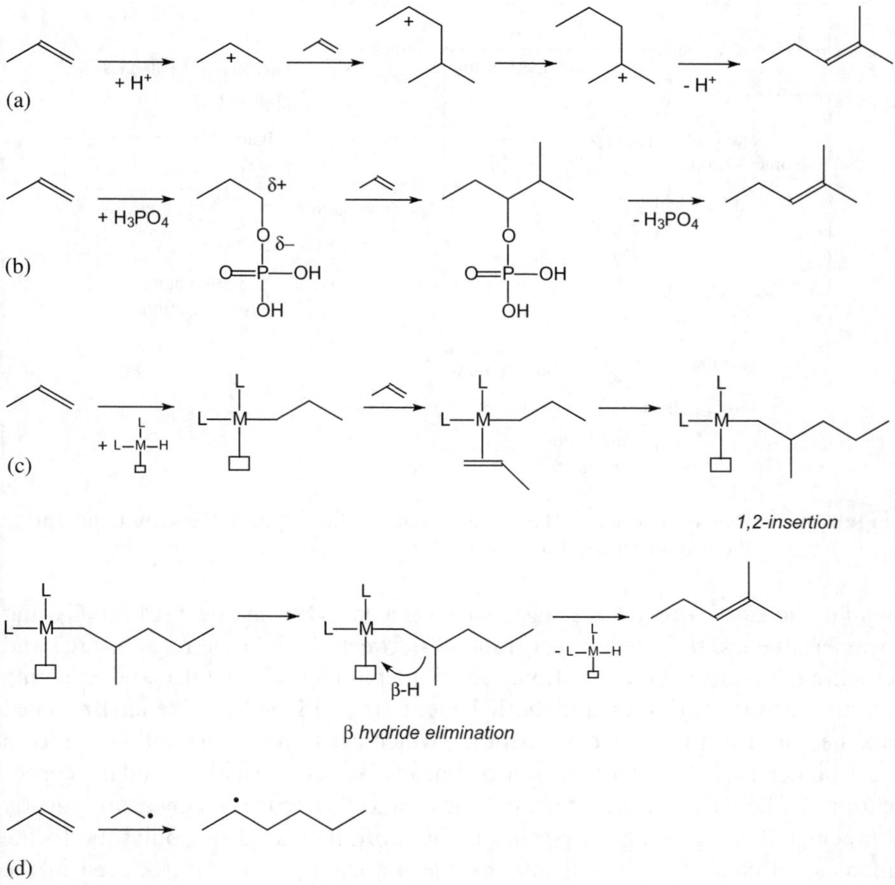

Figure 5.1 Alkene oligomerisation mechanisms. (a) Classic Whitmore-type carbocation mechanism. (b) Ester-based mechanism typical of phosphoric acid. (c) 1,2-Insertion and β-hydride elimination typical of organometallic catalysis. (d) Radical propagation. These simplified descriptions do not reflect the influence of the mechanisms on stereochemistry and branching.

cracking, rather than by direct conversion of the 1-hexene.[16] Benzene selectivity during the acid-catalysed aromatisation of hexene is low.

There is a significant difference between C_6 and lighter alkenes compared with C_7 and heavier alkenes as the feed materials for acid catalysis. In the temperature range 100–300 °C the C_6 and lighter alkenes are susceptible to cracking only after they have been oligomerised. For C_7 and heavier alkenes, the carbon chain is long enough to allow the formation of a secondary carbocation after cracking by β-scission.[17] This has some fundamental implications for optimisation of OLI processes.

During the OLI of C_6 and lighter alkenes, a tradeoff develops between per pass conversion and cracking rate. Because OLI and cracking are reactions in series, at the same temperature and pressure the cracking selectivity

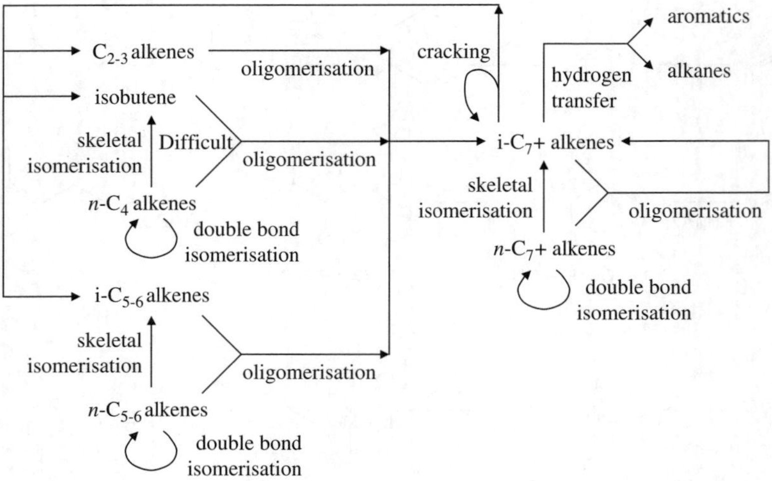

Figure 5.2 Reaction network of acid-catalysed reactions typically encountered during the oligomerisation of alkenes.

would increase with decreasing space velocity. During the OLI of C_7 and heavier alkenes, there is a direct tradeoff between the relative rates of OLI and cracking, because these reactions occur in parallel. Both OLI and cracking require strong acid sites and both benefit from IS of the alkene. Branched alkenes can form tertiary carbocations, which are more stable and hence form at a higher rate than protonation of linear alkenes to yield secondary carbocations. The OLI products from linear and branched alkenes are mostly branched. The cracking propensity is therefore increased not only due to the increase in chain length, but also by the branching that is introduced in the product.[18]

The presence of branching in the feed alkene is also important in determining the relative rates of reactions. This can be illustrated by comparing the conversion of 1-octene and 2,4,4-trimethylpentene (Figure 5.3).[18] At 180 °C, the conversion 2,4,4-trimethylpentene was much more extensive than that of 1-octene, but at 200 °C little difference in conversion could be noted. However, conversion of 2,4,4-trimethylpentene was mainly by OLI and cracking, whereas that of 1-octene was mainly by double bond isomerisation. The OLI of 1-octene took place only after some skeletal isomerisation had taken place and the rate of OLI was much slower for 1-octene than for 2,4,4-trimethylpentene, despite similar overall reaction rates. Branching benefits both oligomerisation and cracking, because the tertiary carbon at the position of branching allows the formation of a tertiary carbocation intermediate.

Skeletal isomerisation of C_5 and heavier alkenes can take place on weaker acid sites than are required for OLI and cracking.[13] Catalysts that are employed for OLI may therefore also serve as catalysts for skeletal isomerisation. This is of specific significance for catalysts used to convert C_5–C_6 alkenes, because it implies that weaker acid sites on the catalyst can be productively used for

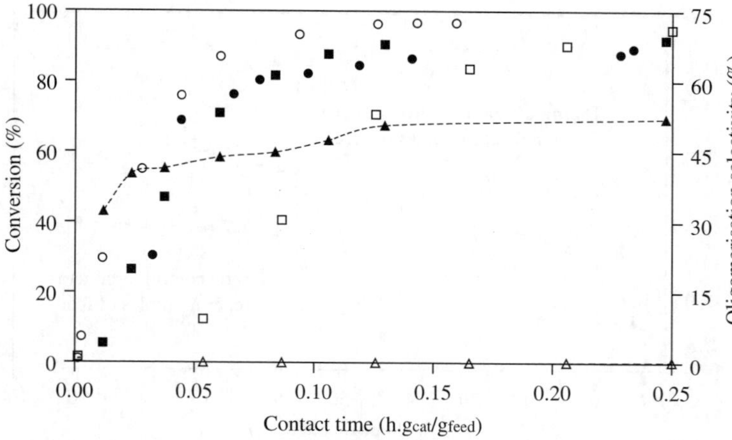

Figure 5.3 Reaction of 1-octene (open symbols) and 2,4,4-trimethylpentene (solid symbols) over a C84/3 solid phosphoric acid catalyst at 3.8 MPa in a batch reactor. The conversion of the two octene isomers at 180 °C (□, ■) and 200 °C (○, ●) are shown, and also the oligomerisation selectivity at 180 °C (△, ▲).

skeletal isomerisation, thereby increasing the rate of OLI on the stronger acid sites, but without increasing the rate of cracking. The same is not true for C_7 and heavier alkenes, because skeletal isomerisation will increase the rate of both OLI and cracking.

This leads to one specific recommendation for the application of OLI in Fischer–Tropsch upgrading. When upgrading C_5–C_6 alkenes by OLI, the catalyst and operating conditions can be selected in such a way that significant skeletal isomerisation takes place. By employing a recycle of the C_5–C_6 material, the per pass OLI conversion can then be used to limit cracking and achieve different ratios of branched naphtha for motor gasoline and branched kerosene for jet fuel production.

The competition between OLI and cracking also affects the carbon number distribution that can be obtained. At lower temperatures, where the reaction is kinetically controlled, it may be possible to produce very heavy products, because cracking is not yet significant. The product typically contains oligomers that are integral multiples of the feed (Figure 5.4).[13] At higher temperatures, the reaction becomes thermodynamically controlled and the carbon number distribution can be equilibrated.

Under kinetically controlled conditions, the catalyst may limit the carbon number distribution by processes such as competitive adsorption, diffusion restrictions or stability of the carbocation intermediate. For example, at temperatures below 200 °C, where cracking is not significant, various solid acid catalysts tested with 1-hexene had a selectivity to dimers (C_{12}) and trimers (C_{18}) in the order of 90%. When these catalysts were tested with 1-octene, the selectivity to dimers (C_{16}) was more than 90%. This indicated that there was a

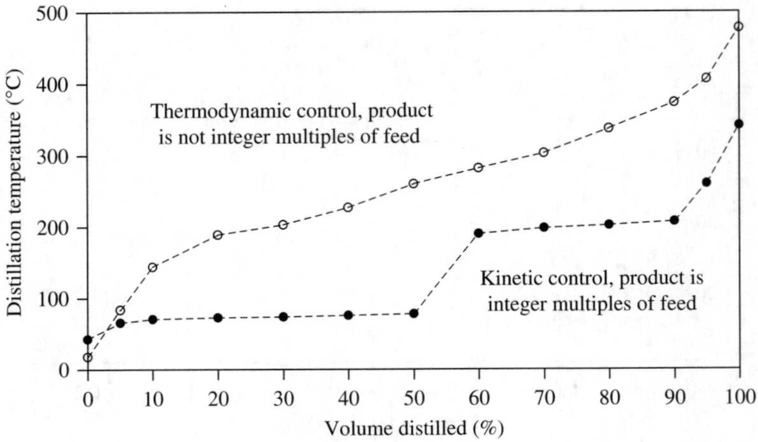

Figure 5.4 Carbon number distribution obtained during 1-hexene OLI over H-ZSM-5 under kinetically controlled conditions (●) and thermodynamically controlled conditions (○).

Table 5.1 Catalyst characterisation data and product selectivity during the oligomerisation of 1-hexene over different solid acid catalysts after 4 h on-stream in a fixed bed reactor at 100 °C, 0.8 MPa and LHSV 1.2 h^{-1}.

Catalyst	*Surface area* ($m^2 g^{-1}$)	*Average pore size (nm)*	*Product selectivity (%)*[a]			
			C_{12}	C_{18}	C_{24}	C_{30}+
SO_4^{2-}/ZrO_2	35	16	78	22	0	0
H-ZSM-5 (Si:Al = 80)	387	4.2	85	9	1	5
H-Y (Si:Al = 2.5)	536	2.9	80	5	4	11
1.3% Cr/H-ZSM-5	–	–	72	16	3	9
1.1% Cr/H-Y	–	–	79	15	3	3
1.1% Cr/H-MCM-41 (Si:Al = 8.2)	725	10	21	17	6	56

[a]Direct comparison of selectivities is not advisable due to differences in conversion.

restriction on the chain length of the product, rather than a restriction on the number of successive alkene dimerisation steps.[13]

The strength and nature of the acid sites also play an important role in determining the carbon number distribution (Table 5.1).[13] In the study reported by de Klerk, most of the heavier oligomers were produced with the H-Y zeolite catalyst, whereas sulfated zirconia catalysts were very selective in producing only lighter than C_{20} material. This was surprising, because it was expected that the large pore zirconia catalysts, rather than the narrow-pore H-Y, would produce the most heavier oligomers. The data suggested that the interaction of

H-Y with the alkene was stronger than with the SO_4^{2-}/ZrO_2, allowing more successive alkene OLI steps to occur before the product could be desorbed. In a study by Keogh and Davis,[19] it was shown that the dimer selectivity over SO_4^{2-}/ZrO_2 catalysts was influenced by the nature of the hexene isomer, and also by the level of conversion. At high conversions the dimer selectivity was high.

The catalyst pore size distribution is important from selectivity and deactivation points of view. It was noted in the literature that the micropores are not essential for the OLI process and that the reaction is mostly catalysed on the large external surface areas.[20] However, it is also known that the constrained pores of zeolites can have a marked effect on the nature of the products. This is illustrated by the difference in the degree of branching in the products obtained from OLI over solid phosphoric acid and H-ZSM-5 (Table 5.2).[21] The open structure of solid phosphoric acid does not limit branching, but the pore-constrained geometry of H-ZSM-5 limits branching.

From the preceding discussion, it is clear that the product composition obtained during OLI is governed by many variables. The mechanism, operating conditions (kinetic *versus* thermodynamic control), nature of the feed (chain length and degree of branching), catalyst geometry and the acid strength distribution of the catalyst all influence the product properties. It is consequently not possible to point out a single type of catalyst that is best for the OLI of Fischer–Tropsch syncrude and, depending on the specific application, different OLI catalysts will be recommended.

5.1.2 Commercial Processes for Oligomerisation

Table 5.3 lists some commercially available processes for the OLI of alkenes together with their intended applications. This list does not include processes devised mainly for the production of chemical commodities or fine chemicals.

The OLI of alkenes for the production of a high octane number gasoline has been practised for several decades. Initially it was limited to the conversion of C_2–C_5 alkenes produced during catalytic cracking.[22,23] The 'Catalytic Polymerisation' (CatPoly) technology of UOP was one of the first solid acid-catalysed alkene OLI technologies to be commercialised.[22,24] The process

Table 5.2 Effect of catalyst structure on the degree of branching, as illustrated by the CH_3 to CH_2 ratio of hydrogenated OLI products obtained from the conversion of Fe-HTFT C_5–C_6 over H-ZSM-5 and Fe-HTFT C_3–C_6 over solid phosphoric acid (SPA).

	Degree of branching (CH_3:CH_2 ratio)	
Product boiling fraction (°C)	*H-ZSM-5*	*SPA*
140–170	0.9	1.4
170–200	0.6	2.0
200–230	0.6	2.4

Table 5.3 List of commercially available alkene oligomerisation technologies that are relevant for fuels refining.

Catalyst	*Technology*	*Supplier*	*Main fuels application*
Solid phosphoric acid	CatPoly	UOP	Motor gasoline, jet fuel
	InAlk	UOP	Motor gasoline
Amorphous silica–alumina	Polynaphtha[a]	IFP/Axens	Diesel fuel
	Selectopol[a]	IFP/Axens	Motor gasoline
Montmorillonite	Octol-A[b]	Hüls/UOP	Motor gasoline
H-ZSM-5 zeolite	MOGD	ExxonMobil	Diesel fuel
	COD	PetroSA	Diesel fuel
H-ZSM-22 or -57 zeolite	EMOGAS	ExxonMobil	Motor gasoline, jet fuel
Acidic resin	NExOCTANE	Fortum Oy	Motor gasoline
Homogeneous nickel	Dimersol G[c]	IFP/Axens	Motor gasoline

[a]IP 811 catalyst; this technology is also available with IP 501, a zeolite-based catalyst.
[b]Available with an Ni-promoted catalyst for butene dimerisation to a more linear product.
[c]Dimersol E for ethene oligomerisation and Dimersol X for butene oligomerisation.

employed phosphoric acid on kieselguhr as catalyst, which is also called a solid phosphoric acid (SPA) catalyst. Three versions of the UOP process were developed that differ mainly in the method of removing heat from exothermic reaction and its operating conditions.[25] Two high-pressure processes were developed based on fixed bed (chamber-type) and tubular fixed bed reactor technology. A low-pressure regenerative-type process was also developed to process the gas from the stabilising unit on a cracking plant. In later years, the 'Indirect Alkylation' (InAlk) technology was introduced, which is essentially the same as the high-pressure fixed bed CatPoly process, but it was designed for more selective, lower temperature conversion of isobutene-rich feed materials.[26]

Isobutene-rich materials can also be converted over acidic resin catalysts, such as polystyrene crosslinked with divinylbenzene.[26–28] This acidic resin-based process selectively converts isobutene, with little *n*-butene OLI taking place.

The Institut Français du Pétrole (IFP) commercialised a series of Dimersol processes that employ a homogeneous catalyst in the liquid phase.[29] The catalyst is of the Ziegler type and comprises of a nickel derivative. An organometallic compound is used to activate the catalyst. Different versions of the Dimersol process have been developed for ethene OLI and for the OLI of C_4 fractions from which isobutene has been removed. A modification of the Dimersol process, called the Difasol process, employs an ionic liquid for biphasic conversion, thereby limiting catalyst loss.[30]

Much research effort in the OLI of alkenes was directed at the use of silica–alumina-based materials as catalysts. This resulted in the development of commercial processes employing activated clays and amorphous silica–alumina materials, *e.g.* the montmorillonite-based Octol process.[31] Crystalline zeolite-type silica–alumina catalysts were later synthesised, and among the zeolites, the MFI-type (ZSM-5) zeolite attracted the most attention. This led to the 'Mobil Olefins to Gasoline and Distillate' (MOGD) process.[10] A modification of this process that has been developed specifically for the conversion of alkenes from

FTS is the 'Conversion of Olefins to Distillate' (COD) process.[32] These H-ZSM-5-based technologies are best suited for distillate production.

The 'ExxonMobil Olefins to Gasoline' (EMOGAS) process is also zeolite based, but it does not make use of H-ZSM-5 as catalyst. ExxonMobil patent applications suggest that the process may employ a zeolite of the MFS type (H-ZSM-57) or TON type (Theta-1/H-ZSM-22). The EMOGAS process was designed for retrofitting solid phosphoric acid units and it is claimed that in the absence of nitrogen bases a catalyst lifetime approaching 1 year can be realised.[33] The carbon number distribution of the product is similar to that of SPA, with little material boiling above 250 °C.[34] However, it is unlikely that the isomer distribution is the same.

It is sometimes claimed that the replacement of older generation OLI catalysts with zeolites is beneficial, but from the subsequent discussion it will be clear that this is not necessarily the case. Each catalyst type has specific advantages and disadvantages that are intrinsically linked to the catalyst and its operating envelope.

5.1.3 Catalysts for Oligomerisation

With the aim of increasing the efficiency and/or of optimising the operation, efforts focusing on the development of more active and selective catalysts for OLI have been reported. Although SPA and zeolitic catalysts have been studied extensively, other types of catalysts have also attracted attention. Table 5.4 illustrates the variety of solids that have been tested as catalysts for OLI.[11] It is evident that the nature of the catalyst has a significant effect on product distribution, a point that was emphasised earlier (Section 5.1.1). Organometallic compounds, such as transition metal (Ti, Ni, Zr and W) complexes and organoaluminium compounds, are also active for various OLI reactions. These materials are not covered in this chapter. A detailed account of OLI over these catalysts was given by Skupinska,[6] with emphasis on the OLI of ethene, propene and higher 1-alkenes.

Alkene OLI has retained a central position in Fischer–Tropsch refining, due to the abundance of gaseous alkenes in the primary FTS product.[35] The subsequent discussion will focus on OLI of Fischer–Tropsch-derived materials, but pertinent studies involving model compounds will also be included. It will be noted that a wide range of experimental conditions have been employed during catalyst evaluations. The operating range of OLI is consequently fairly wide and more limited ranges are applicable to specific catalysts.

5.1.3.1 Solid Phosphoric Acid Catalysts

Production of the high-octane olefinic motor gasoline by the OLI of C_2–C_5 alkenes over solid phosphoric acid (SPA) has been practised commercially since the 1930s. The operating conditions for SPA are usually in the range 150–245 °C, although lower temperatures can be considered for feed materials free

Table 5.4 Product distribution from OLI of propene and butene over various solid acid catalysts.

	Process			Production distribution (mass%)[a]				
Catalyst	*Feed*	*Temperature (°C)*	*Pressure (MPa)*	*Dimer*	*Trimer*	*Tetramer*	*Pentamer*	*Hexamer*
H-ZSM-5	C_3	220	5	18	30	27	14	11
H-Mordenite	C_3	300	5	28	31	23	10	8
H-Mordenite	C_4	250	5	48	40	7	5	–
Solid phosphoric acid	C_3	200	3	9	65	16	10	–
Solid phosphoric acid	C_4	200	3	80	14	6	–	–
Mica–montmorillonite	C_3	150	5	9	25	21	21	24
Ni/mica–montmorillonite	C_3	150	5	8	30	26	15	21
SiO_2–Al_2O_3	C_3	200	3	16	35	27	20	2
Ni/SiO_2–Al_2O_3	C_3	80	3	60	22	13	5	–
Amberlyst 15	C_3	130	5	55	30	10	5	–
Al–tungstophosphoric acid	C_3	230	5	12	44	25	14	5

[a]Although the product distribution suggests integral multiples of the feed, it is unlikely to be the case in all instances.

of C_2–C_3 alkenes.[36] The lower temperature limit is determined by the formation of stable phosphoric acid esters with the alkenes and the upper limit is set by the catalyst deactivation rate, which increases with temperature.

It was found that the feed had little impact on the quality of the olefinic motor gasoline. When ethene was used as feed, the gasoline fraction had a motor octane number (MON) of 82 and a research octane number (RON) of 96.[37] When propene was used as feed, the gasoline fraction had a MON of 81 and a RON of 95.[38] When *n*-butenes were used as feed, the gasoline fraction had a MON of 81,[39] and when a propene–butene mixture was used as feed, the gasoline fraction had a MON of 82.5 and a RON of 97.[25] Many conventional crude oil refineries still make use of SPA-based OLI. It is a very forgiving refinery technology that maintains its high-octane olefinic motor gasoline quality despite fluctuations in the feed.

The main drawback of SPA is its comparatively short catalyst life, around 6 months depending on the operation. Since SPA is not regenerable, this created the impression that SPA has a large environmental footprint. A life-cycle analysis of SPA materials showed that this is not true. Only natural substances (kieselguhr and phosphoric acid) are required during catalyst manufacture.[40] Moreover, the disposal of spent SPA can be equally environmentally friendly. It is possible to neutralise the spent SPA catalyst with ammonia and then employ it as ammonium phosphate for agricultural use.[41] This method of disposal has been practised commercially in South Africa for decades, where the spent SPA from HTFT synthetic fuels production is converted into plant fertiliser.

Historically, the SPA-catalysed conversion of Fischer–Tropsch alkenes focused mainly on mixed C_3–C_4 streams.[1] A similar quality high-octane olefinic motor gasoline could be produced as with alkenes derived from crude oil refining, but unlike a crude oil refinery the motor gasoline from FTS is inherently olefinic. There is a limit to the amount of alkenes that can be blended into a motor gasoline. In the context of Fischer–Tropsch refining, there arose a need to hydrogenate some of the olefinic motor gasoline, which is generally not considered in crude oil refining. On hydrogenation of the product from C_3–C_4 OLI over SPA, the octane number of the motor gasoline decreased considerably. The RON decreased from 94.5 to 63.7 and the MON decreased from 80.9 to 70.6.[42] Initially it was not realised that the feed dramatically affected the degree of branching and thereby the quality of the hydrogenated motor gasoline. When propene and butenes are subjected to OLI separately over SPA and not as a mixture, it is possible to obtain two very different products. SPA has been found to be especially well suited for the upgrading of *n*-butenes to good quality hydrogenated motor gasoline, with straight run FTS-derived butenes giving a hydrogenated motor gasoline with a RON of 86 and a MON of 88.[43,44] Conversely, propene-derived motor gasoline has a very poor hydrogenated octane number, with a RON of less than 50. It is clear that C_3 and C_4 alkenes must be converted separately over SPA if part of the product is to be hydrogenated.

Due to the linear 1-alkene-rich nature of FTS derived naphtha, it has a low octane number despite being very olefinic. This motivated some studies to evaluate the possibility of employing SPA for the OLI of naphtha in order to produce distillate.[45,46] Distillate formation was observed for C_5–C_6 and C_7–C_9 naphtha cuts from FTS, but it was not a very efficient conversion. It was found that the weak interaction of long-chain alkenes with phosphoric acid mechanistically limited distillate production and, as a consequence, SPA was not a suitable catalyst for distillate production. Furthermore, naphtha-range alkene OLI was inhibited by short-chain alkenes, such as propene, which became the main carbocation source. The same was likely to be true for butenes, as indicated in studies with butene and pentene mixtures.[47]

Prinsloo reported that during the OLI of C_3–C_4 alkenes over SPA, the selectivity to diesel range products could be manipulated by careful control of the catalyst hydration through controlling the water content of the feed and the process temperature.[48] However, even for C_3–C_4 mixtures, the propene content played a significant role in determining the ultimate distillate yield from OLI over SPA.[45]

The mechanism over SPA is not described in terms of Brønsted acid catalysis, but it involves the formation of a phosphoric acid ester.[12,49–51] The mechanism is shown in Figure 5.1. Due to the more intimate interaction between the phosphoric acid and the reacting alkene, the reactivity of alkenes is determined not only by the stability of the carbocation, but also by the stability of the phosphoric acid ester. Ethene requires a high temperature to oligomerise over SPA because it forms an ester that is thermally stable to about 200 °C.[50,52] The same is true for propene, which forms an ester that is thermally stable to about 125 °C.[50,53] The *n*-butenes do not form such stable esters and reactions can take place at room temperature,[54] although the stability thresholds for *n*-butene esters are less clearly defined. For *n*-alkenes the ester stability decreases with increasing chain length of the alkene. However, the reactivity sequence of *n*-alkenes does not show a monotonic increase with chain length and *n*-butenes have the highest reactivity, the reactivity of the *n*-pentenes being lower than that of the *n*-butenes.

The C_5 and heavier alkenes show a trend of decreasing olefinic reactivity with increasing chain length. Test work at 160 °C and a total pressure of 3.8 MPa showed a drop in 1-pentene conversion at constant butene conversion with increasing 1-pentene concentration.[43] This suggested that the butenes, rather than the 1-pentene, were more readily protonated and became the primary carbocation source. A reasonable conversion of 1-hexene could be demonstrated at 200 °C and a total pressure of 6 MPa over SPA.[13] However, as in the case of 1-octene,[18] skeletal isomerisation of the 1-hexene was a prerequisite for OLI.[55,56] As expected, 1-octene was considerably less reactive than 1-hexene, because even at 200 °C it was difficult for OLI to take place.[18] All of these results confirm the decreasing reactivity with increasing chain length for C_5 and heavier alkenes.

The reactivity of linear alkenes seems to be a tradeoff between the strength of the phosphoric ester being formed and the time period during which the alkene has an interaction with the acid before desorbing as an alkene again. If the alkene forms an ester with the phosphoric acid that is too strong, the ester is too stable to react with another alkene. If the interaction is too weak, the time period during which the alkene has an interaction with the phosphoric acid is so short that there is a low probability of interacting with another alkene while being polarized, and even the probability of a monomolecular transformation is low. The time period during which the alkene remains polarized is influenced not only by the strength of the ester, but also by the stability of the polarized intermediate. The stability of the intermediate is expected to increase in the same order as carbocation stability: primary < secondary < tertiary.[57] A branched-chain molecule is consequently expected to be more reactive than a linear molecule of the same carbon chain length, because a tertiary carbon results in a more stable polarized intermediate. Other factors may also influence the reactivity of longer chain alkenes over SPA catalysts:

1. Adsorption of alkenes may decrease with increased degree of branching.[58]
2. Longer chain alkenes, which become increasingly apolar, are also less likely to adsorb strongly on the polar catalyst surface.
3. Longer alkenes are also more bulky, with a slower rate of diffusion, and SPA is known to be mass transfer limited.[59]

The mechanism of butene OLI over SPA has some unique features. Industrially the most relevant aspect of the mechanism is the ability of SPA to convert *n*-butenes into trimethylpentenes. This unexpected conversion is responsible for the high octane numbers of hydrogenated motor gasoline from Fischer–Tropsch butene OLI over SPA.[43] The ability of SPA to produce branched products from linear feed is clearly not Brønsted-like behaviour. SPA does not behave like a pure Brønsted acid catalyst, because the OLI mechanism involves the formation of phosphoric acid esters. The products from 1-butene and 2-butenes are different, emphasizing that the intermediate is not a common secondary carbocation, as would be expected from alkene protonation and reaction by the classic carbocation mechanism.[51] It was found that at low temperatures (typically 160 °C and lower) there is a low-temperature skeletal IS pathway involving the rearrangement of a butyl phosphoric acid ester intermediate to produce a trimethylpentene-rich product on dimerisation (Figure 5.5). The trimethylpentene-rich product is especially rich in 2,3,4-trimethylpentene.[60]

Despite double bond isomerisation of the 1-butenes to yield an equilibrated *n*-butene mixture that is rich in 2-butenes, the OLI rate of the 1-butene is much faster than that of the 2-butenes. At low temperatures, most of the OLI occurs *via* 1-butene and ultimately double bond isomerisation is constantly converting the 2-butenes back to 1-butene in order to maintain the *n*-butene equilibrium. This is one of the main reasons for the increase in degree of branching with decrease in temperature. At higher temperatures, the OLI of 2-butenes contributes more to the overall OLI rate and the products from the OLI of 2-butenes are less branched.[51]

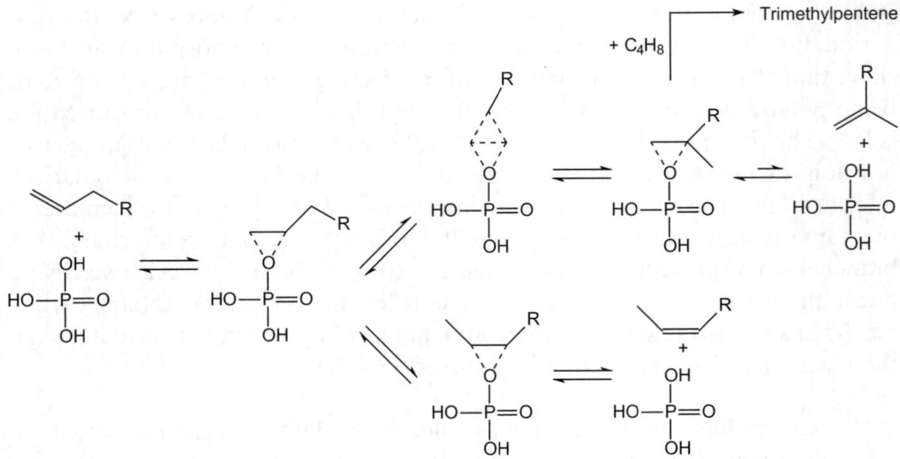

Figure 5.5 Simplified representation of the low-temperature skeletal isomerisation pathway of 1-butene over solid phosphoric acid.

A description of the mechanism on SPA is further complicated by the self-dissociative behaviour of phosphoric acid:[61,62]

$$2H_3PO_4 \rightleftharpoons H_4PO_4^+ + H_2PO_4^- \tag{5.1}$$

$$2H_3PO_4 \rightleftharpoons H_4P_2O_7 + H_2O \tag{5.2}$$

$$3H_3PO_4 \rightleftharpoons H_4PO_4 + H_2P_2O_7^{2-} + H_3O^+ \tag{5.3}$$

These reactions are responsible for creating Brønsted acid sites on the catalyst, which causes reactions to occur *via* both carbocation and ester formation pathways. Empirical evidence suggests that at lower temperatures the ester formation pathway dominates OLI over SPA, but it is not clear whether the more Brønsted-like behaviour at higher temperatures is due to self-dissociation or merely the high rate of ester decomposition.

Temperature is clearly one of the parameters affecting the mechanism and thereby the product distribution, but temperature also affects catalyst hydration and in most studies these effects are not decoupled. In the study by Bethea and Karchmer, the effect of temperature and hydration was decoupled for the OLI of propene.[63] In an analogous study by de Klerk *et al.*, the effects of temperature and hydration were decoupled to investigate the relative contribution of each on the quality of the product for butene OLI (Figure 5.6).[64] Thus, Figure 5.6 confirms that the degree of branching in the OLI product is a strong function of temperature, with low temperature favouring increased branching. Branching is less affected by the hydration level of the phosphoric acid, except at low temperatures, where branching is increased by increasing

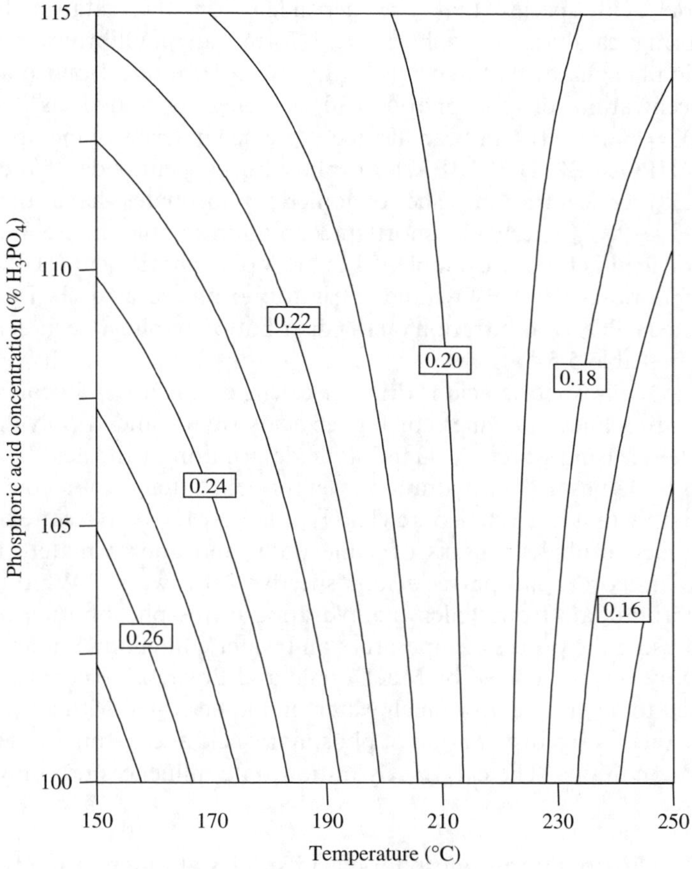

Figure 5.6 Effect of temperature and catalyst hydration on the ratio of trimethylpentenes to total C_8 alkenes in the product from butene oligomerisation over liquid phosphoric acid at 3.8 MPa. The butene feed had an *n*-butene:isobutene ratio of 10:1.

hydration. Similar behaviour was reported for industrial operation with Fischer–Tropsch alkenes.[64]

From Figure 5.6, it is also clear that catalyst hydration affects the mechanism independently. Catalyst hydration refers to the dynamic response of the SPA catalyst to changes in the water partial pressure that affects the phosphoric acid species present on the catalyst surface. One of the peculiarities of SPA catalysts is that they require the feed to contain a small amount of water to maintain catalyst activity.[4,59,65] Oxygenates can also be converted over SPA and many of the SPA-catalysed oxygenate reactions result in the formation of water.[66] This has to be taken into account when using SPA with material from FTS.

The phosphoric acid, which is the active phase, is present as a glassy layer on a quartz or kieselguhr support. The active phase is actually a mixture of phosphoric acid species that are dependent on the catalyst hydration state. During catalysis, phosphoric acid forms an equilibrium mixture of oligomeric phosphates that depends on the concentration of phosphoric acid. The concentration of phosphoric acid is generally stated as % H_3PO_4 (% H_3PO_4 = % P/0.316), but can also be expressed in terms of metaphosphoric acid (% HPO_3 = % H_3PO_4/1.23) or phosphorus pentoxide (% P_2O_5 = % H_3PO_4/1.38) concentration. The oligomeric phosphates have the general formula $H_{n+2}P_nO_{3n+1}$ and the shorthand notation for the different species of phosphoric acid is P*n*; for example, P1 is orthophosphoric acid (H_3PO_4), P2 is pyrophosphoric acid ($H_4P_2O_7$) and P3 is triphosphoric acid ($H_5P_3O_{10}$). The effect of the H_3PO_4 concentration on the distribution of phosphoric acid species is shown in Table 5.5.[64]

When 85% phosphoric acid is dried or calcined, it is not only concentrated, but also forms linear polymers of higher acids (pyro- and tripolyphosphoric acid) while releasing water. A quantitative description of the acid distribution was given by Jameson.[67] A hydration level (or acid 'strength') of 100% H_3PO_4 does not imply that it is 100% pure H_3PO_4, it merely refers to a state of dryness where the active phase consists of some water and approximately 14% and 86% pyro- and orthophosphoric acid, respectively. Brown and Whitt measured the equilibrium data from which the hydration of the phosphoric acid can be estimated using the process temperature and water content at equilibrium.[68] A similar study was performed by MacDonald and Boyack.[69] These calculations can be used to determine how much water should be co-fed with a process feed to achieve the desired distribution of phosphoric acid species under the relevant operating conditions. The catalyst hydration state influences not only catalyst

Table 5.5 Distribution of phosphoric acid species at different levels of phosphoric acid concentration determined by high-performance liquid chromatography combined with ion chromatography (HPLC–IC).

	Phosphoric acid concentration (% H_3PO_4)					
Species	*85*	*100*	*104*	*108*	*115*	*117*
P1	100	76.6	58.9	33.4	5.2	4.2
P2		22.6	38.0	50.5	21.0	14.3
P3		0.8	2.8	13.0	22.3	16.7
P4			0.3	2.6	17.9	15.5
P5				0.5	12.7	12.9
P6					8.7	10.1
P7					5.0	6.9
P8					3.3	5.5
P9					1.8	3.9
P10					1.0	2.9
P11					0.6	2.5
P12					0.3	2.1
P13					0.2	1.8
P14					0.04	0.7

activity, but also the selectivity for reactions such as the OLI of alkenes and the alkylation of aromatics with alkenes.[48,64,70,71]

It should be noted that the crystalline 100% H_3PO_4 consists of only P1.[72] However, in the liquid phase it consists of a mixture of water, P1 and P2 species,[73–76] as can also be seen from Table 5.5 and mentioned previously. In earlier studies, this observation was ascribed to the persistence of the hydrate, such as $(H_3PO_4)_2 \cdot H_2O$,[61,62] but it is more likely due to the self-dissociation behaviour of phosphoric acid [Equation (5.2)].[77,78] Although it is known that oligomeric phosphates can exist as linear polyphosphate or cyclic metaphosphate species,[79,80] it was shown that in the range 94–112% H_3PO_4, which includes the concentration range of industrial interest for catalytic applications, only linear phosphoric acid species are found.[25,74,76]

Catalyst hydration is also important for the structural integrity of the SPA catalyst. If the catalyst is over-hydrated, swelling of the catalyst is observed, associated with a rapid increase in pressure drop. This happens as a result of the softening of the catalyst, which ultimately causes disintegration. Excess of water will also reduce acid viscosity and cause the loss of acid from the catalyst.[65] When the SPA catalyst is under-hydrated, the catalyst loses activity, becomes brittle and may also disintegrate.

Due to the impact of hydration on SPA, the hydrolysis behaviour of SPA is of interest. Hydrolysis of 115% H_3PO_4 in deionised water caused the heavier phosphoric acid species to break down rapidly, and after 3 h at 80 °C only P1–P3 species were present (Figure 5.7).[64] It was found that P3 is the preferred intermediate hydrolysis product rather than P1, contrary to the hydrolysis mechanism proposed before.[81]

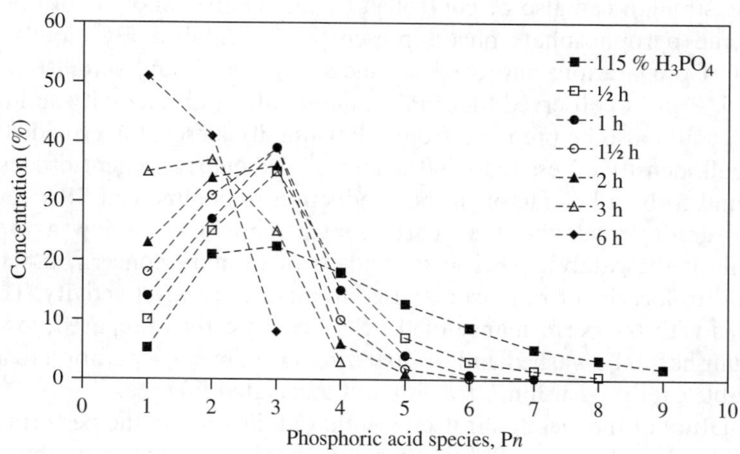

Figure 5.7 Hydrolysis of 115% H_3PO_4 in water at 80 °C, showing the composition of the phosphoric acid species over time.

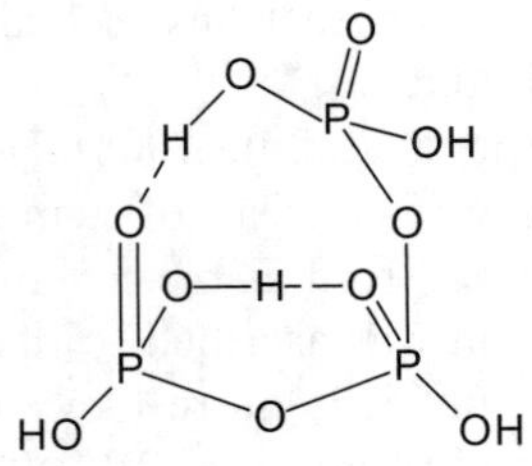

Figure 5.8 Stabilisation of the phosphoric acid P3-species ($H_5P_3O_{10}$) by intramolecular hydrogen bonding.

Converting P3 to P2 and P1 took much longer than hydrolysis of the heavier phosphoric acid species. The observation of slow P3 hydrolysis is in agreement with those of Bell,[82] who also observed the formation of P3 during the hydrolysis of hexametaphosphate. A plausible explanation is that P3 is stabilized due to intramolecular hydrogen bonding. The formation of strong hydrogen bonds between P–OH and P=O units in the molecule creates a P3-containing cage (Figure 5.8), which protects the P3 fragment from further hydrolysis. With the formation of the stable P3 species, catalytic activity is decreased.[83]

The mechanical properties of SPA, and specifically the crushing strength of the catalyst particles, is a key parameter determining the performance of SPA catalysts. This topic was the focus of the study of Coetzee *et al.*[40] Inadequate crushing strength may result in the structural breakdown or collapse of catalyst particles while the catalyst is still active, thereby necessitating premature catalyst replacement. Traditionally, the mechanical strength of SPA catalysts has been improved by the use of additives, such as Fuller's earth, as binders. The crushing strength can also be controlled by the relative amounts of the silicon ortho- and pyrophosphate phases present in the catalyst. By modifying the method of preparation, improved hardness, robustness and stability could be achieved. Prinsloo observed that SPA catalysts of suitable activity and particle strength could also be prepared from a low-quality kieselguhr provided that it had a bulk density of less than 300 $kg\,m^{-3}$.[84] The quartz content of kieselguhr was found to be a key factor in the production of commercial SPA catalysts. For low-grade kieselguhr, the quartz content should be as low as possible, especially if the catalyst particle strength is of primary concern, otherwise a high acid-to-kieselguhr ratio can be used to increase catalyst activity. To avoid problems with the performance of the SPA catalyst thus prepared, low-grade kieselguhr has to be washed and/or quartz removed by air separation to achieve a desirable catalyst crushing strength and associated lifetime.

The nature of the kieselguhr itself also has an impact on the performance of an SPA catalyst. It is assumed that the catalysis is not affected by the support (kieselguhr), which is reportedly not catalytically active.[53] Liquid phosphoric acid may therefore safely be employed as a model catalyst to study the mechanistic behaviour of SPA. In fact, there is a significant body of literature that suggests that the kieselguhr type mainly influences activity by mass transport resistance.[83,85–87] However, not all studies found the kieselguhr support to be inactive.[88] It has been noted that the nP_2O_5–$mSiO_2$ system may form heteropolyacids and that as hydrated compounds, hydrogen ions may be formed.[89] The hydrolysis chemistry of the H_3PO_4–SiO_2 system may therefore play some role in generating catalytic activity.[90] Furthermore, kieselguhr is a silica-rich material, but it is not a pure silica. The concentration of metal impurities in kieselguhr varies depending on the geological deposit (Table 5.6).[91] The main impurities are Al and Fe, with one or more of Na, K, Mg, Ca and Ti that may also be present. Some of these metals and their phosphates are catalytically active.[92] Observed differences in the reactivity of differently supported phosphoric acid catalysts have been ascribed to the formation of catalytically active phosphates.[93] Considering the influence of kieselguhr type as a mass transport effect only may consequently be an over-simplification. In the literature this issue is still unresolved.

The OLI of alkenes over SPA has a limited carbon number distribution as a direct consequence of its mechanism. The weak interaction of long-chain alkenes with SPA inherently limits the chain length of the product. SPA is consequently an excellent catalyst for OLI to make motor gasoline and jet fuel, but not diesel fuel. However, the carbon number distribution can be manipulated to increase distillate range products by decreasing catalyst hydration (increasing the acid strength), decreasing space velocity and increasing temperature. This holds true for propene as feed material (Figure 5.9),[48] and also for butenes (Figure 5.10).[64] At very low levels of hydration, the motor gasoline-to-distillate ratio becomes insensitive to temperature. The aromatics content of the product increases with increasing temperature and phosphoric acid concentration, but aromatics remain a minor product even at high temperature and high catalyst hydration. At 115% H_3PO_4, 250 °C and 3.8 MPa, less than 2% of aromatics were found in the product from butene OLI over SPA.[64] At low

Table 5.6 Composition of the mineral content of different kieselguhr (diatomaceous earth) types to give an indication of the variation in composition.

	Mineral matter (mass%)							
Kieselguhr deposit	SiO_2	Al_2O_3	Fe_2O_3	Na_2O	K_2O	MgO	CaO	TiO_2
Alterschlirf, Hesse, Germany	94.4	1.0	2.8	0.6	0.4	0.3	Trace	0.5
Auxillac, Auvergne, France (white)	95.5	1.6	2.3	Trace	Trace	0.3	0.2	0.1
Pit River, CA, USA	98.2	1.1	0.6	–	–	–	0.1	–
Pope's Creek, MD, USA	84.5	3.6	3.4	–	–	5.8	2.7	–
Richmond, VA, USA	84.6	11.0	3.3	–	–	0.8	0.3	–
Toome, Ireland	87.7	7.8	2.2	0.3	–	0.8	1.2	–
Unterlüss, Hannover, Germany (white)	96.1	2.0	0.4	0.7	0.4	0.2	Trace	0.1
Wilmont Wharf, VA, USA	85.8	7.0	2.4	1.0	1.1	1.1	0.4	1.1

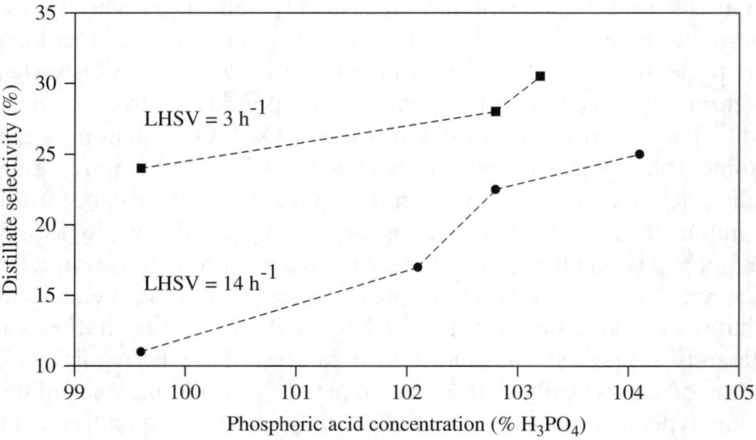

Figure 5.9 Effect of catalyst hydration and space velocity on the distillate selectivity in the product from propene oligomerisation over SPA at 180–200 °C and 3.8 MPa.

temperatures, the hydration level of the phosphoric acid has little influence on hydrogen transfer.

The nature of the OLI product in combination with temperature also plays a role in limiting the carbon number distribution. When the alkene contains an isobutene-like fragment in its structure, it is susceptible to selective cracking (depolymerisation). Depolymerisation of 2,4,4-trimethylpentene has been reported to occur even at 100 °C over SPA.[18] This form of depolymerisation was historically used as a method to determine the trimethylpentene content of OLI products.[39,94] This reaction clearly affects both the carbon number distribution and product quality. Distillate produced from propene is less susceptible to depolymerisation than distillate produced from butene OLI.

5.1.3.2 *Heteropolyacid Catalysts*

There are numerous heteropolyacid (HPA) compounds, but for catalysis it is primarily the Keggin type that are of importance. The acid strength of the three most acidic types of HPA catalysts are $H_3PW_{12}O_{40} > H_4SiW_{12}O_{40} \approx H_3PMo_{12}O_{40}$. In contrast to Keggin-type HPA structures, other structures are thermally less stable and cannot be employed for reactions at temperatures above 150 °C.[95]

The use of HPA catalysts has been investigated for the OLI of propene,[96] isobutene[97,98] and mixtures of *n*-butenes and isobutene.[99] Isobutene can be converted at low temperatures; for example, Burrington *et al.* described a slurry phase process operating at −5 °C for the production of a lubricant additive from isobutene.[97] At higher temperatures the products are less heavy.

Only one of these studies focused on material derived from FTS. Propene OLI was conducted over various HPA catalysts to investigate the use of HPA

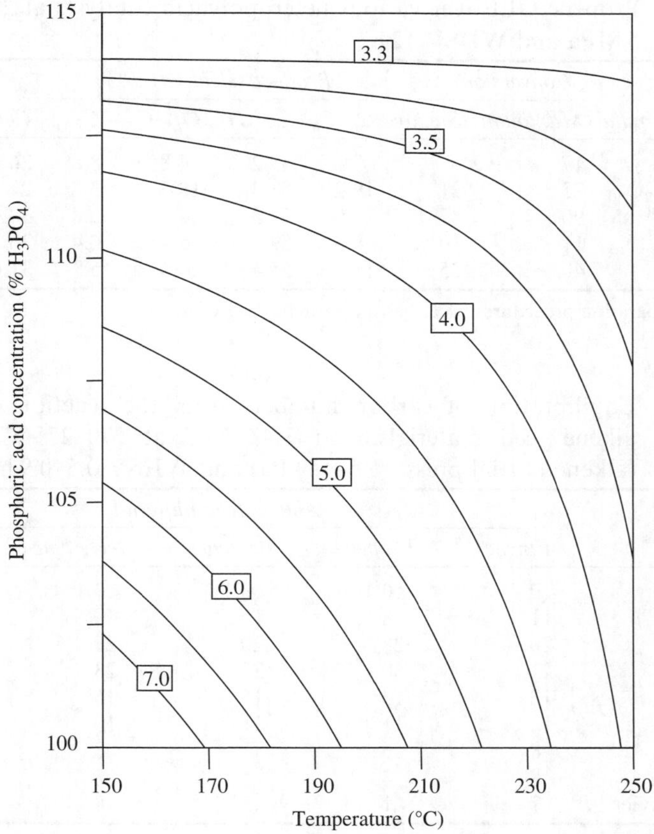

Figure 5.10 Effect of temperature and catalyst hydration on the ratio of naphtha (<174 °C) to distillate (>174 °C) in the product from butene oligomerisation over liquid phosphoric acid at 3.8 MPa.

catalysts for distillate production (Table 5.7).[96] Conversion decreased when the FTS-derived propene was not dried before use.

5.1.3.3 Zeolitic Silica–Alumina Catalysts

Although various types of silica–alumina-based zeolites have been studied for alkene OLI, the pentasil zeolite ZSM-5 (MFI) is the best known commercial OLI catalyst. As was shown by O'Connor and Kojima,[100] the performance of this catalyst can be influenced by the methods of preparation and pretreatment, and also by conditions applied during testing.

The ZSM-5 catalyst has a three-dimensional structure with sinusoidal pores (5.1 × 5.5 Å) and straight pores (5.4 × 5.6 Å). The shape-selective property of this catalyst, resulting from its pore size, ensures a low degree of branching.[101,102] By preventing the formation of bulky hydrocarbons and coke precursors, the

Table 5.7 Propene OLI over various heteropolyacid catalysts at 220–230 °C, 5 MPa and WHSV 12 h^{-1}.

Catalyst formula	*Conversion (%)*		*Product selectivity (%)*				
	Maximum	*Steady*	*C_6–C_8*	*C_9–C_{14}*	*C_{15}–C_{20}*	*C_{21+}*	*Cetane number*
$H_3PW_{12}O_{40}$	17	–	7.2	49.2	34.8	8.8	26
$(NH_4)_3PW_{12}O_{40}$	22	21	19.2	58.3	18.7	3.8	22
$AlPW_{12}O_{40}$[a]	90	87	25.9	57.3	14.1	2.8	38.4
$FePW_{12}O_{40}$	11	10	8.8	59.5	25.4	6.4	36
$H_4SiW_{12}O_{40}$	40	25	19	55.4	20.4	5.2	31.6

[a]Various preparation procedures used, best performance selected.

Table 5.8 Equilibration of carbon numbers from the reaction of various alkene feed materials over H-ZSM-5 at 270–275 °C, 0.1 MPa (alkene partial pressure 5–15 kPa) and WHSV 0.5–0.9 h^{-1}.

Product	*Alkene feed material*				
	Ethene[a]	*Propene*	*Pentenes*	*1-Hexene*	*1-Decene*
C_2	0	<0.1	<0.1	<0.1	<0.1
C_3	11	8	10	9	4
C_4	20	28	20	20	13
C_5	21	30	27	23	26
C_6	13	13	15	16	20
C_7	12	11	11	10	17
C_8	8	6	7	8	8
C_9	8	3	5	6	7
C_{10} and heavier	7	1	5	8	5

[a]Based only on converted ethene; product contained 47.5% ethene.

deactivation of ZSM-5 catalysts is significantly diminished. The Si:Al ratio in the ZSM-5 catalyst is another parameter that can be used to manipulate activity and selectivity.[101] Ultimately the properties of the ZSM-5 catalyst must be balanced with the operating conditions and product requirements.

The chemistry and catalysis of alkene OLI over H-ZSM-5 has been studied extensively, with the pioneering work of Garwood clearly showing its equilibration properties at high temperature (Table 5.8).[103] At low temperature, typically below 230 °C, H-ZSM-5 catalyses OLI with limited cracking, resulting in the formation of oligomers that are multiples of the monomer. At higher temperatures, the carbon number distribution of the product is equilibrated (Figure 5.4). In the temperature region where the feed is 'equilibrated', the process is insensitive to the carbon number distribution of the alkenes in the feed. The operating conditions (temperature and pressure) and product recycling can then be used to determine the product distribution.[104]

Over H-ZSM-5 catalysts, higher conversions can be achieved at higher temperatures and higher partial pressure of alkenes in the feed. Higher temperatures also favour parallel reactions, such as cracking, copolymerisation and disproportionation, but these reactions are not necessarily favoured by an increase in pressure. The response of OLI over ZSM-5 to changes in the operating conditions provides significant flexibility to the process, and the product distribution can be varied from mainly naphtha to mainly distillate production by adjusting the operating parameters (Table 5.9).[11] Distillate can be maximised using moderate temperatures (200–220 °C) at a total pressure of around 5 MPa, whereas naphtha can be maximised by increasing the temperature to 300 °C and lowering the total pressure to about 3 MPa.

Oxygenates are known to reduce catalyst activity,[21,105] but this does not preclude the use of H-ZSM-5 with FT feed material.

Fuel properties from commercial-scale operation of the OLI of alkenes on H-ZSM-5 using the MOGD and COD processes are shown in Table 5.10.[32,104,106–108]

Table 5.9 Product yield from the conversion of a C_3–C_6 feed (82% alkenes, 15% alkanes, 1.5% aromatics and 1.8% oxygenates) over H-ZSM-5 in naphtha (gasoline) mode and distillate mode.

	Product yield (mass%)	
Products	*Naphtha mode*	*Distillate mode*
C_1–C_3	4	1
C_4	5	2
C_5–165 °C naphtha	–	15
>165 °C distillate	–	82
C_5–200 °C naphtha	84	–
>200 °C distillate	7	–

Table 5.10 Fuel properties of products from OLI of C_3–C_4 alkenes from fluid catalytic cracking in the MOGD process and OLI of C_3–C_6 Fischer–Tropsch-derived material in the COD process.

Fuel property	*MOGD*	*COD*	
Motor gasoline (unhydrogenated)			
RON	92[a]	81–85	85
MON	79[a]	74–75	75
Density @ 15.6 °C ($kg\,m^{-3}$)	730	–	738
Alkene content (mass%)	94	–	94
Aromatic content (mass%)	2	–	2
Diesel fuel (hydrogenated)			
Cetane number	52–56	52–54	51
Density @ 15.6 °C ($kg\,m^{-3}$)	779	787[b]	801[b]
Viscosity @ 40 °C (mPa s)	2.5	2.55	–
Distillation, ASTM D86 (°C)			
IBP	166	198	229
T50	236	245	–
T90	342	320	323
FBP	–	358	361

[a]Better octane numbers than in the COD process due to isobutene-rich feed material.
[b]Density reported at 20 °C.

These properties compare well with those of fuels obtained by once-through operation due to the equilibrating nature of OLI over ZSM-5.[21] The distillate has a good cetane number and excellent cold flow properties have been reported for the kerosene fraction (freezing point – 60 °C).[109] It is clear that H-ZSM-5 is primarily suited for distillate production. The naphtha fraction from ZSM-5 OLI of Fischer–Tropsch materials has poor transportation fuel properties, especially considering that it is an olefinic motor gasoline component.

The quality of the motor gasoline obtained from alkene OLI over H-ZSM-5 increases with catalyst age. The olefinic RON obtained from 1-hexene OLI over H-ZSM-5 improved from 66 to 80 after 30 OLI reaction–regeneration cycles.[21] An analogous improvement was reported by Minnie,[106] who found that the RON and MON values of the olefinic motor gasoline improved from 68.0 to 80.0 and from 71.4 to 84.5, respectively. These values were obtained from the pilot plant-scale (0.7 kg of catalyst) OLI of a C_3–C_6 FTS feed on a commercial H-ZSM-5 catalyst at 210–253 °C, 5.6–5.7 MPa and space velocity 0.5 h^{-1}. As the catalyst becomes more deactivated, there are fewer strong acid sites available that are responsible for OLI and cracking, whereas double bond IS and skeletal IS are less affected. The C_5–C_6 alkenes in the feed are consequently isomerised, rather than oligomerised, thereby increasing the octane number of the product.

The OLI of light alkenes such as ethene, propene and isobutene was investigated using templated and non-templated ZSM-5 zeolites.[110] For the latter, the conversion of ethene was significantly greater than that over the templated ZSM-5, whereas the opposite order of activity was observed for propene and isobutene. For the non-templated zeolite, the Na-exchanged samples were more active than the H-form. The product distribution indicated that the oligomers underwent transformations such as alkylation, IS and cracking. The testing was conducted at 350 °C in a flow of nitrogen.

The OLI of propene over H-AlMFI zeolite using a sub-atmospheric pressure of the reactant (13 kPa absolute) at 200–550 °C showed increased conversion with temperature, reaching a maximum at 300 °C.[111] Butenes and hexenes were the major products. The data suggested that most of the products are dimers and cracked products from the trimers to yield pentenes and butenes. The overall conversion could be explained by primary dimerisation, followed by disproportionation of carbocations within the C_6–C_9 range. Hydrogen transfer reactions also occurred. At 550 °C the conversion was dominated by cracking of dimers, yielding an almost the equimolar mixture of C_2 and C_4 products.

Apart from the large body of literature dealing with H-ZSM-5, there are also studies dealing with OLI over other zeolites, including H-Y (FAU), H-Mordenite (MOR), H-A (LTA), H-Beta (BEA), H-Offerite (OFF) and H-Omega (MAZ).[112–117]

A modified form of ZSM-5 zeolite has been used as catalyst in the NESCO process, whereas the Exxon EMOGAS process employs ZSM-22 (TON) as catalyst.[118] The zeolitic type of catalyst used in the Shell process is characterized by a high flexibility because of its capability to convert ethene under conditions similar to those for higher alkenes. These processes are suitable for upgrading alkenes from FTS, although they were developed and demonstrated for the conversion of alkenes produced during the cracking of various hydrocarbon streams and gaseous by-products from FCC.

The zeolite ferrierite (FER) is well known as a selective catalyst for the skeletal IS of butene to isobutene, having a limited selectivity for OLI.[102] However, one of the suggested mechanisms for butene skeletal IS is by the formation of the C_8 dimer, which is then cracked to produce isobutene. The migration of C_8 alkenes is restricted by the shape selectivity of ferrierite. This disfavours the formation of branched alkenes heavier than C_8 and favours cracking reactions instead. Under operating conditions where the cracking rate is lower, one may employ ferrierite as a selective catalyst for butene dimerisation. In the case of pentene, skeletal IS is the dominant reaction over ferrierite and few heavier than C_5 products are formed. Similarly, Tiitta *et al.* reported that for hexene, more dimerised products were formed on Beta-zeolite and Y-zeolite than on ferrierite.[119] These results are understandable, because skeletal IS of C_5 and heavier alkenes can readily take place by monomolecular rearrangement, which is more difficult for butene.

Despite the strong acidity of some zeolite catalysts, they are not often used for OLI at temperatures below 200 °C. This is mainly due to the rapid deactivation of most zeolites under such conditions. During a comparative study of 1-hexene and 1-octene OLI that involved, amongst other, the zeolites ZSM-5, Y-zeolite and Omega, catalyst activity was almost completely lost after 10 h on-stream at temperatures in the range 100–200 °C.[13] It was found that the fairly rapid deactivation of all zeolite catalysts in this study was due to the formation of heavy oligomers that are difficult to remove from the catalyst surface. At higher temperatures, where cracking becomes significant, these heavy oligomers are cracked and thereby removed from the surface to rejuvenate the catalyst.

5.1.3.4 Amorphous Silica–Alumina Catalysts

The most obvious difference between amorphous silica–alumina (ASA) and zeolite catalysts used for OLI is the less pore-constrained geometry of ASA, which is by definition not crystalline. There are other differences also that allow ASA to yield a very different product from OLI. These include its apparent lower acid strength, high hydrogen transfer propensity[109] and a reaction mechanism that is somewhat different to the classic Whitmore-type carbocation mechanism. The latter is evidenced by its *cis*-selective nature for double bond IS and the differences in products obtained from the OLI of 1- and 2-butene (Table 5.11).[120] The blending RON of 1-butene-derived dimers approached that of isobutene-derived dimers and there is a clear analogy with SPA-catalysed OLI of butenes.

It has been found that ASA catalysts work well with Fischer–Tropsch feeds, including oxygenate-containing feeds.[121,122] The distillate thus produced has a higher density (810 kg m^{-3}; much needed in FT refining) than any of the other

Table 5.11 Dimerisation of butenes over amorphous silica–alumina (85% SiO_2, 15% Al_2O_3) at 120 °C, 3.5 MPa and WHSV 8 h^{-1}. Product quality is expressed in terms of its blending research octane number (BRON), which is also a measure of the degree of branching.

Feed material	*Conversion (%)*	*BRON*
1-Butene	85	138
2-Butenes	35	110
Isobutene	95	145

Table 5.12 Performance of different Fischer–Tropsch-derived feed materials during OLI over ASA at 180 °C and LHSV 0.5 h^{-1}.

	Feed material		
Description	*C_5–105 °C SLO*	*C_3–C_6 HTFT*	*C_7–C_{10}*[a]
Feed properties			
Oxygenate content (%)	1–4	<0.01	0.05
Alkene content (%)	85	85	92
Oligomerisation			
Alkene conversion (%)	72–74	97	–[b]
Distillate yield (%)	52–55	65–67	52–60
Unhydrogenated naphtha			
RON	74–76	92–94	78–82
Unsaturation (g Br per 100 g)	44–66	82–114	44–62
Hydrogenated distillate			
Cetane number	37	28–29	29–30
Density (kg m^{-3})	810	810–816	809–810
Viscosity (mPa s)	2.5–2.8	2.8–3.4	3.5–3.6

[a]Naphtha fraction from C_3–C_4 HTFT OLI over SPA.
[b]Not calculated because the feed and product carbon number ranges overlap.

OLI catalysts. The hydrogenated distillate also has good cold flow properties, but with a cetane number of only 28–30. The naphtha properties are feed dependent and short-chain alkenes yield a better quality motor gasoline (RON = 92–94, MON = 71–72) than H-ZSM-5. Similar catalyst cycle lengths and regenerability as H-ZSM-5 have been demonstrated in service as an alkene OLI catalyst, making ASA-based OLI technology as environmentally friendly, based on catalyst use, as H-ZSM-5-based technology.

Typical properties of naphtha and distillates obtained from the OLI of alkene-containing feed from FTS over ASA are shown in Table 5.12.[122] Once-through operation with C_7–C_{10} feed resulted in a distillate yield of 53–60%. When the motor gasoline fraction from the OLI of C_7–C_{10} feed was recycled, the distillate yield increased to 63–67%. The distillate yield on a fresh feed basis was insensitive to the recycle ratio used and it was similar to the 65–67%

distillate yield obtainable with C_3–C_6 feed under comparable conditions. To increase distillate production from C_3–C_6 olefinic feed, recycling of the gasoline fraction was considered, but only a moderate gain in distillate yield based on the fresh feed was observed.

During the OLI of C_7–C_{10} olefinic feed over ASA,[121] it was not possible to maintain a constant distillate yield over time. The distillate yield typically decreased by about 10% every 50–60 days. This was attributed to the formation of viscous products and the deposition of coke on the catalyst which gradually blocked the active sites or access to the active sites. Over time, the operating temperature had to be increased to maintain constant conversion. The increase in temperature in turn decreased the maximum distillate yield that is thermodynamically possible. Both of these effects contributed to the decrease in the distillate yield with time on-stream.

There is also a fair amount of interest in making use of the more structured silica–alumina catalysts, such as MCM-41, for alkene OLI.[123–128] These catalysts have large pores and are not geometrically constraining, but have not yet been applied industrially for this purpose.

Some natural clay materials can also be employed as silica–alumina catalysts and montmorillonite specifically has been investigated as a catalyst for 'polyalphaolefin' (PAO) synthetic lubricant production, an application that is well matched with FTS. Acid-activated montmorillonite exhibited better OLI activity than several zeolites when C_{12}–C_{18} alkenes were subjected to OLI over the clay catalysts at 150–180 °C.[129] The dimer:trimer ratio in the products decreased with increasing conversion and alkenes with an internal double bond were more reactive towards OLI than linear 1-alkenes. The aluminium nitrate-treated clay was particularly active for the OLI of C_{14} alkenes. Montmorillonite and Al^{3+}-, Zr^{4+}- and H^+-exchanged montmorillonite clays, when evacuated at high temperatures, exhibited a high activity for the OLI of 1-decene.[130] The following order in activity was established: montmorillonite-H > montmorillonite-Zr > montmorillonite-Al > montmorillonite-K10 > montmorillonite-Na. The activity order was in line with the change in acidity of the clays.

Montmorillonite has also been used as a catalyst in the Huls Octol process.[31] For chemical applications, the Octol B catalyst is employed, which is a nickel-promoted montmorillonite that yields a more linear product. For fuels applications, the Octol A catalyst that gives a more branched product is preferred. The addition of nickel to the Octol B catalyst introduces a different reaction mechanism, namely 1,2-insertion and β-hydride elimination, which implies that more than one mechanism is operative in parallel.

The use of nickel to modify the properties of silica–alumina has been extensively studied for the OLI of ethene,[131–135] propene[136,137] and butene.[138,139] In the context of FTS, it is worthwhile pointing out that these catalysts are sensitive to water and deactivation has been reported when NiO/SiO_2–Al_2O_3 catalysts adsorb as little as 0.5% moisture.[131] This is contrary to the behaviour of silica–alumina catalysts, which are activated by the addition of small amounts of water.[140–142]

5.1.3.5 Silico-aluminophosphate Catalysts

Silico-aluminophosphate (SAPO) catalysts are not often considered for OLI. One of the few studies is that by Vaughan *et al.*, who studied the OLI of propene using various formulations of SAPO-11 (Table 5.13).[143]

Among three unmodified SAPO-11 catalysts, the best activity was exhibited by the pelleted form, compared with the extruded and powdered forms. The results show that the incorporation of Ni, Co and Fe into SAPO-11 decreased the yield. In these cases, the metals were added by impregnation. This resulted in catalyst deactivation due to increased diffusional resistance. The performance of Mn-SAPO-11 was similar to that of SAPO-11. Both mild steaming and severe steaming had a dramatic effect on the yield increase, similar to reducing diffusional resistance. Silanising and acid washing had adverse effects on SAPO-11 activity.

When studying SAPO-5 (AFI), SAPO-11 (AEL) and SAPO-34 (CHA) for the skeletal IS of butene, Yang *et al.* observed that the porosity of catalysts influenced the extent of OLI of butene.[144] Thus, the OLI was important for a large pore SAPO-5 catalyst. The medium-pore SAPO-11 favoured double bond IS and skeletal IS, whereas the small-pore SAPO-34 restricted the OLI reactions and favoured the formation of small over more bulky isomers.

5.1.3.6 Sulfated Zirconia Catalysts

Sulfated zirconia (SO_4^{2-}/ZrO_2) and some other sulfated metal oxides are solid superacid catalysts and therefore possible replacements in processes employing liquid acid catalysts for the OLI of alkenes. Studies with sulfated zirconia were focused mainly butene dimerisation[145,146] and the OLI of hexene and heavier alkenes.[13,147,148]

Table 5.13 Propene OLI over various SAPO-11 (AEL)-based catalysts.

Catalyst	Maximum conversion (%)	Liquid yield (g_{cat}^{-1})[a]	Product distribution (mass%)		
			Dimer	Trimer	C_{12+}
SAPO-11 (powder)	78	702	56.3	22.3	21.2
SAPO-11 (extrudate)	84	416	57.5	28.9	13.6
SAPO-11 (pellet)	95	1368	65	27	8
SAPO-11 (mild steaming)	93	1674	53.7	27.7	18.6
SAPO-11 (severe steaming)	89	1762	70.3	19.5	10.1
SAPO-11 (silanised)	33	12	69.8	23	7
SAPO-11 (acid washed)	55	35	47.9	24.4	27.7
Ni–SAPO-11	60	16	56.2	30.4	13.4
Fe–SAPO-11	60	43	60.4	26.2	13.4
Co–SAPO-11	65	47	73.3	20.8	5.8
Mn–SAPO-11	59	680	54.9	20.9	24

[a]Liquid yield is defined as the product mass collected from the period of maximum conversion to half the catalyst lifetime.

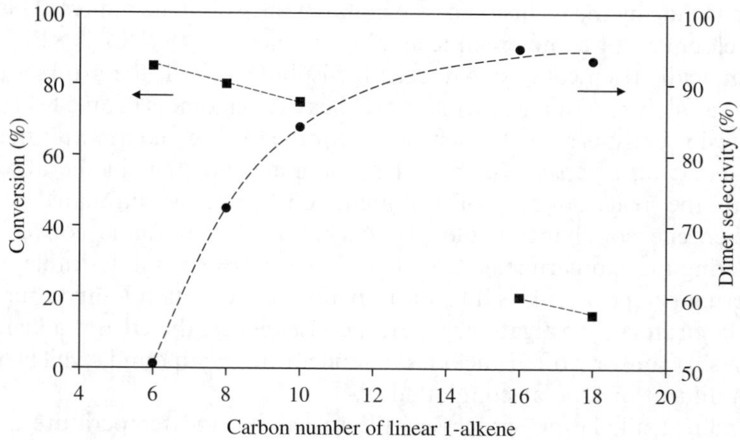

Figure 5.11 Conversion (■) and dimer selectivity (●) during the OLI of linear 1-alkenes over sulfated zirconia. Reactions were conducted in the slurry phase, with a 10% catalyst concentration for 8 h at 180 °C and autogenous pressure.

It was indicated earlier that sulfated zirconia (SO_4^{2-}/ZrO_2) exhibited better activity and stability than several zeolites during the OLI of 1-hexene and 1-octene at 100 °C.[13] Only dimers and trimers were produced over SO_4^{2-}/ZrO_2, compared with heavier products produced over zeolitic catalysts (Table 5.1). This was ascribed to accessibility, since fouling by heavy oligomers rapidly deactivated the catalysts when operating at such low temperatures. Likewise, mainly dimers and trimers were found when operating in the temperature range 120–180 °C,[147] but when operating at room temperature Keogh and Davis reported heavier OLI products from hexene.[148] The reported data also suggest that SO_4^{2-}/ZrO_2 is not a good catalyst for PAO production, since activity for alkene OLI and selectivity to heavier products decrease with increasing mass of the alkene (Figure 5.11).[147]

5.1.3.7 Acidic Resin Catalysts

Acidic resin catalysts became popular for the production of high-octane motor gasoline when it became clear that the use of methyl *tert*-butyl ether (MTBE) as an oxygenated fuel component would be banned in some regions.[27,149] MTBE is produced by the etherification of isobutene with methanol over an acidic resin catalyst and isobutene dimerisation is a side-reaction in this process. The selectivity for isobutene dimerisation can be increased by increasing the isobutene-to-methanol ratio in the feed and it is not difficult to see how an MTBE unit can be converted into an isobutene dimerisation unit. The use of acidic resin catalysts for isobutene dimerisation therefore had a twofold aim, namely to reuse the existing MTBE refining infrastructure (same catalyst, same operating conditions and partly the same feed) and to address the octane shortfall in motor gasoline by the inclusion of the high-octane isobutene dimers. This led to the development of commercial technologies, such as NExOCTANE.[28]

When acidic resin catalysts are used for isobutene OLI, the product usually contains a mixture of dimers, trimers and heavier oligomers (Table 5.14).[150–152] Acidic resin catalysts are extremely active for the OLI of alkenes with the C=C double bond on a tertiary carbon. Haag pointed out that in the absence of moisture the reaction rate of isobutene OLI over a sulfonated styrene–divinylbenzene copolymer could approach that of enzymatic conversion.[150] Considering the exothermic nature of OLI, this is clearly not desirable from an engineering perspective; it is difficult to maintain a constant temperature, even during laboratory investigations.[153] As a consequence, the OLI of alkenes over acidic resin catalysts often includes a moderating compound, which may be either a diluent or a polar compound.

In 'indirect alkylation' processes, the selectivity and temperature are controlled by making use of diluents or less reactive alkenes as moderating compounds.[26,154] For example, the reaction can be moderated by *n*-butenes, which are less reactive than isobutene, and also by butanes that are unreactive under the process conditions. Generally, isobutene in a mixed C_4 feed reacts selectively, with little *n*-butene conversion that takes place in parallel.[26,28] In order to codimerise *n*-butenes with isobutene to produce 3,4,4-trimethylpentenes, which are also high-octane motor gasoline components, a higher reaction temperature is required. Higher temperature processes, such as the Bayer process,[154] have the added advantage that a complete C_4 cut can be processed, albeit at reduced selectivity to the dimer.

Polar compounds interact strongly with acidic resin catalysts due to the polar nature of the sulfonic acid groups. The selectivity and temperature can therefore also be controlled through the judicious addition of a polar solvent, such as alcohols,[151,153,155,156] or water.[157] The advantage of using polar compounds to moderate OLI is that they change the acid strength of the resin catalyst by solvating the acid groups.[158]

Although oxygenates are capable of many acid-catalysed side-reactions, it was reported that typical Fischer–Tropsch oxygenate classes mainly inhibited reaction over acidic resin catalysts, which is desirable for OLI. Only a few side-reactions were noted on Amberlyst 15 at 70 °C and 0.4 MPa.[159] The main drawback of using an acidic resin-catalysed process in a Fischer–Tropsch refinery is the lack of isobutene. The application of acidic resin catalysts for alkylate-type production in a Fischer–Tropsch context has been evaluated previously.[44]

5.1.3.8 Homogeneous Catalysts

Alkene OLI by the Dimersol process from IFP/Axens is currently one of very few refinery technologies where homogeneous organometallic catalysis is applied industrially.[8] The OLI reaction is catalysed by a nickel-based Ziegler-type catalyst and proceeds by 1,2-insertion and β-hydride elimination

Table 5.14 Isobutene OLI over acidic resin catalysts in the absence of polar compounds to moderate the reaction.

	Operating conditions				*Product distribution (mass%)*[a]		
Catalyst	*Temperature (°C)*	*Pressure (MPa)*	*Space velocity (h^{-1})*	*Conversion (%)*	*Dimer*	*Trimer*	*Tetramer*[b]
Amberlyst 15 (dry)	60	1	3600	58	52	40	8
Amberlyst 15 (dry)	16	1	180	89	33	57	10
Commercial resin[c,d]	80	1.5	Batch	97	24	71	5
Dow XUS-40036.01	85	Near atm	1.9	72.1	40.7	53.7	5.6
Dow XUS-40036.01	105	Near atm	1.9	64.5	42.4	50.6	7.9
Nafion-H	85	Near atm	1.9	54	29.4	62.5	8.1
Nafion-H	105	Near atm	1.9	52	46.9	50	3.1

[a]Products may include some cracking products, dimer = C_5–C_8, trimer = C_9–C_{12} and so forth.
[b]May include some heavier products.
[c]Sulfonated styrene–divinylbenzene resin.
[d]Feed is isobutene (47 mass%) in alkane mixture.

(Figure 5.1). There are a number of variants of the Dimersol process, each tailored to a specific feed and product combination:[160]

1. Dimersol E for the OLI of ethene and C_2–C_3 mixtures in FCC off-gas to produce olefinic motor gasoline. This type of technology was applied industrially at the HTFT refinery of Sasol in Secunda, South Africa, in order to convert excess ethene to motor gasoline. It was originally installed as a risk-mitigation option to avoid flaring of ethene, but this unit is no longer in use.[2]
2. Dimersol G for the OLI of propene and C_3–C_4 alkene mixtures to an olefinic motor gasoline component.[161,162]
3. Dimersol X for butene dimerisation to produce octenes with a low degree of branching for the manufacture of plasticiser alcohols.[163,164]

Because the technology makes use of a homogeneous organometallic catalyst system, it is sensitive to any impurities that will complex with the nickel. Among others, it is sensitive to dienes, alkynes, water and sulfur, which should not exceed 5–10 $\mu g\, g^{-1}$ in the feed.[30,163] It is possible to compensate for deactivation by feed impurities by increasing the catalyst dosing, but this will increase the catalyst cost.

After OLI, the catalyst must be removed from the reaction product by a caustic wash. In a more recent version of this technology, called Difasol, the catalyst is contained in an ionic liquid phase, which makes catalyst separation easier.[30,165] The Difasol process generates less caustic effluent than the Dimersol process. In a lifetime test conducted over a period of 5500 h, it was found that the nickel catalyst consumption in the Difasol process was only 10% of that in the Dimersol process, and the co-catalyst consumption was half.[30] The Difasol technology has been piloted successfully with Fischer–Tropsch alkenes.

Nickel is not the only metal active for alkene OLI. There is a significant body of literature on organometallic alkene OLI. One specific application that has attracted much interest in relation to FTS is the selective trimerisation and tetramerisation of ethene over chromium-based catalysts.[166,167]

The older literature abounds with accounts of alkene OLI over liquid acid catalysts, such as sulfuric acid[168] and phosphoric acid.[49,169] Although sulfuric acid is still used for aliphatic alkylation of isobutane with alkenes in many conventional crude oil refineries, it is not a mainstream OLI catalyst. Historically, liquid phosphoric acid was used industrially for OLI of FTS alkenes,[5] but in this role it has since been replaced with solid phosphoric acid (Section 5.1.3.1). Another liquid-phase catalyst that has been investigated for the OLI of alkenes from FTS is boron trifluoride (BF_3). Linear 1-alkenes was oligomerised over BF_3 to produce PAO lubricants.[170]

5.1.3.9 Other Catalyst Types

Nickel-based heterogeneous silica–alumina and homogeneous catalysts (Sections 5.1.3.4 and 5.1.3.8) have already been discussed. In addition to these catalyst classes, Ni has been added to many other catalysts and supports in order to improve OLI of alkenes. Among others, OLI investigations have been reported on the use of different Ni salts on various silica and alumina supports,[171] Ni/Al_2O_3 and its phosphorus-promoted analogues,[172] $NiSO_4/TiO_2$–ZrO_2[173] and $Ni/SO_4^{2-}/Al_2O_3$.[174]

Cai *et al.* reported that $NiSO_4/Al_2O_3$ was active for OLI of propene; a propene conversion of 98% with 55–88% dimer selectivity could be obtained at 30 °C, 2.5 MPa and LHSV 2 h^{-1}.[175] In subsequent work on the dimerisation of ethene over $NiSO_4/Al_2O_3$, it was reported that the nickel loading and temperature of calcination influenced the catalyst activity.[176] Maximum activity was achieved at an Ni loading of 5–10% and calcining temperatures of 500–600 °C. The experiments were conducted in a closed circulating system and the optimal operating temperature for OLI was 50 °C. It was observed that the catalyst could be poisoned by the addition of a base (NaOH), indicating that the acid sites were necessary for OLI and that reaction did not proceed purely by Ni catalysis. The catalyst could also be deactivated by CO, but the activity could be restored simply by evacuation of the catalyst. Poisoning by CO indicated that Ni also played a role in the OLI catalysis. Cai also reported on other combinations of $NiSO_4$ as catalysts and concluded that OLI proceeded both over the acid and metal functions of these catalysts.[177]

The use of aluminium chloride ($AlCl_3$) as catalyst for alkene OLI is of historical importance in a Fischer–Tropsch context. Linear 1-alkenes produced from the cracking of alkanes derived from FTS have been oligomerised on an industrial scale in the slurry phase over $AlCl_3$ to produce synthetic lubricants.[5,178]

Ionic liquids have been recognised as a useful medium for alkene OLI. Various systems have been investigated, some bearing a resemblance to $AlCl_3$. For example, Yang *et al.* conducted the OLI of isobutene over an ionic liquid catalyst containing $FeCl_4^-$ and $Fe_2Cl_7^-$.[179] The study was carried out in an autoclave. The conversion of isobutene was above 83 mass% and selectivity to the dimer and trimer was better than 75%. Such conversions were observed for a mole ratio of $FeCl_3$ to $[(C_2H_5)_3NH]Cl$ in the range 1.2:1–2:1. The addition of CuCl to the ionic liquid increased the conversion and selectivity to dimers and trimers. The selectivity reached 90% at a mole ratio of CuCl to $[(C_2H_5)_3NH]Cl^- \cdot 1.5FeCl_3$ of 0.25:1. The reaction pathway of isobutene OLI catalysed by iron(III) chloride ionic liquids was explained in terms of the Whitmore carbocation mechanism.

5.1.4 Comparison of Commercial Oligomerisation Catalysts

The fuel properties from OLI of Fischer–Tropsch-derived feed materials over commercial OLI catalysts are compared in Table 5.15.[21,43,44,121] It is evident that the values are dependent on the OLI catalyst, feed, operating conditions and whether the product was hydrogenated or not. The differences between products from these catalysts can be related to two molecular properties: degree of branching and degree of cyclisation.

Table 5.15 Comparison of selected product properties obtained from OLI of Fischer–Tropsch alkenes over solid phosphoric acid (SPA), amorphous silica–alumina (ASA) and H-ZSM-5 zeolite catalysts.

	SPA			*ASA*	*H-ZSM-5*
Property	*C_3 feed*	*C_4 feed*	*C_3–C_5 feed*	*C_3–C_6 feed*	*C_3–C_6 feed*
Naphtha:distillate[a]	75:25	85:15	80:20	35:65	35:65
Olefinic motor gasoline					
RON	95–97	95–97	95–97	92–94	81–85
MON	81–82	81–82	81–82	71–72	74–75
Hydrogenated motor gasoline					
RON	<50	86–88	64–80[b]	~75	–
MON	–	86–88	70–80[b]	–	–
Diesel fuel					
Cetane number	–	–	<35	29–30	52–54
Density ($kg\,m^{-3}$)	–	–	750–760	~810	~790
Viscosity at 40 °C (mPa s)	–	–	1.0–1.2	~2.8	~2.6
T10 (°C)	–	–	160–180	180–190	200
T90 (°C)	–	–	190–200	330–350	320

[a]Naphtha (<177 °C) to distillate (>177 °C) ratio from once-through conversion; the distillate yield can be increased by recycling naphtha.
[b]Very dependent on the feed composition and operating conditions.

The octane number of motor gasoline increases with increasing degree of branching within a specific boiling range. Conversely, within a specific boiling range the highest cetane number of diesel fuel is obtained with linear alkanes. Branching decreases the cetane number, but improves the cold flow properties of diesel fuel. The pore-constraining geometry of the ZSM-5 zeolite limits the degree of branching during OLI (Table 5.2). The products from H-ZSM-5 OLI are therefore less branched than those obtained over SPA and ASA catalysts. As a consequence, the lowest RON value of motor gasoline and the highest cetane number of diesel fuel were obtained over the H-ZSM-5 catalyst.

The degree of cyclisation affects the density of the fuel. Density generally also increases with increasing molecular mass. In order not to skew the comparison by changes in carbon number distribution, similar boiling fractions must be considered. When the OLI products in the same boiling fraction from SPA and ZSM-5 were compared, the density and viscosity of these products were similar.[21] Despite the difference in degree of branching, OLI products from SPA and ZSM-5 are mainly acyclic aliphatic hydrocarbons. The low viscosity of SPA distillate noted in Table 5.15 is a consequence of its limited product carbon number distribution. The OLI distillate from ASA has a higher density due to the higher amount of cyclic material compared with that from SPA or ZSM-5. For similar boiling range products, the viscosities of ASA- and ZSM-5-derived OLI are comparable, indicating that viscosity is not significantly affected by cyclisation.

The selection of an OLI catalyst to convert alkenes from FTS is determined by the product properties that are desired. Although the comparison was restricted to SPA, ASA and ZSM-5, niche applications can be indicated for some of the other catalyst types. For example, it has been reported that homogeneous nickel-based catalysts were employed commercially for ethene OLI from FTS and acidic resin catalysts may be considered if the alkene feed is rich in isobutene.

5.1.5 Radical Oligomerisation

Radical OLI does not require a catalyst. Free radicals can be generated either by thermal treatment of hydrocarbons or by various free radical initiators. In the former case, the relative strengths of the C–C and C–H bonds dictate that the radical generation will be governed by the rupture of C–C bonds, which have a lower bond dissociation energy. It has been generally observed that the cleavage of the C–C bonds begins at about 300 °C. Therefore, thermally initiated radical OLI can be attempted only above 300 °C, unless a free radical initiator is employed.

The predominance of linear 1-alkenes in primary products from FTS suggests that these materials may be suitable feeds for radical OLI. This option was explored by de Klerk,[180] who used three alkene-rich feed materials from FTS, namely a C_4, a C_5/C_6 and a C_7/C_{14} fraction. The experiments were carried out in a continuous fixed bed reactor at 300–400 °C and 1–18 MPa. It was observed that pressure was the most important parameter, as indicated in Figure 5.12. The effect of pressure was particularly evident in the case of the C_4 fraction. Oxygenates which were present in every feed fraction were readily converted, hence their adverse effect on OLI was not evident.

Radical-initiated polymerizations have been applied commercially in the polymer industry. Radical initiators usually possess one weak bond (peroxide O–O or disulfide S–S) which can be readily cleaved, either thermally or

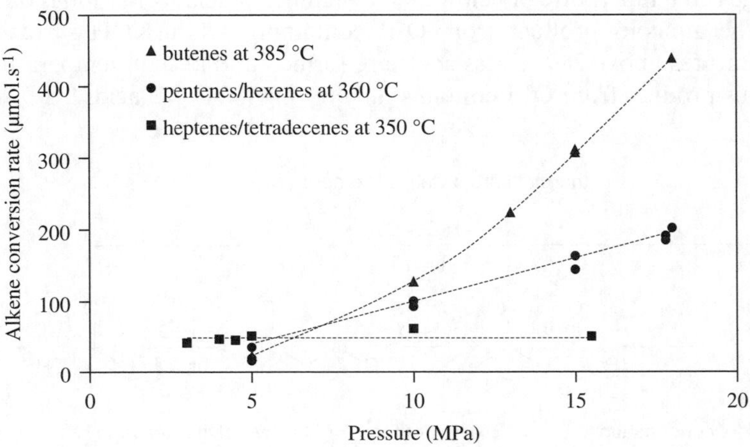

Figure 5.12 Effect of pressure on the rate of radical OLI of linear 1-alkene-rich Fischer–Tropsch C_4 (▲), C_5/C_6 (●) and C_7/C_{14} (■) fractions.

photolytically. The same approach can be applied to OLI, provided that the reaction can be terminated at the right stage to obtain the product of interest. The advantage of using a radical initiator is that radical OLI can be conducted at lower temperatures and pressures than required for thermal OLI without a radical initiator. Di-*tert*-butyl peroxide (DTBP), one of the best known radical initiators, was used by Cowley to study the OLI of 1-octene and 1-pentene derived from FTS.[181] The study was conducted in the temperature range 100–200 °C and at 1–2 MPa. These conditions coincide with the activation temperature range of the DTBP. The products of radical OLI of alkenes aided by a radical initiator were less branched than can be obtained by catalytic OLI. Such alkenes with a very low degree of branching find application as plasticiser alcohols, detergent alcohols, PAO lubricants and high cetane number distillates.

5.1.6 Carboxylic Acid Formation Over Acid Catalysts

From a transportation fuel point of view, one of the most important side-reactions that can occur during the OLI of material from FTS is the acid-catalysed conversion of carbonyl compounds to carboxylic acids. Thus, although the feed may contain no carboxylic acids, the ketones in the feed can be converted into carboxylic acids. The chemistry of this reaction is shown in Figure 5.13.

This conversion has been observed at OLI conditions on SPA,[46,66,182] ASA,[122,183] H-ZSM-5[21,105,106,184] and various other acid catalysts.[185,186] As the reaction temperature is increased, the conversion of carbonyls by other reaction pathways than to carboxylic acids reduces overall carboxylic acid formation, while conversion of the carboxylic acids to other products is increased.[187]

Two distinct operating regimes are found during industrial OLI with oxygenate-containing Fischer–Tropsch feed over H-ZSM-5.[106] At a weighted average bed temperature of below 280 °C significant acid formation is observed, with the aqueous product from OLI containing 1.1 mg KOH g^{-1} acids. At temperatures above 280 °C less acids are formed and at high temperatures the aqueous product from OLI contains only 0.1 mg KOH g^{-1} acids.

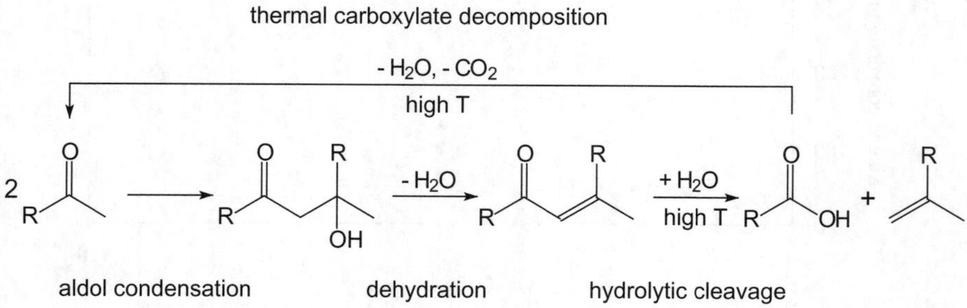

Figure 5.13 Acid-catalysed interconversion of carbonyl compounds and carboxylic acids.

Therefore, even at high conversion temperatures it is possible that the OLI product from FTS-derived feed may contain carboxylic acids. This has some implications for the post-processing and/or utilisation of the products from the OLI of oxygenate-containing feed materials. When OLI of ketone-containing material from FTS is considered, the material of construction of the OLI unit and units downstream from OLI must make provision for the change in corrosion behaviour due to the presence of short-chain carboxylic acids in the product.

5.1.7 Catalyst Deactivation During Oligomerisation

The causes of catalyst deactivation during OLI are similar to those of other catalytic reactions occurring during the upgrading of primary FTS products and conventional crude oil.[188] Some specific deactivation problems are due to the oxygenates present in Fisher–Tropsch syncrude. Due to the polar nature of many oxygenate classes, they interact strongly with polar catalytic surfaces and it is expected that the nature of this interaction will be different from that of the less polar hydrocarbons. Another unavoidable source of deactivation over acid catalysts is the formation of heavy oligomers, carbonaceous deposits and coke on the catalyst surface. In addition to these two deactivation mechanisms, which will be discussed in some detail, there are of course many others. For example, basic compounds that neutralise the acidic sites of OLI catalysts will cause catalyst deactivation.

5.1.7.1 Oxygenate-related Deactivation

One of the important differences between Fischer–Tropsch syncrude and conventional crude oil is the high content of oxygenates present in the former. The light naphtha cuts from HTFT, such as the condensate from cryogenic separation and the C_5–C_6 cut from the stabilised light oil (SLO), contain ketones as the main oxygenate class, with little alcohols and esters and no detectable carboxylic acids. In heavier naphtha and distillate cuts (C_7 and heavier), other oxygenate classes, such as alcohols and esters, also become significant. This difference in oxygenate distribution is not caused by the FTS process, but by the way in which the material is fractionated and stabilised (see Section 4.2.1).

Oxygenates inhibit alkene OLI by preferential adsorption on the active sites. The oxygenates may also lead to side-reactions that result in inhibition or deactivation. The effect of water on catalysts such as SPA, acidic resin and sulfated zirconia is pronounced and water may easily form by various acid-catalysed oxygenate reactions.[66,159] Carbonyl compounds have an especially rich chemistry. Acid-catalysed aldol condensation and aromatisation of carbonyl compounds can lead to the formation of carbonaceous deposits that may cause catalyst deactivation. Oxygenate-related deactivation of OLI catalysts is therefore catalyst dependent, as will be illustrated by SPA and ASA as examples.

Oxygenates in the feed to SPA catalysts result in inhibition of OLI and at high levels they may undermine the mechanical strength of the SPA catalyst, leading to catalyst disintegration. The extent of inhibition is determined by the nature of the oxygenates. For example, ketones have the least effect, with only mild inhibition of alkene OLI reactions, whereas alcohols, ethers and esters may cause significant inhibition of alkene OLI.[66] The inhibiting effects of oxygenates were confirmed by comparing butene OLI of oxygenate-free and oxygenate-rich feeds.[46] The conversion of butenes in the latter feed was always lower. Moreover, for the oxygenate-rich feed, the yield of oligomers was lower due to increased hydration of SPA by reaction water. With increasing hydration, more orthophosphoric acid is formed, which is a weaker acid than pyrophosphoric acid. In spite of high catalyst hydration, butene was still reactive for OLI, but the acid strength was too weak for an effective interaction with the heavier alkenes.

The OLI of Fischer–Tropsch alkenes over ASA was affected differently by the presence of oxygenates. When OLI of an oxygenate-free feed is compared with that of a feed containing 1–4% oxygenates, similar products could be obtained, but the rate of catalyst deactivation was higher with oxygenate-containing feed.[122] The adverse effect of oxygenates during alkene OLI over ASA was attributed to a higher rate of coke deposition as result of carbonyls in the feed being converted to aromatics. This reduced the operation cycle of ASA catalysts. However, some beneficial effects were also observed, such as an increase in reaction rate due to reaction water from oxygenate conversion.

5.1.7.2 Deactivation by Carbonaceous Deposits

The design of the MOGD process for the OLI of light alkenes to liquid fuels over the ZSM-5 zeolite incorporates three fixed bed reactors in series. In this case, the first reactor contains the most deactivated catalyst and the third reactor the least deactivated catalyst.[119] A fourth reactor that is off-stream at any particular time undergoes oxidative regeneration before being reconnected to the system to replace the reactor with the most deactivated catalyst. This reactor arrangement illustrates the effect of catalyst deactivation by coke deposition during OLI on process design. Other causes of catalyst activity loss cannot be ruled out,[189] but such activity loss is generally not recovered by oxidative regeneration. During operation, the rate of catalyst deactivation may also be decreased by increasing the dissolving capacity of the fluid,[190,191] although this is not always practical.

After the first oxidative regeneration of silica–alumina catalysts, an increase in activity has been reported for ASA catalysts,[121] and also H-ZSM-5 catalysts.[192,193] This has been ascribed to hydrodealumination and the formation of highly active extra-framework alumina species.[194–196] The hydration of ASA catalysts is also known to affect catalyst activity,[197] suggesting that the formation of water vapour during regeneration might also be responsible for the increased initial activity observed after regeneration.

For the OLI of alkenes at high temperature, ZSM-5 zeolite is usually the catalyst of choice, because deactivation by coke deposits is limited compared with most other zeolite catalysts. The formation of polynuclear aromatic coke on ZSM-5 zeolite is inhibited by its small pore structure.[198,199]

General reaction schemes for coke formation on acidic catalysts during the conversion of alkenes have been proposed.[200,201] The deactivation pathway depends somewhat on the alkene. For example, during the OLI of 1-hexene over ZSM-5, the coke formation begins with the dehydrogenation of the reactant to give cyclopentadienes and their dimers, which further undergo gradual conversion to indanes, indenes, tetralins and naphthalenes. The naphthalenes are ultimately converted to tricyclic aromatic structures, which are the main constituents of coke. On the other hand, during the OLI of ethene over USY zeolite, the coke formation begins with the production of the dimer carbocation ($C_4H_9^+$), which subsequently reacts to give *n*-butenes in parallel with hydrogen transfer/cracking reactions to give methane and propene.[202] Another portion of the carbocations is converted to higher molecular weight species that are unable to desorb from the catalyst. With time on-stream, the aromaticity of the heavy adsorbed species increases and the coke becomes more refractory.

The temperature at which the carbonaceous deposits are formed is very important. Catalysts deactivated by carbonaceous deposits during OLI at low temperature could readily be regenerated by controlled oxidation.[13] The deposits in this study were not analysed, but the ease of regeneration suggested that the carbonaceous deposits were not refractory and mostly aliphatic in nature. Heavy oligomers were detected in the product and low temperature and high pressure favoured OLI rather than hydrogen transfer and aromatisation.

No aromatic compounds were reported in the product after 1-hexene OLI over H-ZSM-5 at 290 °C and 5 MPa.[101] Even at lower pressure and 300 °C, the primary reactions were found to be IS, OLI and cracking.[203] The residue retained by H-ZSM-5 during the reaction of 1-hexene at atmospheric pressure and 320 °C contained some naphthalenes and polycyclic aromatics,[201] indicating that with increasing temperature the carbonaceous products change from more aliphatic to more aromatic in character. The nature of the catalyst determines the temperature threshold where this transition occurs. In a study with 1-hexene over USY zeolite, heavy aromatics started to form at 180 °C.[204] The formation of aromatics over USY zeolite is to be expected, because USY has a much higher hydrogen transfer activity than OLI activity compared with ASA and H-ZSM-5.[109]

A high alkane selectivity is often indicative of hydrogen transfer and possibly the formation of aromatic coke. However, this is not always the case. For example, Pater *et al.* reported high hexane selectivity during the OLI of 1-hexene at 200 °C and 5 MPa, but could not identify hydrogen-deficient molecules in the product.[205] This was likely due to hydrogen transfer from the diluent, *n*-hexane, which is not inert. A 5% conversion of *n*-hexane was reported over H-ZSM-5 at 270 °C, 4.8 MPa and LHSV 1 h^{-1}.[13]

5.2 Isomerisation and Hydroisomerisation

Linear hydrocarbons dominate the product spectrum from FTS. Isomerisation (IS) and hydroisomerisation (HIS) and are among the most important reactions for adjusting the properties of *n*-alkanes and *n*-alkenes, without changing their chain length. In fact, both IS and HIS are important in the production of motor gasoline, jet fuel and diesel fuels from Fischer–Tropsch syncrude.

The isomerisation studies conducted in the presence of excess H_2 over catalysts with metal sites for HYD/deHYD are referred to as HIS and the products are by implication branched alkanes. Isomerisation of alkenes in the absence of H_2, or with little H_2 only to limit catalyst deactivation, is referred to as IS and the products are branched alkenes. In the case alkene IS, double bond IS may take place as part of the overall mechanism, but the subsequent discussion will focus on skeletal IS.

The carbon chain length of the feed determines the type of IS/HIS catalyst and the associated operating conditions of the IS/HIS process. The acid strength needed for HIS is less than that for HCR, but the reaction intermediate is the same. Thus, HIS of light alkanes (butane, pentane and hexane) can be conducted over catalysts with strong acidic sites, whereas that of longer chain alkanes out of necessity has to be conducted over weaker acid sites to reduce HCR in competition with HIS. It is convenient to define three classes of IS/HIS:

1. *C_4 hydrocarbon isomerisation.* The HIS of *n*-butane to methylpropane (isobutane) is performed to provide feed for aliphatic alkylation units. Aliphatic alkylation is an important source of high-octane paraffinic motor gasoline. Likewise, the IS of *n*-butene to methylpropene (isobutene) is important as feed for indirect alkylation and etherification to produce high-octane motor gasoline. Mechanistically, the isomerisation of C_4 has to be considered as a separate class. In the classic sense, a C_4 carbocation intermediate cannot rearrange skeletally *via* a monomolecular pathway to a branched C_4 without forming a primary carbocation. This has resulted in considerable debate in the literature on the mechanism of butene isomerisation.[206–213]
2. *C_5–C_6 hydrocarbon isomerisation.* The light straight run (LSR) naphthas from FTS and conventional crude oil are both rich in *n*-alkanes and can be hydroisomerised to increase the octane number. The product from HIS of C_5–C_6 naphtha is often referred to as 'isomerate' in the refining industry and it is a motor gasoline blending component. The IS of C_5–C_6 alkenes is less common and is generally found in conjunction with etherification to produce high-octane fuel ethers for use in motor gasoline. Mechanistically, C_5 and C_6 carbocation intermediates can readily rearrange *via* a monomolecular pathway to produce branched products. At the same time, the carbocation intermediates have a low tendency to crack, since this would involve the formation of a secondary carbocation intermediate.[17]

3. *C_7 and heavier hydrocarbon isomerisation.* The ability to hydroisomerise *n*-heptane is of specific interest to all fuel refiners, since it would provide an efficient refining pathway for upgrading this difficult to refine molecule. It is highlighted separately, since C_7 is the lowest carbon number aliphatic material with a significant cracking propensity.[17] As such, it illustrates the tradeoffs involved in HIS of C_7 and heavier hydrocarbons when HCR becomes a meaningful side-reaction. In general, the HIS of longer chain *n*-alkanes is important to improve cold flow properties (Figure 5.14). This may include HIS of kerosene-range material to make it suitable for jet fuel, HIS of distillate for diesel fuel and HIS of heavier material for lubricating oil production. Catalytic dewaxing is a special class of heavier alkane HIS and will be discussed in more detail in Chapter 6.

To various extents, acid-catalysed cracking/HCR reactions occur in parallel with IS/HIS or as consecutive reactions after IS/HIS. Cracking and HCR reactions require a catalyst with a higher acid strength than is required for IS/HIS. The acidity in IS/HIS catalysts must be regulated to prevent excessive formation of unwanted gaseous by-products and coke. At the same time, the acid strength needed for isomerisation depends on the chain length. Thus, the isomerisation of C_4 hydrocarbons requires very strong acidic sites, but the acid strength required for isomerisation of longer chain hydrocarbons is lower. A bifunctional catalyst that exhibits good activity for HIS requires optimisation of the acid sites. For example, an acidic catalyst with predominantly medium-strength and weak acid sites may exhibit a high activity for HIS, whereas its activity for HCR may be rather low. It is therefore not surprising to find that different catalysts have very different HIS performances (Table 5.16).[214]

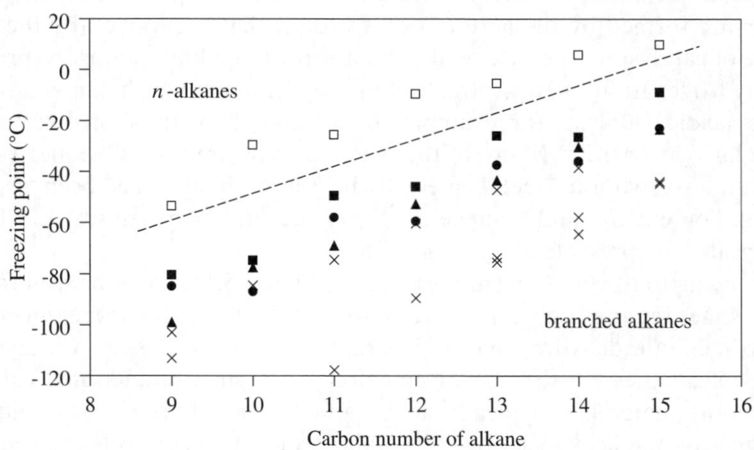

Figure 5.14 Freezing points of linear (□), 2-methyl (■), 3-methyl (●), 4-methyl (▲) and dimethyl (×) branched C_9–C_{15} alkanes.

Table 5.16 Product distribution from HIS of *n*-octane over bifunctional Pt-promoted acidic catalysts at 6.9 MPa and H_2:*n*-octane feed ratio of 16:1. All selectivity data reported at 30% conversion.

Property	*HY*	*ZSM-5*	*SAPO-5*	*SAPO-11*	*ASA*
Temperature (°C)	257	260	304	331	371
Selectivity (mass%)					
Total branched octanes	96.8	56.6	49.3	94.8	96.4
Dibranched octanes	12.0	1.8	9.0	2.3	8.5
Product ratios					
2-Methylheptane:3-methylheptane	0.71	1.54	0.46	1.07	0.67
(Propene + pentenes):3-methylheptane	0.64	2.10	0.86	1.00	0.95

However, it is also clear that the acid strength is not the only factor that determines the catalyst selectivity. This will be discussed in the next section on the mechanism.

5.2.1 Mechanism of Isomerisation

There is a significant volume of information on various aspects of the mechanism of IS/HIS available in the literature. Some of the studies originated more than 50 years ago.[214–222] Essential information can also be found in several textbooks. Because the focus of this review is on upgrading of the primary products from FTS, only a cursory account of the mechanism, as it applies to IS/HIS, is given. Whenever appropriate, reaction networks are included and discussed as part of the interpretation of experimental observations.

On acid catalysts, the mechanisms of skeletal IS/HIS of alkanes and alkenes involve the formation of a carbocation as the initial step. For alkenes, carbocations are formed by the addition of a proton that is supplied by the acidic surface of catalyst. In the case of alkanes, the proton addition must be preceded by dehydrogenation or by abstraction of a hydride ion which can be accepted by the acidic catalyst, for example, by combining with a proton to yield molecular hydrogen.[223] Many of the reaction steps are reversible and even on acid catalysts without metal sites, hydrogen exchange has been reported between butene-d_8 and butene-d_0.[224] A general mechanism of hydro-isomerisation is presented in Figure 5.15

According to the mechanism presented in Figure 5.15, it can be seen that the IS of alkenes requires only an acid catalyst, but the IS of alkanes requires metal sites to facilitate dehydrogenation in addition to the acid sites. Catalysts for IS/HIS of alkanes are therefore bifunctional. On such bifunctional catalysts, the support material is typically acidic and the metal sites are provided by impregnating the acidic support with an appropriate metal, often platinum.

The carbon chain length affects the way in which the carbocation intermediate isomerises on the catalyst surface. In this respect, the mechanism involving C_4 hydrocarbons differs from that for C_5 and heavier hydrocarbons,

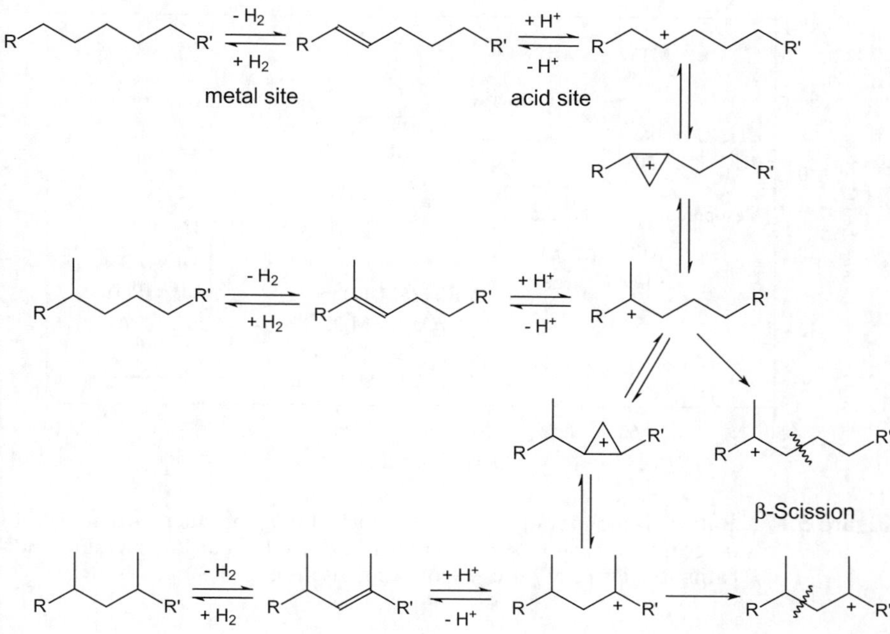

Figure 5.15 Mechanism of hydroisomerisation.

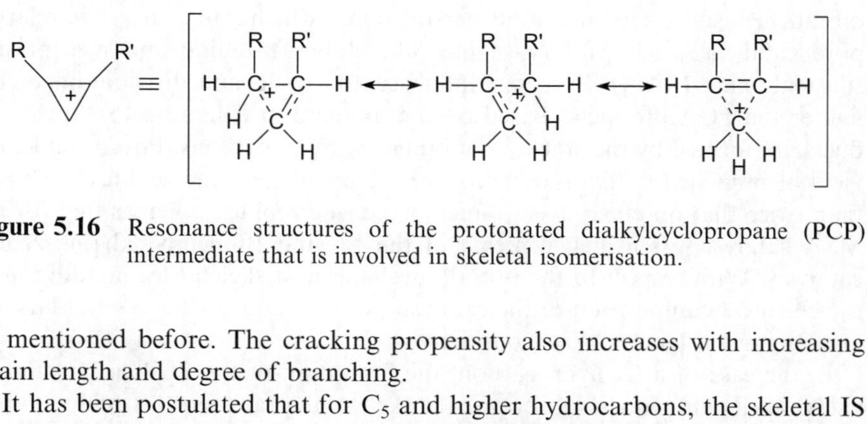

Figure 5.16 Resonance structures of the protonated dialkylcyclopropane (PCP) intermediate that is involved in skeletal isomerisation.

as mentioned before. The cracking propensity also increases with increasing chain length and degree of branching.

It has been postulated that for C_5 and higher hydrocarbons, the skeletal IS involves the rearrangement of a classical secondary carbocation into a protonated dialkylcyclopropane (PCP).[224–229] Evidence for the involvement of a PCP-based rearrangement was provided by carbon isotope experiments.[224,226] As was pointed out by Sie,[228,229] such a transformation has a low energy barrier, because of the existence of several resonance structures of the PCP species, shown in Figure 5.16. The resonance stability is sufficient to compensate for the strain of the three-membered ring structures.

Information about the mechanism of skeletal IS and the role of the catalyst structure can be deduced from the reaction products. In Table 5.16, it was

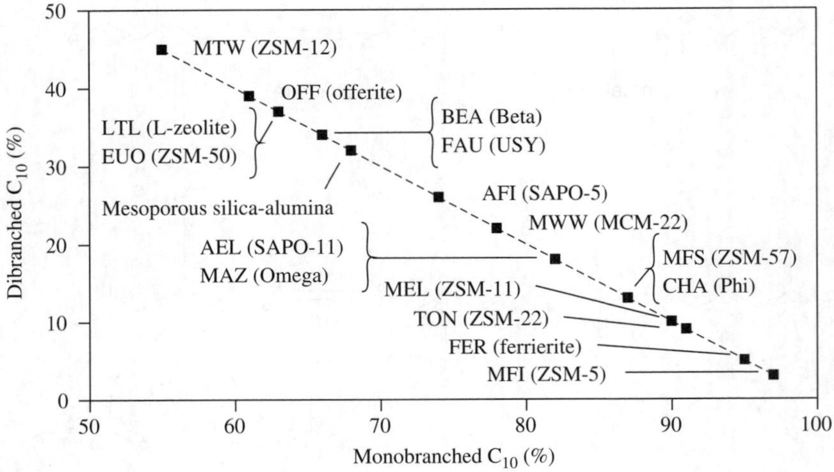

Figure 5.17 Ratio of monobranched to dibranched C_{10} products from HIS of *n*-decane over various Pt-promoted acid catalysts at low pressure, indicating the effect of pore size on isomerisation.

shown that catalyst activity and indirectly catalyst acid strength are not the only factors influencing isomerisation. Figure 5.17 shows the importance of the catalyst structure on the yield of mono-branched *versus* dibranched isomers of *n*-decane.[230,231] It is evident that ZSM-5 catalyst results in a very low yield of dibranched isomers because of its narrow pores, which is in contrast with large-pore catalysts, such as Beta-zeolite, ultrastable Y-zeolite and mesoporous silica–alumina (MSA). Pore size influences the yield and distribution of the lighter than C_{10} products. Based on the mechanism (Figure 5.15), such products are formed by the cracking of branched hydrocarbons. For example, the yield of branched C_5 hydrocarbons over 12-membered ring zeolites was more than twice that observed over 10-membered ring zeolites, whereas that for the MSA catalyst was in between that of the 12- and 10-membered ring zeolite catalysts. With respect to the overall mechanism of skeletal IS, in addition to the chemical composition of the catalyst, porosity and/or shape selectivity are also important factors to be considered.

In the case of a C_4 hydrocarbon, the formation of the PCP carbocation is energetically unfavourable, because the skeletal rearrangement would require a primary carbocation intermediate.[229] It is therefore not surprising that Brouwer and Oelderik reported a significant difference between the skeletal IS of C_4 and of C_5 and heavier hydrocarbons.[232] In the presence of a superacid, the latter were rapidly converted to branched isomers, whereas the C_4 conversion was low.

Assuming some similarity between the carbocations formed from *n*-butane, for example by by hydride abstraction and *n*-butene by protonation, the involvement of the PCP intermediate in skeletal IS may be excluded unless intimate contact of the C_4 cation with the catalyst surface stabilises the transition state intermediate. This suggests that the mechanism of the skeletal IS of butene and butane may be strongly influenced by the chemical composition of the catalyst. It was reported that for *n*-butane, skeletal IS proceeded most efficiently on strongly acidic centres (superacids) and it occurred at relatively low temperatures.[102,233–235]

Apparently, skeletal IS can proceed by both monomolecular and bimolecular mechanisms, which lies at the root of the debate about the mechanism mentioned previously. The operating conditions, nature and deactivation state of the catalyst influence the mechanism and may favour one pathway over another. This is illustrated by the different observations in the literature.

Mooiweer *et al.* observed that a large amount of C_3 and C_5 products formed simultaneously with isobutene during the IS of butene.[102] This indicated the involvement of bimolecular reactions. Houžvička and Ponec reported similar product trends (Figure 5.18),[236] which were also found by others.[237] While using ferrierite, Meriaudeau *et al.* found products indicative of a bimolecular reaction mechanism on the fresh catalyst, but indicative of a monomolecular reaction on the coke deposited catalyst.[238] These observations were in agreement with the results published by Guisnet *et al.*[239] This was attributed to the modification of catalyst porosity by deposited coke. Čejka *et al.* observed that on a CoAlPO-11 catalyst very selective monomolecular formation of isobutene took place, whereas on ferrierite a large amount of isobutene was formed *via* the bimolecular mechanism.[240] This could be reconciled on the basis of differences in restrictive transition state selectivity. For example, for ferrierite, the three-dimensional channel system allowed the formation of dimers in addition to the monomolecular mechanism. The monomolecular mechanism prevailed on CoAlPO-11, because of the one-dimensional elliptical channels. Therefore,

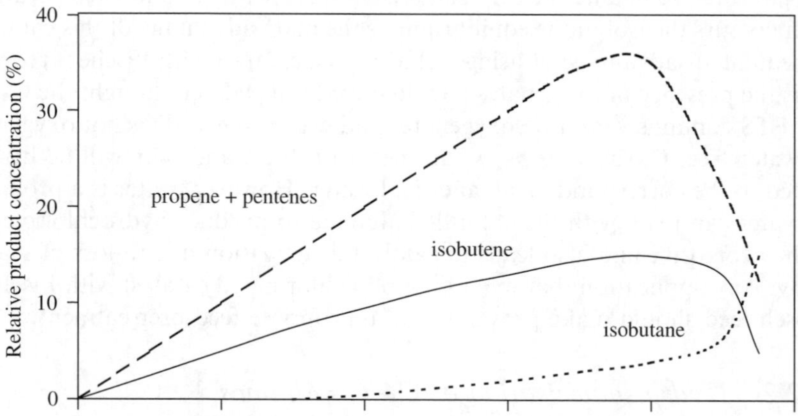

Figure 5.18 Skeletal isomerisation of *n*-butene at 350 °C over different H-ZSM-5 catalysts with Si:Al ratios varying from 25 to 1000 all resulted in the same trend.

shape selectivity of catalysts is an important factor in controlling the IS mechanism.

When the butene reacts with the catalyst surface by formal bonding, skeletal rearrangement becomes possible without the formation of a primary carbocation. Direct and indirect evidence for the formation of such a formal (σ-bond) during butene skeletal IS has been presented for a number of catalysts, including alumina,[241] phosphoric acid[242] and coked ferrierite.[243] There is a clear analogy with the low-temperature SI pathway reported during butene OLI over SPA,[51] as shown in Figure 5.5.

The mechanism may be influenced by the operating conditions. It stands to reason that operation at low partial pressure of alkenes would favour a monomolecular over a bimolecular pathway, while the converse is true for high alkene partial pressure. During HIS, the partial pressure of alkenes on the catalyst is determined by the H_2 pressure and the operating temperature. The rate and equilibrium of alkane deHYD to produce alkenes may be slowed with increasing pressure of H_2 and decreasing temperature.

5.2.2 Commercial Processes for Isomerisation

5.2.2.1 Hydroisomerisation of Butane

The principal technology for IS of *n*-butane to isobutane is the chlorinated Pt/Al_2O_3-catalysed Butamer process of UOP,[244] which operates at 180–220 °C, 1.5–2.0 MPa, space velocity 2 h^{-1} and hydrogen-to-hydrocarbon ratio 0.5–2.0:1. There are generally two reactors, the first operating at a higher temperature to increase the reaction rate and the second at a lower temperature to improve the isobutane equilibrium concentration. The ability of chlorinated Pt/Al_2O_3 to catalyse butane skeletal isomerisation efficiently at a low temperature, which favours the isobutane equilibrium, is the main advantage of this catalyst.

The main disadvantage of using a chlorinated catalyst with Fischer–Tropsch feed is the presence of oxygenates and dissolved water. Even thought the C_4 cut from FTS contains very little oxygenates and water, syncrude is not oxygenate and water free. Oxygenates *per se* are not a problem and they will be hydrogenated to the corresponding alkanes and water. However, water is a problem. The water can react with the chlorided alumina to produce hydrochloric acid, which is corrosive and also leads to catalyst deactivation due to loss of strong acidity. Any application that makes use of a chlorinated catalyst with Fischer–Tropsch feed should make provision for appropriate feed pretreatment.

5.2.2.2 Hydroisomerisation of C_5–C_6 Alkanes

There are three main classes of catalysts that are currently used for C_5–C_6 alkane HIS, namely chlorinated Pt/Al_2O_3 (*e.g.* UOP I-8/I-80, Procatalyse IS 612 and Albemarle AT-20), Pt/MOR (*e.g.* Süd-Chemie Hysopar, Procatalyse IS 632 and UOP HS-10) and Pt/SO_4^{2-}/ZrO_2 (*e.g.* UOP LPI-100, and Süd-Chemie Hysopar-SA).[244–251] The chlorinated Pt/Al_2O_3 catalysts have similar requirements, advantages and drawbacks to those already listed for butane HIS. The main advantage is the low operating temperature (120–180 °C), which favours the IS equilibrium. The main drawback, apart from the environmental concern related to the use of a chlorinated system, is the water sensitivity of the catalyst.[250] These catalysts are also sulfur sensitive, but that is not a drawback when dealing with FTS.

On the opposite end of the spectrum is the Pt/MOR zeolite catalysts, which are much more resistant to sulfur and water, and very long catalyst lifetimes (more than 10 years) have been reported. The main drawback of Pt/MOR catalysts is that they require an operating temperature of 250–280 °C, which is much higher than that required by either chlorinated Pt/Al_2O_3 or Pt/SO_4^{2-}/ZrO_2 catalysts. The use of Pt/MOR is less favourable in terms of the IS equilibrium and should preferably be employed with technologies that include an *n*-alkane recycle. When an *n*-alkane recycle is employed, the final product quality is not determined by the per pass equilibrium conversion.

5.2.2.3 Isomerisation of C_4–C_5 Alkenes

Considering the linear alkene-rich nature of the products from FTS, skeletal IS of alkenes is of interest. For practical applications, ferrierite is by far the most selective catalyst for high-temperature IS of *n*-butene.[102,252] The operating temperature of butene SI is typically 350 °C and higher. For commercial processes, cycle lengths of the order of 500 h have been reported.[253]

The skeletal IS of *n*-pentene is more facile than that of *n*-butene. Commercial technologies are available using different catalysts, such as ferrierite (Lyondell),[253] acidic molecular sieves (UOP)[254] and alumina (IFP/Axens).[255] Oxygenates may influence the operation of these technologies. For example, the strong adsorption of oxygenates reduces the operating window of the acidic molecular sieve-based UOP Pentesom process.[256] For alumina-based processes, the operating conditions have to be optimised when there are oxygenates in the feed, although oxygenates are not necessarily detrimental.[257]

5.2.3 Catalysts for Isomerisation

It has been pointed out that IS/HIS processes can be divided into three classes based on the feed: C_4, C_5–C_6 and C_7 and heavier. The feed must be matched to the catalyst type, as will be apparent from the discussion, but in this section the discussion is organised by catalyst type.

A wide range of catalysts have been developed and tested for IS and HIS of *n*-alkanes and *n*-alkenes. The literature abounds with studies involving different combinations of active metals (*e.g.* Pt and Pd) with silica–alumina materials, especially zeolites, but also amorphous silica–aluminas and active clays. Silico-aluminophosphate (SAPO) catalysts, molecular sieves and sulfated zirconia (SZ)-based catalysts have also been studied. In addition, active metals supported on different acidified supports, such as fluorided and chlorinated

Al_2O_3, SiO_2 and various carbon supports, were employed in several studies. Activity determination involved both model compounds and real feeds. The studies in which different types of catalysts were tested under identical conditions are of particular importance for comparison of catalyst performance.[258,259] Experimental studies have been reported with hydrogen pressures ranging from atmospheric up to about 6 MPa, and with temperatures in the range 100–400 °C. In order to promote monomolecular reactions and isomerisation in particular, some investigations were performed with the hydrocarbons being diluted in an inert carrier gas, mostly nitrogen.

Linear alkanes and alkenes, typical of those found in primary products from FTS, have been employed in most studies involving the development and testing of catalysts for IS/HIS reactions. This is a natural consequence of the objective transformation, namely the conversion of linear hydrocarbons into branched hydrocarbons. Although the IS/HIS studies generally did not have applications involving material from FTS as the aim, they are nevertheless directly relevant.

The literature dealing with IS/HIS of long-chain hydrocarbons, which are in a semi-solid or solid form under ambient conditions, such as waxes, will be dealt with in Chapter 6. There is inevitably some overlap with the literature on the HIS of C_7 and heavier material. Here the focus will be mainly on gas-phase HIS of lighter hydrocarbons, with only limited coverage of liquid-phase HIS.

5.2.3.1 *Zeolitic Silica–Alumina Catalysts*

Some catalyst development for skeletal IS of light alkenes (C_4 and C_5) focused on ZSM-5 and ZSM-11 zeolites with different contents of alumina and boron.[259] Incorporating boron into the zeolite framework reduced acidity. This increased IS selectivity, while decreasing the selectivity for acid-catalysed side-reactions, namely cracking, OLI and hydrogen transfer. It was observed that the activity, selectivity and stability of the H-ZSM-5 zeolite can also be optimised by varying the Si:Al ratio, particle size and substitution (*e.g.* Fe for Al).[260,261]

The skeletal IS of *n*-butene (in N_2 at 350 °C) was conducted over a series of M-ZSM-22 (M = Al, Ga and Fe) catalysts with the different Si:M ratios and particle sizes.[262] For catalysts having similar composition and particle size, the activity increased with increasing acidity, Al > Ga > Fe, whereas the opposite trend was observed for the isobutene selectivity. For Ga- and Fe-ZSM-22, better than 80% selectivity to isobutene at 50% butene conversion was achieved.

The Mg^{2+} cation-exchanged ZSM-22 exhibited the best performance during the IS of *n*-butene (in He) compared with cations such as H^+, Mn^{2+}, Cu^{2+} and Ca^{2+}.[263] The activity order was proportional to the ratio of Lewis acidity to Brønsted acidity. The addition of boron and phosphorus had mainly adverse effects on the activity and stability of the catalysts. Steaming Mg-ZSM-22 decreased the catalyst acidity.[264] The conversion of 1-butene decreased rapidly and the selectivity to isobutene increased with increasing steaming temperature.

The effect of He as diluent was compared with that of H_2 on the IS of *n*-butene to isobutene over naturally occurring clinoptilolite zeolite at 450 °C.[265] The conversion was improved by replacing He with H_2 due to diminished coke formation in the presence of the latter. As one would expect, the selectivity for skeletal IS in H_2 decreased with the increased conversion.

In a comparative study on *n*-butene IS,[266] the yield of isobutene obtained over ferrierite was much higher than that obtained over the zeolites SAPO-11 and H-mordenite. The large-pore H-mordenite, which is commercially employed for C_5–C_6 HIS, showed a low conversion and low selectivity for IS of *n*-butene. Ferrierite exhibited excellent IS activity for *n*-butenes and *n*-pentenes,[260] as noted during the discussion on the IS mechanism. The same was found by Onyestyak *et al.* for *n*-hexene IS, but the catalyst deactivated rather rapidly.[267] Over ferrierite at 300 °C with a 10% 1-hexene in nitrogen mixture, the decrease in conversion with time on stream was quite evident, although at the same time, the selectivity to IS was increasing with decreasing conversion as one would expect.

The studies on skeletal IS of *n*-hexene have highlighted the shape selective properties of ZSM-5 zeolite.[268–271] Abbot *et al.* compared ZSM-5 and Y-zeolites and observed a double bond shift in 1-hexene for both zeolites.[268,269] The *cis*-2-hexene-to-*trans*-2-hexene ratio was nearer to equilibrium on ZSM-5 zeolite than on Y-zeolite. Also, the relative rate of skeletal IS was higher on the former catalyst. The amount of cracking and OLI products was smaller on ZSM-5 zeolite than on Y-zeolite. For the same reaction, ZSM-5 zeolite was reportedly much more active than Pd/SAPO-11, SAPO-11 and mordenite.[270] At comparable conversion (80–85%) selectivity to branched hexenes increased in the order ZSM-5 < Beta < SAPO-11.[271]

Zeolites such as Y, Beta, ZSM-22 and ferrierite were compared in the study of Tiitta *et al.* using *n*-hexenes as feed.[119] In this study, ferrierite had the highest selectivity for the formation of branched hexenes, and Y-zeolite was the least active. Also, more dimer products were formed on Beta- and Y-zeolites than on ferrierite and ZSM-22. To deal with the excessive coke formation during alkene IS, Sandelin *et al.* designed a continuous circulating fluidised bed system comprising a reactor and a regenerator.[272] Using this system, the alkene-rich C_4–C_6 fraction was successfully isomerised over ferrierite.

The H-ZSM-5 zeolite and Ni/H-ZSM-5 catalysts prepared by impregnation of the former were tested for the transformation of 1-hexene in the temperature range 160–400 °C and an H_2 pressure of 0.4–2.0 MPa in a continuous flow microreactor.[273] Below 220 °C, the conversion was dominated by double bond IS, whereas skeletal IS only became evident at higher temperatures. The addition of Ni enhanced the formation of aromatics, branched alkanes and C_{12} hydrocarbons (dimerisation). Above 350 °C, aromatics were the major products. Over the Ni/H-ZSM-5 catalyst, increasing H_2 pressure decreased the yield of aromatics, cycloalkanes and alkenes, but it had little effect on the yield of branched alkanes (due to HIS) and C_{12} hydrocarbons (due to OLI). Liu *et al.* studied the HIS of *n*-hexane in a continuous flow reactor (230 °C, 1.5 MPa and H_2:*n*-hexane = 8:1) and observed that a further increase in the activity and selectivity of the Ni/H-ZSM-5 catalysts can be achieved by the addition of Mo and phosphorus.[274] The optimum combined effect of Ni, Mo and phosphorus was attained at contents of 1.0, 2.0 and 1.5 mass%, respectively.

During the conversion of heptane, in both H_2 and N_2, the Ni/H-ZSM-5 catalyst was much more active than a Co/H-ZSM-5 catalyst.[275] This was evident particularly in H_2 (2 MPa). For the same amount of metals, the activity of Pt/H-ZSM-5 was much higher than that of the other two catalysts. With respect to the activity and selectivity, and also catalyst stability, conditions were much more favourable in H_2 than in N_2. Under conditions identical with those used for the HIS of heptane, the activity of the Ni-containing zeolites for the HCR and HIS of *n*-octane, 2,5-dimethylhexane and 2,2,4-trimethylpentane decreased in the sequence Ni/H-ZSM-5 $>>$ Ni/H-Beta $\approx$ Ni/H-MOR.[276] However, the selectivity for HIS of *n*-octane and 2,5-dimethylhexane was the highest over Ni/H-Beta and the lowest over the Ni/H-ZSM-5 catalysts.

During the preparation of a Pd/HZSM-5 catalyst, Canizares *et al.* observed that Pd dispersion could be controlled by pH.[277] A strong acidity was attained at low pH because of the partial exchange of Na^+ with protons. In addition, dealumination increased the density of strong acid sites. The catalysts were tested for the HIS of *n*-butane. Because of the strong acidity present, the yield of isobutane was increased. With the aim of increasing mechanical strength, the Pd/H-ZSM-5 catalyst was combined with a binder.[278] This decreased the density of strong acid sites due to solid ion exchange between zeolite protons and binder sodium. As a consequence, the conversion of *n*-butane decreased. However, this was compensated for by an increase in selectivity to isobutane.

The relative contributions of acidity, deHYD/HYD and the metal–support interactions in HIS were illustrated by the work of Zhang *et al.* (Figure 5.19).[279]

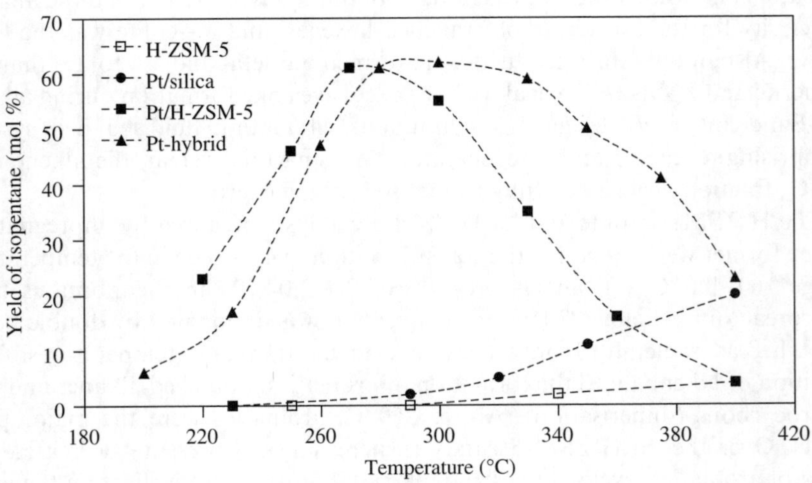

Figure 5.19 Yield of methylbutane (isopentane) from HIS of *n*-pentane over H-ZSM-5 (□), Pt/SiO_2 (●), Pt/H-ZSM-5 (■) and Pt-hybrid (▲) catalysts at 100 kPa, 0.2 mol h^{-1} g_{cat}^{-1} and H_2:*n*-pentane molar ratio 9:1.

The addition of Pt to H-ZSM-5 zeolite resulted in a significant enhancement in conversion and selectivity during HIS of *n*-pentane. However, the best performance was exhibited by a Pt hybrid catalyst prepared by co-grinding four parts by weight of the H-ZSM-5 zeolite with one part of Pt/SiO_2 and pressure moulding the mixture into granules. The Pt hybrid catalyst showed higher HIS activity over a wider range of temperature and pressure. This was attributed to the regeneration of Brønsted sites and stabilisation of the $C_5H_{11}^+$ intermediate aided by hydrogen spillover. In this process, the gaseous H_2 was dissociatively adsorbed on the noble metal and subsequently spilled over on to the zeolite.[280]

The nature of the metal promotion has an influence on both HIS conversion and catalyst sensitivity to feed contaminants. During the HIS of hexadecane at 320 °C and an H_2:hexadecane ratio of 400, the overall conversions over H-ZSM-5 and Ni/H-ZSM-5 catalysts were 35 and 46%, respectively.[281] For Pd/H-ZSM-5, the conversion approached 99%. With sulfur in the feed, the conversions over H-ZSM-5 and Ni/H-ZSM-5 increased, whereas the conversion over Pd/H-ZSM-5 decreased. The conversions over Ni/H-ZSM-5 and Pd/H-ZSM-5 were both around 53%.

Bifunctional Ni-Pd/HY zeolite catalysts containing 0.1–0.5 mass% Ni and 0.1 mass% Pd prepared by incipient wetness impregnation were used for the HIS of *n*-octane between 200 and 450 °C at atmospheric pressure.[282] It was found that Ni addition up to 0.3 mass% to the Pd/HY zeolite increased *n*-octane conversion and HIS selectivity. At the same time, the yield of cracked products was decreased. Above 0.3 mass% of Ni, the conversion decreased and the yield of cracked products increased. It was reported that bimetallic catalysts were more selective towards the formation of dibranched isomers having a higher octane number than mono-branched isomers.

Three paraffinic naphtha fractions of variable compositions were used by Ramos *et al.* to study the HIS activity of a Beta-zeolite agglomerated with bentonite.[283] The experiments were conducted in an autoclave at temperatures of 290–390 °C and 1 MPa pressure using an H_2:feed ratio of 14:1. The HIS activity was measured by determining the ratios of branched to linear C_6–C_8 hydrocarbons in the feed and product. The highest HIS conversion was observed for the naphtha having the highest content of *n*-alkanes. Under the conditions employed in this study, most of the aromatics were converted to cycloalkanes.

In a series of Pt/HY catalysts tested by Giannetto *et al.* for the HIS of heptane, the yield of branched products increased with increase in platinum loading and increase in Si:Al ratio (Figure 5.20).[284] In the range of Pt loadings from 0 to 1.5% and Si:Al ratios from 3:1 to 35:1, little improvement was found beyond a platinum loading of 1% and an Si:Al ratio of 9:1. The ratio of single to multi-branched products decreased with increasing conversion. The authors suggested that an ideal HIS catalyst should have one metallic (Pt) site for 10 acidic sites. As can be seen from Figure 5.20, the best catalysts had a low HCR activity and a low yield of gaseous products. Since HIS of *n*-heptane is more demanding in this respect, such a catalyst should also be well suited for HIS of *n*-pentane and *n*-hexane.

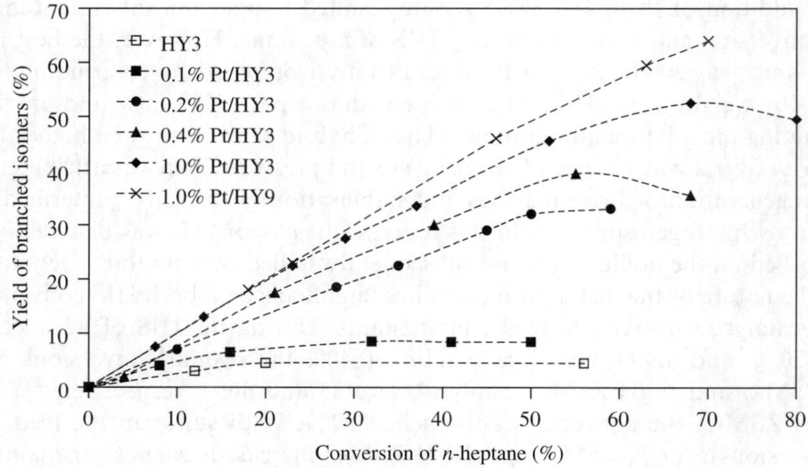

Figure 5.20 Yield of branched isomers during HIS of *n*-heptane over HY-zeolite (FAU) catalysts with different platinum loadings and different Si:Al ratios. The HY3 catalysts had an Si:Al ratio of 3:1, whereas the HY9 catalyst had an Si:Al ratio of 9:1.

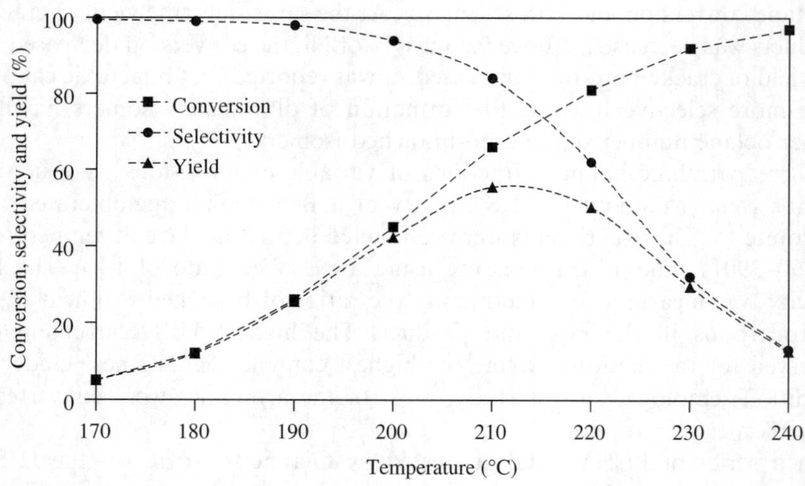

Figure 5.21 Typical relationship between conversion (■), selectivity (●) and yield (▲) to branched isomers during the HIS of C_7 and heavier hydrocarbons. Shown is the HIS of *n*-heptane over Pt/H-Beta at an H_2:*n*-heptane ratio of 7.5:1.

The relationship between conversion and selectivity for HIS for *n*-heptane shown in Figure 5.21[285] is typical of that for HIS of C_7 and heavier alkanes. As the temperature is increased, the conversion increases, but as the conversion increases, the selectivity to branched isomers decreases due to the increased contribution of cracking. Hence the total yield of the C_7 isomers passed through a maximum at about 210 °C. The yield and selectivity to branched isomers increased with increasing partial pressure of H_2. At the same time, the selectivity to multi-branched isomers exhibited a slight decrease with increasing H_2 partial pressure. Increasing amounts of Pt in the HIS catalysts had a beneficial effect on the selectivity and yield of isomers, but it had little effect on conversion. This supported the interpretation that conversion is dependent on acid-catalysed IS of alkenes that are in deHYD/HYD equilibrium with the alkanes, whereas the selectivity is dependent on how quickly the branched alkene can be hydrogenated before it can undergo β-scission and is cracked to form lighter products. The Pt/H-Beta zeolite used by Wang *et al.*[285] for HIS of *n*-heptane showed good stability at 220 °C over a 78 h test period.

A detailed study on the HIS of *n*-pentane, *n*-hexane and *n*-heptane was conducted by Chao *et al.* that involved Pt/H-Beta and Pt/H-MOR catalysts with Si:Al ratios from 5 to 112 and a Pt loading of 0.5%.[221] The catalysts tested therefore had a wide range of acid site densities. The objective of the study was to maximise HIS of *n*-alkanes into branched alkanes, while suppressing cracking activity (Table 5.17). It is evident that the most acidic catalysts were very active for HCR, whereas their selectivity for HIS was limited. However, maximum HIS activity occurred at temperatures where HCR was still not very evident. With a further increase in temperature the HIS activity passed through a maximum and HCR became the main contributor to the overall conversion. The trend for all catalysts was similar to that shown in Figure 5.21. The HIS activity could be increased and HCR activity could be decreased by reducing zeolite acidity through ion exchange, substituting H^+ ions for Mg^{2+} ions.

Similar findings were reported for HIS at elevated pressure. The content of the framework alumina influenced the HIS activity of Pt/MOR catalysts employed for HIS of *n*-hexane and *n*-octane at 220 °C and 2 MPa.[286] The hydroconversion increased almost linearly with the increase in content of the framework aluminium to a maximum and then abruptly decreased.

Table 5.17 Influence of acid site concentration on *n*-heptane HIS over different Pt/H-Beta and Pt/H-MOR catalysts in a flow reactor at near atmospheric pressure and using an H_2:*n*-heptane ratio of 18:1.

Catalyst[a]	*Si:Al ratio*	*Acidity (mmol g^{-1})*	*Maximum HIS*[b] *Yield (%)*	*Maximum HIS*[b] *Temperature (°C)*
Pt/H-Beta	11	0.8	73	210
Dealuminated Pt/H-Beta	78	0.174	75	270
Pt/H-MOR	5	1.4	15	210
Pt/H-MOR	18	0.4	57	250
Dealuminated Pt/H-MOR	37	0.3	61	240
Pt/H-MOR	112	0.13	57	270
Dealuminated Pt/H-MOR	112	0.106	55	280

[a]0.5 mass% Pt loading with different Si:Al ratios.
[b]Approximate yield and temperature values; experiments conducted at 10 °C intervals.

A commercial Pt/MOR catalyst was evaluated for the direct conversion of 1-pentene into isopentane with applications in Fischer–Tropsch refining in mind.[287] The Pt/MOR catalyst was selected due to its water tolerance, which is needed for processing feed from FTS without pretreatment. Despite the high alkene partial pressure in the feed, catalyst deactivation was limited over the test period that involved evaluation at different operating temperatures in the range 200–270 °C at 2 MPa with H_2:1-pentene molar ratios in the range 3:1–5:1.

A series of Pt/HY zeolite catalysts containing 1% Pt were prepared by progressive dealumination with $SiCl_4$ and used for the HIS of *n*-decane.[288] With increasing degree of dealumination, the number of Brønsted sites decreased. This reduction in the number of acid sites was compensated for by increased catalytic efficiency of the remaining acidic sites. At low overall conversion, methylnonane and ethyloctane were the only isomers produced. The yield of the latter decreased with increasing dealumination, but increased with increasing conversion.

It has been reported that the Pt dispersion has a pronounced effect on the activity of Pt/H-ZSM-5 catalysts employed for the HIS of *n*-heptane.[289] The best dispersion could be achieved during calcination at 350 °C. Over this catalyst, the HIS selectivity improved relative to HCR. The HIS selectivity was increased with increasing number of accessible Pt atoms. Methylhexane was the main HIS product. In addition, small amounts of 2,2- and 2,4-dimethylpentane were formed. The experiments were conducted in a flow reactor at 250 and 350 °C using an H_2:*n*-heptane ratio of 9:1.

Alvarez *et al.* prepared a series of Pt/HY catalysts with varying metal to acid site ratios and used them for the HIS/HCR of *n*-decane.[233] The balance between the metal and acid functions influenced the transformation of *n*-decane. A low cracking conversion in favour of HIS was observed over catalysts with high metal-to-acid site ratios. On the other hand, light products formation was favoured over catalysts with low metal-to-acid site ratios.

It has been generally observed that the HIS activity of MOR-based catalysts may be optimized by promoters, conditions of preparation and various pretreatments.[290] For Pt/MOR catalysts, the activity can be controlled by the amount of Pt and the conditions applied during its addition to MOR. The significantly better HIS/HCR activity of Pt/MOR compared with H-MOR catalysts was further improved when the Pt/MOR was used as part of a composite catalyst with Pt/Al_2O_3. This parallels the observations with other hybrid catalysts (Figure 5.19). The experiments were carried out at 200–500 °C and 1.5 MPa H_2 pressure using C_7–C_{12} *n*-alkanes as feed. For the composite catalyst, hydrogen spilled over from Pt/Al_2O_3 on to Pt/MOR, which resulted in a reduction of coke precursors from the catalyst surface. When unpromoted H-MOR was used as catalyst for alkane isomerisation with N_2 as co-feed, rapid catalyst deactivation took place.[291]

A Pt/HMOR catalyst was compared with a Pt/mazzite catalyst for the HIS of *n*-hexane at 250 °C and 4.8 MPa H_2.[292] With both catalysts, methylpentanes and 2,3-dimethylbutanes, and also a small amount of 2,2-dimethylbutane, were the primary products. The Pt/mazzite catalyst was found to be more active than the Pt/MOR catalyst.

Bifunctional Pt/MCM-22 catalysts showed features of both 10- and 12-membered ring zeolites during HIS of *n*-decane at 190–250 °C, 3 MPa and an H_2:*n*-decane ratio of 8:1.[293] Based on the yield of multi-branched C_{10} isomers, which is impeded by 10-membered ring structures, Pt/MCM-22 behaved more like a 12-membered ring zeolite. However, the distribution of the mono-branched isomers also indicated the involvement of 10-membered ring structures. These observations suggest that a larger space is available in the void volume of MCM-22 compared with zeolites such as ZSM-5. At lower pressure, during the HIS of *n*-hexane over Pt/MCM-22 at 230 °C and 0.1 MPa, the product distribution was dominated by methylpentanes, giving a methylpentanes: dimethylbutane ratio of >5.[294] As expected, the overall conversion decreased with increasing H_2:hexane ratio.

5.2.3.2 Silica–Alumina Catalysts

Various forms of ASA exhibit activity during the IS/HIS of hydrocarbons. Unpromoted ASA catalysts are active for the IS of alkenes and many of the earlier studies have been summarised in a review by Dunning.[216] Conversion can typically take place at fairly low temperatures. However, without metal promotion ASA catalysts are not active for IS of *n*-alkanes. It was reported that in the temperature range 100–150 °C, branched alkanes can be isomerised over unpromoted ASA, but not *n*-alkanes.[295]

The activity of unpromoted ASA for IS could be improved by the addition of water.[295] The activating effect of water on ASA catalysts have been noted before (Section 5.1.3.4), with small amounts of water increasing catalyst activity, which passes through a maximum with increasing water co-feeding, before decreasing at higher water content in the feed. This has implications for IS of Fischer–Tropsch-derived feeds on account of their oxygenate content.

When ASA materials are promoted with a metal, they become active for HIS of alkanes. Corma *et al.* prepared a mesoporous silica–alumina (MSA) and used it for the HIS of *n*-decane in a continuous fixed bed system between 250 and 300 °C.[231] This catalyst was compared with various other catalysts to illustrate the effect of catalyst structure on isomer distribution (Figure 5.17). In this study, the effects of Pt content and temperature on *n*-decane conversion and HIS selectivity were also investigated. It was evident that an increase in temperature resulted in an increase in conversion, whereas selectivity exhibited the opposite trend, similar to that shown in Figure 5.21. In this study, the Pt supported on ASA catalysts exhibited higher selectivity than the catalysts based on USY zeolites. The least selective USY catalysts gave the highest conversions. Compared with the ASA-based catalysts, the conversion difference decreased with increasing temperature. The superior selectivity of the MSA catalyst was attributed to its moderate acidity and mesoporosity. The latter favoured diffusion of the C_{10} branched isomers, thus preventing their cracking.

The inclusion of metal sites on a silica–alumina material can also help to prevent deactivation during alkene IS. In the absence of oxygenates, Fischer–Tropsch-derived pentenes were readily isomerised at 260 °C, 1.7 MPa and at an H_2:pentene ratio of 2:1 over a metal-promoted non-zeolitic molecular sieve achieving a catalyst cycle length of around 10–12 months.[256]

The HIS of *n*-octane over Ni–$WO_3/SiO_2–Al_2O_3$ was investigated by Rezgui and Guemini.[296] The HIS selectivity increased with increasing Ni loading and reached a plateau at about 15% Ni. The best results, namely 69% HIS selectivity at 33% conversion, were obtained with a catalyst containing 15% Ni and 10% W, respectively. The HIS activity was also influenced by H_2 pressure and SiO_2:Al_2O_3 ratio of the catalyst.

Shi and Shen used a sulfided Co/Co-MCM-41 catalyst to study the skeletal IS of 1-hexene, where a 5% Co-MCM-41 catalyst was used as support for additional Co loading.[297] The additional 5% Co on 5% Co-MCM-41 catalyst exhibited stronger surface acidity than the Co-MCM-41 without additional Co loading. The surface acidity was greatly enhanced upon sulfidation. Thus, the sulfided catalyst possessed strong surface acidity. At 300 °C, more than 60% of the 1-hexene feed was skeletally isomerised.

5.2.3.3 *Alumina Catalysts*

Halogenated platinum-promoted alumina catalysts form an important class of industrially applied catalysts for the HIS of alkanes. The dominant catalyst type is chlorinated Pt/Al_2O_3 and these catalysts have already been discussed (Section 5.2.2). Detailed studies describing the modification of both Pt and Al_2O_3 by chlorination with CCl_4 indicated that the Pt is chlorided and the Al_2O_3 develops strong Lewis acid sites, but no Brønsted acidity.[298] The chlorination also caused redistribution of Pt on the surface, which at high Pt loadings led to some Pt sintering rather than increased dispersion.

During HIS over chlorinated Pt/Al_2O_3, some activity is lost and continuous addition of a chloroalkane is necessary to maintain catalyst activity. Two pathways were suggested for alkane IS, one involving strong Lewis acidity [Equation (5.4)] and the other involving the creating of Brønsted acidity through the action of weakly bound HCl [Equation (5.5)]:[299]

$$\mathrm{Al-O-(\,)-O-AlCl_2 + RH \rightarrow Al-O-(R^+)-O-(AlHCl_2^-)} \qquad (5.4)$$

$$\mathrm{-AlCl_2 + HCl \rightarrow -AlCl_3^- H^+} \qquad (5.5)$$

Kinyakin *et al.* studied the HIS of *n*-hexane over chlorinated Pt/Al_2O_3 over a range of operating conditions to develop an accurate kinetic description of the reaction.[300] It was found that good HIS activity could be maintained for liquid- and gas-phase reactions, as long as sufficient hydrogen was present in each phase. The results of their study is typical of chlorinated Pt/Al_2O_3 catalysts (Table 5.18), indicating that high HIS activity and selectivity can be

Table 5.18 Effect of temperature on the HIS of *n*-hexane over a chlorinated Pt/η-Al_2O_3 catalyst in a fixed bed flow reactor at 2 MPa, LHSV 1.5 h^{-1} and H_2:*n*-hexane ratio 1:1. The catalyst contained 3 mass% Pt and was chlorinated with CCl_4 to a level of 9.7 mass% chloride.

			Product distribution (mass%)[a]			
Temperature (°C)	*Conversion of n-C₆ (%)*	*Selectivity for HIS (%)*	*n-C_6*	*2-MP, 3MP + 2,3-DMB*	*2,2-DMB*	$\sum C_2$–C_5
130	51.5	99.2	48.5	45.1	5.5	0.4
140	62.1	99.4	37.9	53.8	8.0	0.4
150	70.2	97.7	29.8	55.3	13.3	1.6
170	70.4	97.2	29.6	53.9	14.5	2.0
180	76.0	91.7	24.0	54.0	15.7	6.3

[a]*n*-C_6 = *n*-hexane, 2-MP = 2-methylpentane, 3-MP = 3-methylpentane, 2,3-DMB = 2,3-dimethylbutane and 2,2-DMB = 2,2-dimethylbutane.

achieved at low temperature. Increasing the pressure over the range 0.6–3.5 MPa caused an increase in both conversion and selectivity for HIS. It was also noted that the rate of 2,2-dimethylbutane formation was about 20 times faster than that of reverse IS and in their kinetic analysis the reverse reaction could be ignored.

As in the case with chlorination, fluorination decreases the temperature at which Pt/Al_2O_3 catalysts become active for the HIS of alkanes. The degree of fluorination, and also the nature of the alumina, influence activity. It has been reported that Pt/Al_2O_3 catalysts based on γ-Al_2O_3 are about 25–30 °C more active than similar catalysts based on η-Al_2O_3.[301] With fluoridation, an activity gain of around 40–45 °C per 5% F was achieved over the range 0–10% for the HIS of alkanes.

Alumina modified by fluorination was also active for skeletal IS of *n*-butene at 350 °C (5% *n*-butene + N_2).[302] The mechanism of IS was influenced by the degree of fluorination. The low F content alumina favoured a monomolecular mechanism involving Lewis acid–base pairs. For severely fluorinated alumina, Brønsted acid sites developed, which favoured the bimolecular mechanism, namely dimerisation–IS–cracking. It was observed that Brønsted acid sites could be developed on alumina by severe fluorination.[303] Houžvička *et al.* compared the fluorinated alumina with chlorinated alumina, phosphated silica and SZ during the skeletal IS of *n*-butene under identical conditions, as noted above.[261] The selectivity of the fluorinated alumina was much lower than that of the other catalysts. This was evident particularly at a high *n*-butene conversion. The chlorinated alumina catalysed the HIS of *n*-butane *via* the bimolecular mechanism involving a C_8 intermediate.[304]

Non-halogenated metal-promoted alumina catalysts can also be used for the HIS of alkanes, but require much higher operating temperatures. For example, Pt/Al_2O_3 and Pd/Al_2O_3 catalysts were evaluated for HIS of *n*-hexane and at temperatures below 290 °C the conversion was less than 10%.[305]

Sulfur is detrimental to catalyst activity and extensive poisoning of the HIS sites of a Pt/Al_2O_3 catalyst at H_2S concentration of 30 $\mu g\, g^{-1}$ has been reported.[306] At the same H_2S concentration, the poisoning effect on a Pt/Beta zeolite was less evident, indicating that sulfur poisoning is not purely metal site related.

5.2.3.4 Silico-aluminophosphate Catalysts

The complexity involved in the preparation of the SAPO-based HIS catalysts has been reported.[307] As in other bifunctional catalysts, both the acidic function provided by the SAPO material and deHYD/HYD function of the noble metals have to be optimally balanced. It is desirable that the SAPO materials are free of any contaminants. The choice of initial reagents for synthesis and conditions of hydrothermal treatment of reaction mixtures have an impact on the activity of the catalyst, as does the thermal treatment under oxygen or hydrogen. It should be noted that not every noble metal in combination with SAPO yields acceptable HIS activity and selectivity.

The IS of light alkenes has been investigated by several researchers using SAPO-11 and MeAPO-11, where Me refers to Co, Zn and Mn.[308–310] The high selectivity of SAPO-11 for isobutene was attributed to its medium-strength acidity. The selectivity to isobutene could be further increased by potassium ion exchange of the SAPO-11. The established trends indicated that the IS selectivity increased with decreasing acidity, whereas the same change resulted in a decrease in the overall conversion. Thus, for more acidic catalysts, more *n*-butene underwent cracking reactions. Catalyst selectivity comparison, of course, has to be made at similar conversion to be meaningful. Good selectivity towards skeletal rearrangement of *n*-butene to isobutene was observed for the MeAPO-11 catalysts. For example, at 400 °C the selectivity of MnAlPO-11 approached 80% at about 50% conversion. The catalysts exhibited good stability over a 24 h test period. For comparison, zeolites ZSM-22 and ferrierite were also included in the study of Yang *et al.*[309] These catalysts exhibited much higher conversions than the SAPO-11- and AlPO-11-based catalysts, but the selectivity of ZSM-22 and ferrierite for skeletal IS was lower and high yields of cracking products were obtained. Čejka *et al.* found that CoAlPO-11 was very selective for HIS of *n*-butene to isobutene (10% *n*-butene in N_2, 347 °C, near atmospheric pressure and WHSV 4.5 h^{-1}).[240] They concluded that most of the isobutene over CoAlPO-11 catalyst was produced by monomolecular skeletal IS. However, for ferierrite, at least 30% of isobutene was formed *via* dimerisation followed by IS and cracking.

Wei *et al.* investigated the conversion of *n*-butane to isobutene.[311] Rather than HIS, the reaction conditions (300–350 °C and H_2:*n*-butane ratio 2:1) were selected to promote alkane dehydrogenation, followed by skeletal IS to yield the branched alkene as product. The catalysts tested included molecular sieves such as SAPO-5, SAPO-11, SAPO-34 and AlPO-11 in combination with Pd. Among these catalysts, Pd/SAPO-11 exhibited the highest activity and selectivity for isobutene. This was attributed to the medium-strong acidity and suitable pore geometry of this catalyst. The dehydrogenation of the alkane was also observed over Pd/SAPO-5 and Pd/SAPO-34 catalysts, but skeletal IS of the butenes was suppressed. The eight-membered ring pore opening in Pd/SAPO-34 may have sterically inhibited skeletal IS. The 12-membered ring pore opening in Pd/SAPO-5 did not pose similar steric constraints and the low selectivity of this catalyst for skeletal isomerisation was attributed to its low acidity.

Several studies revealed that SAPO-5, SAPO-11, SAPO-31 and SAPO-41 exhibited good selectivity for HIS of *n*-alkanes, yielding mostly mono-branched isomers.[312–316] This was attributed to their moderate acidities and suitable shape selectivity. Among several SAPOs, the suitability of SAPO-11 for the HIS of longer chain *n*-alkanes has also been reported.[315,316] The SAPO-11 crystal has the AEL structure and consists of non-intersecting elliptical 10-membered ring pores. Such a structure ensures that a significant amount of multi-branched isomers can be produced inside the channels. These isomers can then diffuse out more easily because of the elliptical pore opening with sufficiently large diameter (0.64 nm). The active sites for HIS are located near pore mouths on the external surfaces.[317]

Campelo *et al.* compared the Pt/SAPO-11 catalysts with the Pt/SAPO-5 catalysts during the HIS of *n*-hexane and *n*-heptane using a microcatalytic pulse reactor at 400 °C and 0.3 MPa.[312] The catalysts contained 0.5% Pt. SAPO-5 has 12-membered ring pores with cylindrical channels of 0.80 nm diameter, which is larger than the 10-membered elliptical ring pores of SAPO-11. Some variation in properties can also be introduced by employing different conditions during catalyst preparation.[318] For the overall conversion of both *n*-hexane and *n*-heptane, the Pt/SAPO-5-based catalysts were more active than the Pt/SAPO-11 catalyst. The Pt/SAPO-11 catalyst was more selective for conversion to mono-branched isomers. In other studies on the HIS of *n*-hexane, *n*-heptane and *n*-octane over SAPO-5- and SAPO-11-based catalysts, Campelo *et al.* showed that to a large extent, the size of pores determines the selectivity.[319,320] The difference between the selectivity of SAPO-5 and SAPO-11 could be attributed to a slow migration of alkene intermediates in the channel of the latter and the steric constraints at the pore mouths. These authors also observed that with increasing chain length of the *n*-alkane, the selectivity for HIS decreased for SAPO-5, whereas it increased for SAPO-11. The reaction paths for HIS of *n*-alkanes on SAPO-5 differed from those on SAPO-11. The selectivity patterns of these SAPOs were interpreted in terms of a series of reaction pathways that incorporated both confinement effects and shape selectivity factors.[321] For example, the mono-branched isomers from the HIS of dodecane over Pt/SAPO-11 at 300–400 °C and 0.3 MPa consisted of only methylundecanes and contained no ethyldecanes.[322]

A series of the SAPO-11-based catalysts were prepared and tested by Zhang and co-workers for the HIS of *n*-heptane in a fixed bed microreactor at 340 °C and 0.5 MPa.[323,324] The pretreatment of catalysts involved either direct reduction in H_2 or oxidation in air followed by flushing with N_2 and reduction under a flow of H_2. For a Pt/SAPO-11 catalyst containing 0.4% Pt and pretreated by direct reduction, the overall conversion decreased from 60.5 to 55.9%, whereas the selectivity increased from 46.5 to 58.0% after the H_2:*n*-heptane ratio was increased from 5.5:1 to 11.5:1. For the Pt/SAPO-11 catalyst pretreated by oxidation–reduction, the selectivity exceeded 70% at an H_2:*n*-heptane ratio of 11.5:1. This demonstrated that the method of catalyst pretreatment can have an effect on its selectivity for HIS.

The activity of Pt/SAPO-11 catalysts can also be influenced by the method of preparation. This was demonstrated by comparative testing of three Pt/SAPO-11 catalysts, each prepared by a different method.[317] The catalysts were tested using hexadecane as feed in a continuous flow fixed bed reactor at 8 MPa H_2. The method of preparation influenced the number and strength of acidic sites. The HIS activity was related to the number of Brønsted acid sites, whereas HCR activity was related to the number of strong acidic sites. The conversion, selectivity and yield relationship followed a similar trend to that shown in Figure 5.21. Thus, the maximum yield of isomers occurred at about 350 °C for all SAPO catalysts. Above 350 °C, the yield of isomers decreased and the yield of cracking products increased. Methylpentadecanes were the dominant isomers. When a static hydrothermal method was employed to synthesise nanosized SAPO-11, the catalyst samples prepared by this method had a larger specific surface area and a larger external surface area. These high surface area Pt/SAPO-11 catalysts exhibited better selectivity for HIS than the catalysts prepared by conventional methods.

Ultradispersed SiO_2 prepared by reacting $SiCl_4$ with O_2 in a capacitively coupled plasma was employed for the synthesis of SAPO-31 by Zubkowa *et al.*[325] The Pd-supported catalysts prepared using this SAPO-31 exhibited a higher selectivity for the HIS of *n*-heptane (15% in H_2) at 300 °C than a reference Pd/SAPO-31 catalyst. Ageing the catalyst prepared from ultradispersed SiO_2 by storage at room temperature over several weeks had no detrimental effect on the catalyst activity and stability.

Sinha *et al.* gave detailed accounts of the performance of SAPO-11 and SAPO-31 during HIS of *n*-hexane, *n*-octane and *n*-hexadecane.[326] The SAPOs were synthesised from either aqueous or non-aqueous solutions. After drying and calcining, the catalysts were loaded with Pt using a wet impregnation method to obtain 0.5% Pt. The experiments were carried out in a continuous downflow reactor between 275 and 375 °C, near atmospheric pressure and using a H_2:hydrocarbon molar ratio of 5:1. The SAPOs prepared from non-aqueous media were more active due to a larger number of acidic sites being present on the catalysts. The Pt/SAPO-31 was more active than Pt/SAPO-11, but the ratio of multi-branched to mono-branched isomers was consistently greater for the Pt/SAPO-11-based catalyst. Above 80% conversion of *n*-hexadecane, the selectivity to multi-branched isomers exceeded that to mono-branched isomers over all of the catalysts. However, the same was not found for *n*-hexane or *n*-octane. The results published by Liu *et al.* on the HIS of dodecane over Pt/SAPO-11 at 4 MPa indicated that mono-branched isomers were never produced at a greater rate than the multi-branched isomers.[327] The method of catalyst preparation and Pt content were similar for the catalysts prepared by

Sinha *et al.*[326] and Liu *et al.*[328] However, the difference in chain length of the feed and operating pressure were sufficient to affect the isomer distribution at high conversion.

The general trend for mono-branched and multi-branched isomer selectivity as a function of conversion is illustrated by the HIS of *n*-hexadecane over Pt/SAPO-11 (Figure 5.22).[329] Huang *et al.* compared the performance of a Pt/SAPO-11 catalyst during the HIS of *n*-hexadecane with that of Pt containing ZSM-5, H-Beta and MCM-22 catalysts. As noted earlier, the better HIS activity and selectivity of Pt/SAPO-11 was attributed to its weak acidity. At low conversion, mono-branched isomers dominated the product selectivity, but when the conversion exceeded 90%, the yield of multi-branched isomers rapidly increased (Figure 5.22). Reportedly this is consistent with the reactions occurring on the external surface rather than in the pores of the Pt/SAPO-11 catalyst.

Multi-branched isomers are more susceptible to HCR and even over a mildly acidic SAPO-11 this was well illustrated by HIS of *n*-hexadecane alone and in a mixture with 2,6,10,14-tetramethylpentadecane over Pd/SAPO-11.[330] When the feed contained a highly branched alkane, conversion was dominated by cracking to gaseous products. When the feed contained only an *n*-alkane, the HIS selectivity was markedly increased. Over Pd/SAPO-11, the HIS selectivity was 70% at 97% conversion.[330] Over Pt/SAPO-11 and under similar conditions, the HIS selectivity was 85% at 94% conversion.[331]

A study that had some commonality with feed from FTS involved the HIS of sunflower oil, which contained more than 10 mass% of oxygen.[332] The approach taken by Hancsok *et al.*[332] was to hydrotreat the sunflower oil first,

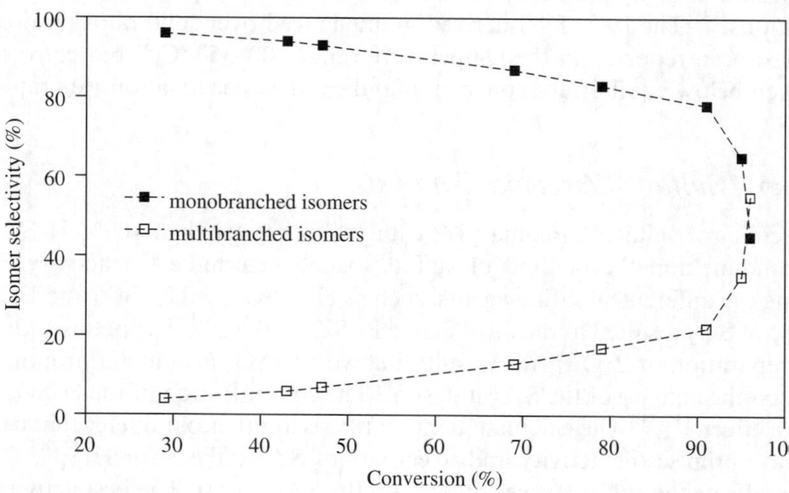

Figure 5.22 Typical relationship between conversion and selectivity to mono-branched (■) and multi-branched (□) isomers during HIS of long-chain alkanes. Shown is the HIS of *n*-hexadecane over Pt/SAPO-11.

which reduced the oxygen content to less than 0.05 mass%, before HIS of the product. The hydrotreated sunflower oil contained more than 90% *n*-octadecane and was hydroisomerised over a hydrothermally synthesised SAPO-11 catalyst promoted with 0.2–1.0% Pt. The best conditions for HIS of the hydrotreated sunflower oil over 0.5% Pt/SAPO-11 were 320–330 °C, 5–6 MPa and 300 normal m^3 H_2 per m^3 of liquid feed. Under these conditions, the cetane number of the product exceeded 88 and the product had excellent cold flow properties.

5.2.3.5 Phosphate and Phosphoric Acid Catalysts

Among phosphates, boron phosphate has attracted attention as a potential catalyst for the IS of butenes.[333] The early methods that were used for the preparation of such catalysts from a mixture of boric acid and phosphoric acid produced catalysts with a limited surface area. It was observed that the surface area could be increased by employing alkyl derivatives of boric acid as starting material.[92] Further improvements in catalyst performance could be achieved by promoting the boron phosphate catalysts with silicon.[334] More than 20 combinations of BPO_4 with either silicon or aluminium were prepared and tested. The silication of BPO_4 increased the surface area and the stability of the catalyst and resulted in a substantial improvement in catalyst activity.

Phosphoric acid on silica was active for the IS of *n*-butene (5% butene in N_2) at 450 °C and the product distribution approached equilibrium.[242] With the most active catalyst (65 mass% P_2O_5), the relative concentration of isobutene was 42%, whereas the by-products (isobutane, propene and pentenes) amounted to about 5%. Catalysts with a very high content of P_2O_5 increased the formation of by-products. Several zeolites were compared under identical conditions.[335] The IS of FTS-derived *n*-butene feed over solid phosphoric acid has also been reported in the temperature range 300–350 °C,[51] but conversion was well below equilibrium conversion and catalyst deactivation was rapid.

5.2.3.6 Sulfated Zirconia Catalysts

The acidity of sulfated zirconia (SZ) catalysts is greater than 100% H_2SO_4.[336] The amount and the method of sulfate loaded determine the activity.[337,338] Among the different sulfating agents, such as H_2S, SO_2, $(NH_4)_2SO_4$ and H_2SO_4, the use of SO_2 resulted in the most active Pt/SZ catalyst.[339] The pH used during the preparation of ZrO_2 prior to sulfating with H_2SO_4 also had a pronounced effect on the activity of the SZ catalyst.[340] The concentration of the H_2SO_4 also had an effect.[341,342] The calcining temperature is another parameter that can be used to optimise the activity and selectivity of SZ catalysts for HIS.[343] Calcination affects the sulfur species present on the SZ catalyst. The best activity for HIS of *n*-hexane over Pt/SZ was obtained after calcination in the temperature range 530–605 °C.[344] Below this temperature range, a high concentration of sulfur species such as S^{4+} and S^{6+} coexisting over the surface of the amorphous material was observed. After calcining in the optimal temperature range, only S^{6+} was detected. Given a specific method of catalyst preparation, the amount of sulfate in the SZ catalyst can be directly related to the activity.[345,346] However, there is a limit, and a plateau is reached at a sulfur concentration of 0.5–0.8 atoms nm^{-2}, after which further sulfate loading has no beneficial effect on activity.[347]

It has been observed that to various extents other acid-catalysed reactions occur in parallel with HIS. Considering that SZ is a superacid, side-reactions requiring strong acidity may occur in addition to those side-reactions generally expected, such as OLI and cracking. For example, SZ is active for alkylation of alkane–alkene mixtures.[348–350]

Unpromoted SZ is active for alkane and alkene IS. Table 5.19 compares the performance of the SZ with that of H-Beta during the IS of *n*-butane in the temperature range 150–350 °C.[351] It is clear that the SZ is considerably more active than H-Beta. Both catalysts are active for the alkylation of isobutane with 2-butene. Conversions higher than 60 and 90% were achieved at 0 and 50 °C, respectively. An active catalyst for *n*-butane IS could also be prepared by activation of ZrO_2 with SF_4.[352] The activation resulted in a larger number of Brønsted acid sites compared with the original ZrO_2. When TiO_2 was treated in a similar way, the resulting catalyst was less active than the catalysts based on ZrO_2.

Despite the activity of unpromoted SZ for the IS of alkanes, in the absence of H_2 SZ catalysts became deactivated with time on-stream because of coke deposition.[353,354] The coke deposition can be controlled by the addition of H_2 to the reactant stream.[355] The catalyst life of the SZ can be further extended by the addition of a metal promoter, such as Pt.[356] Views regarding the role and form of the Pt metal in these catalysts are unclear, although some information suggests that after calcining at high temperatures, most of the Pt is in a metallic form.[357,358]

Keogh *et al.* studied the effect of the amount of Pt added to SZ on the conversion of *n*-hexadecane.[359,360] The overall conversion of *n*-hexadecane reached a maximum at about 0.6 mass% of Pt (Figure 5.23) and a further increase in Pt loading did not improve the activity. The yields of HIS and cracking products remained constant with increasing Pt content over the range 0.6–5.0 mass%. However, the average carbon number of the cracked products

Table 5.19 Activity comparison of sulfated zirconia and H-Beta (Si:Al = 15) for the isomerisation of a mixture of *n*-butane in nitrogen.

			Product selectivity (%)		
Catalyst	*Temperature (°C)*	*Conversion (%)*	C_3	*Iso-*C_4	C_5
SO_4^{2-}/ZrO_2	150	21.4	10.7	89.2	Trace
	250	25.8	22.9	77.0	Trace
H-Beta	250	4.9	20.4	79.6	–
	350	40.6	36.9	49.7	13.4

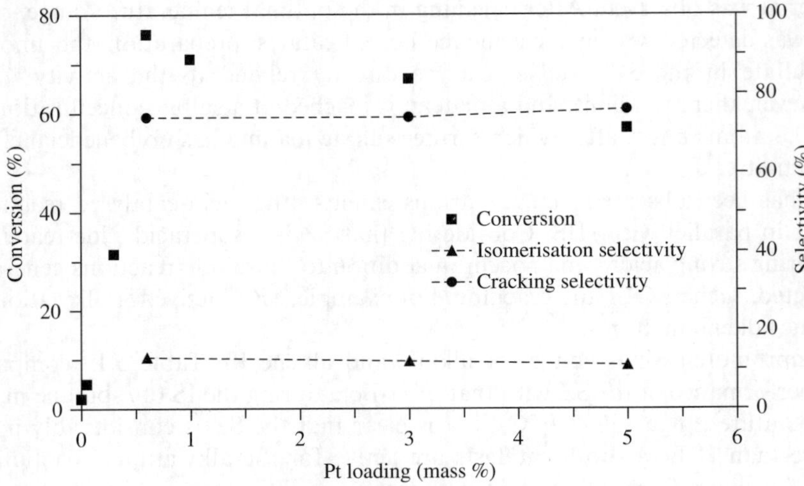

Figure 5.23 Influence of Pt loading on the performance of sulfated zirconia as catalyst for the HIS of *n*-hexadecane in an autoclave at 150 °C and about 3 MPa of H_2.

was influenced by the change in Pt content. At constant temperature, the conversion of *n*-hexadecane increased with increasing H_2 pressure. The increased conversion was mostly accounted for by an increase in the C_5–C_9 fraction. The distribution of the cracked products was asymmetric, suggesting that HCR was not ideal and that the mechanism over SZ deviated from the conventional mechanism. At a 5% Pt loading, one would expect to observe a high HIS selectivity and ideal HCR depending on the temperature.

A similar observation was made by Venkatesch *et al.* using C_7 and heavier alkanes.[361] The addition of alkenes to the feed inhibited the HCR/HIS of alkanes, suggesting that over Pt/SZ catalysts the mechanism may not involve metal-catalysed dehydrogenation of alkanes to alkenes as a necessary first step for protonation to take place. The Pt/SZ catalysts seem to have strong enough acidity to allow direct protonation of the alkane, with rearrangement possibly taking place *via* a pentacoordinated carbon intermediate. This suggestion was also supported by the trends in the effect of H_2 pressure on conversion.

The combination of Pt with a pure SZ and/or with SZ supported either on Al_2O_3 or SiO_2 was investigated with the aim of anchoring Pt on the support rather than on SZ.[362] These catalysts were tested for the HIS of *n*-octane at 300 °C, 1.5 MPa, LHSV 4 h^{-1} and an H_2:*n*-octane ratio of 6:1. The catalyst consisting of Pt dispersed on SZ/SiO_2 exhibited the highest activity and stability.

When Ni- and Pt-promoted SZ catalysts were compared for the HIS of *n*-butane at 300 °C, it was found that Pt was a much better promoter than Ni.[363] The beneficial effect of the promoters resulted from enhanced hydrogen activation to reduce catalyst deactivation by coke deposition and in this respect Pt was better than Ni. However, it was noted that the same beneficial effect of Pt addition to SZ on the HIS of *n*-butane was not observed for SZ promoted with Pd.[364]

The performance of the Ni/SZ and Pt/SZ catalysts for HIS of *n*-butane was compared in a flow of either He or H_2.[365] For the Pt/SZ catalyst, the conversion to isobutane in He reached a maximum at about 140 °C, whereas in H_2 the conversion was very low. However, above 200 °C the conversion in H_2 abruptly increased because of the diminished coke deposition on the catalyst. These effects were more evident over Pt/SZ than Ni/SZ,[365] thereby supporting the conclusion of Yori and Parera.[363] At 206 °C in a flow of H_2, the rate of HIS of *n*-pentane over Pt/SZ was almost seven times greater than HIS of *n*-butane.[366] After increasing the H_2 pressure from 0.22 to 0.61 MPa, the rate of HIS of *n*-butane decreased markedly compared with little decrease in the HIS of *n*-pentane. These observations were contrary to those for *n*-hexadecane, where an increase in H_2 pressure at the same temperature increased the conversion.[359,360]

Under conditions similar to those employed in other studies,[363–366] it was found that the activity of Pt/SZ for HIS of *n*-butane was significantly greater than that of Pt/H-MOR.[367]

Several studies on SZ catalysts by Grau and co-workers reiterated the conclusion reported in many studies on different HIS catalysts, namely that good catalysts for HIS must have good IS activity and mild cracking activity.[368–372] For example, the acid strength required to isomerise *n*-octane to branched octanes is low. The addition of SO_4^{2-} to ZrO_2 and its subsequent calcination at 620 °C produced a solid acid with a high percentage of strong acid sites that are responsible for deep cracking and the production of light gases. The promotion of ZrO_2 with tungstate anions and calcination at 800 °C generated a milder acidity than SZ.[370] The addition of Pt increased the acidity and the yield of light alkanes.[368] In the Pt-supported catalysts, the crystalline structure of ZrO_2 influenced the acid and metal properties of the catalyst. Pt supported on tetragonal ZrO_2 had a lower dehydrogenation activity than that of Pt supported on monoclinic ZrO_2, whereas the opposite effect was observed for IS.

Sulfated zirconia-based catalysts that were modified with Fe and Mn displayed high activity for the IS of *n*-butane in argon at 450 °C.[235] Compared with the unmodified catalysts, the reaction was several orders of magnitude faster even at room temperature.[373] Because the acid strength of the surface sites on modified and unmodified SZ catalysts was similar, the high activity was ascribed to the presence of Fe and Mn and specifically the ability of the metals to produce butene from butane. The product distribution indicated that IS took place *via* a bi- (or multi-) molecular mechanism. The addition of Fe and Mn also resulted in a significant enhancement in the activity and stability of the SZ catalyst during the hydroconversion of *n*-pentane.[339,374] These catalysts nevertheless deactivated within hours. The Fe-promoted catalyst exhibited the highest activity. The Mn-promoted catalyst showed the longest induction period before reaching maximum conversion. The Fe- and Mn-promoted SZ catalysts also catalysed disproportionation reactions, giving isobutane as a major product in addition to small amounts of hexanes, propane and isopentane.

5.2.3.7 Tungstated Zirconia Catalysts

Tungstated zirconia (TZ)-based catalysts, particularly those promoted with Pt, have attracted attention as catalysts for the HIS/HCR of *n*-alkanes. These catalysts are generally less acidic than equivalent sulfated zirconia catalysts (Section 5.2.3.6), making them better suited for HIS.

Platinum-promoted tungstated zirconia catalysts exhibited a high activity for HIS.[375–377] After calcination at 730–830 °C, these catalysts were more active for the HIS of *n*-heptane than many acidic zeolites. As in the case of SZ catalysts, the performance of Pt/TZ catalysts can be influenced during preparation. The method of Pt loading and calcination combined with reduction affect the catalyst activity.[378]

Yori and co-workers investigated several TZ-based catalysts prepared by different methods for the HIS of *n*-butane.[379,380] The focus was on the effect of Pt on the catalyst activity and selectivity. For TZ catalysts, the conversion of *n*-butane to isobutane required at least 0.6% of Pt. This corresponds to the Pt for maximum activity reported by Keogh and co-workers for SZ (Figure 5.23).[359,360] The absence of activity at low loadings of Pt was attributed to the strong interaction of Pt with TZ. Experiments conducted in a continuous fixed bed reactor at 300 °C and near atmospheric pressure of H_2 found that the most active catalyst for butane HIS was 0.4% Pt/ZrO_2, with 70% conversion and high selectivity to isobutane.

In the study of Busto *et al.*, the acid and metal function of the bimetallic Pt–Pd/TZ catalyst was controlled by varying the W content and calcination temperature.[381] The highest activity and stability for the conversion of *n*-decane were obtained for a catalyst with 15% W that was calcined at 700 °C. All catalysts produced a high RON, typically between 75 and 95, with a low yield of light gases. Coke formation occurred on the Lewis acid sites. Thus a correlation between the amount of Lewis acidity and the amount of carbon deposition could be established.

5.2.3.8 Other Catalysts

A novel catalyst comprising a caesium hydrogen salt of 12-tungstophosphoric acid (TPA) promoted with Pt was compared with Pt-promoted H-ZSM-5 and SZ catalysts for the HIS of *n*-pentane and *n*-hexane in the presence of a small amount of H_2 at 180 °C.[382] The Pt/TPA catalyst exhibited the highest activity for conversion of both feed materials. The deactivation rate of the Pt/TPA catalyst was rather low, because of the moderate and uniform strength of the acid sites. The activity and selectivity could be further increased by employing a mechanical mixture of the Pt/TPA catalyst with Pt/Al_2O_3.

Hino and Arata also found that the activity of the sulfated oxides such as TiO_2, Al_2O_3 and Fe_2O_3 could be enhanced by mechanical mixing with Pt/ZrO_2.[383] A similar effect was observed when Pt/Al_2O_3 was mechanically added to a TPA/ZrO_2 catalyst.[384] Indeed, it was shown before that mechanical mixtures of properly selected catalysts can exhibit significantly enhanced activity and selectivity for HIS of hydrocarbons compared with the individual solids.[379]

When TPA was supported on ZrO_2, the resulting catalyst was active for skeletal IS of 1-butene.[385] A 5–25% yield of isobutene could be obtained at 400 °C with an H_2:1-butene ratio of 1:1. Catalysts prepare using ZrO_2 only and TPA/SiO_2 were not active for skeletal IS of 1-butene.

Supported tungsten oxide (WO_3/Al_2O_3) exhibited a high activity and selectivity for the IS of *n*-butene.[384] Based on this observation, an extensive investigation of this reaction was undertaken by Benitez *et al.*, who prepared and tested a series of catalysts with different W loadings.[386] Conversion was evaluated with pure *n*-butene and *n*-butene diluted with N_2 in a fixed bed flow system at 380 °C. The products contained no C_1 and C_2 products. The IS took place by a bimolecular mechanism, as evidenced by a C_3:C_5 ratio of 1.0 in the products. The *n*-butene conversion and isobutene yield increased with increasing W content, both reaching a maximum at about 7 mass% W in the catalyst.

The activity of WO_3/TiO_2 and TiO_2 were compared at 420 and 450 °C in a microreactor using pure 1-butene feed.[387] Significant activity for skeletal IS and double bond migration were observed over WO_3/TiO_2. Some cracking and HYD also occurred. Moreover, the presence of aromatic structures in the coke on the catalyst confirmed the occurrence of aromatisation by hydrogen transfer. With time on-stream, the selectivity to products other than butenes decreased due to catalyst deactivation. Under the same conditions, the activity of TiO_2 was lower only during the early stages of reaction. With increasing time on-stream, the activity difference became small, because less coke was deposited on TiO_2 than on the WO_3/TiO_2 catalyst.

Akhmedov *et al.* used the metal vapour deposition method during the preparation of Ni catalysts supported on MgO, Al_2O_3 and ZSM-5 zeolite.[388] This method ensured a high dispersion of the metal on the supports. The catalysts were evaluated for the conversion of *n*-heptane at 190–220 °C and a near atmospheric pressure of H_2. In comparison with the Ni/ZSM-5 catalyst, the HIS activities of the Ni/MgO and Ni/Al_2O_3 catalysts were rather low. For these catalysts, the product distribution revealed C_1–C_3 and linear C_4–C_6 products, with hardly any isomers. Over the Ni/ZSM-5 catalyst, isobutane and branched heptanes accounted for more than 70% of products. Branched C_5 and C_6 isomers were not present. This work again illustrated the importance of acidity for HIS. The catalyst support materials with little acidity catalysed only hydrogenolysis, which is associated with the Ni.

The Mo oxycarbide catalyst prepared by the oxidation of molybdenum carbide exhibited a high selectivity for HIS of *n*-heptane at 350 °C and 0.65 MPa.[306,389] The branched C_7 products were dominated by monomethylhexanes, such as 2-methyl- and 3-methylhexane, which were close to their equilibrium ratio. The C_7 selectivity was affected by H_2 pressure. This catalyst was compared with Pt/Beta-zeolite and Pt/Al_2O_3 catalysts. The Mo oxycarbide was more resistant to sulfur poisoning than the Pt-supported catalysts. The MoO_3 modified with carbon was resistant to poisoning at both 30 and 120 μg g^{-1} S in the feed.

Ruthenium catalysts were prepared using different supports, namely graphite, activated carbon, SiO_2 and Al_2O_3, and were used for the HIS/HCR of *n*-hexane.[390] Ten catalysts with different pretreatments were evaluated in a continuous flow fixed bed reactor operated at 477 °C, near atmospheric pressure and an H_2:*n*-hexane molar ratio of 5.3:1. For all the carbon-supported catalysts the yields of cracking product (C_1–C_5) were much greater than that of the branched C_6 isomers. Over the Ru/SiO_2 and Ru/Al_2O_3 catalysts, the yields of branched C_6 isomers approached 58 and 54%, respectively, but the catalysts deactivated at a faster rate. In a related study conducted under similar conditions, Pt on activated carbon was employed for HIS of *n*-heptane.[391] In this case, hydrogenolysis and dehydrogenation to C_2–C_6 alkanes and alkenes were the dominant reactions, while HIS and aromatisation proceeded at comparable, but lower rates. The product distribution from this study is ascribed mainly to metal site catalysis, due to the absence of acidic sites on the activated carbon surface.

Catalysts based on natural vermiculite, pillared with an Al–Ce hydroxide solution and promoted with 1 mass% Pt, were evaluated for the HIS of decane at 150–300 °C using an H_2:decane ratio of 375:1.[392] The volume of the pillaring solution corresponded to 12 mmol (Al + Ce) per gram of clay. Different Al:Ce ratios were used and little change in conversion or HIS selectivity was observed up to an Al:Ce ratio of 10:2. In this range, the catalyst activity and selectivity, and also the carbon number distribution, approached those obtained with Pt-promoted 10- and 12-membered ring zeolites. However, when the Al:Ce ratio was increased to 8:4, the catalyst activity declined and the temperature required for maximum HIS selectivity increased from 210 to 260 °C.

5.2.4 Catalyst Deactivation During Isomerisation

The comments made about deactivation during OLI (Section 5.1.7) are equally applicable to catalyst deactivation during IS/HIS. In the case of unpromoted acid catalysts, the deactivation mechanisms during IS and OLI are the same, but in the case of metal-promoted catalysts the metal sites modify the deactivation behaviour somewhat.

5.2.4.1 Oxygenate-related Deactivation

Most of the studies on the IS/HIS of hydrocarbons have paid little attention to the effect of oxygenates. However, the primary products from FTS always contain oxygenates in concentrations ranging from trace amounts to percentage levels. Water and oxygenates affect to various extents all reactions occurring during upgrading. In the case of HIS and IS, water and oxygenates competitively adsorb and even modify the surface structure of acidic catalysts. The modifying effects will vary from catalyst to catalyst. Of commercial relevance is the inability to employ chlorinated Pt/Al_2O_3 catalysts with Fischer–Tropsch feed unless feed pretreatment reduces the water and oxygenate content to limit dechlorination of the catalyst surface with its associated deactivation.

The inhibiting effect of strong adsorption of oxygenates on IS/HIS catalyst performance has been documented by Cowley,[256] who conducted a study on the HIS of pentenes from FTS using a commercially available metal-promoted non-zeolitic molecular sieve catalyst (UOP Pentesom). One feed containing oxygenates and the other oxygenate-free were investigated under identical conditions. When an oxygenate-free feed was employed, stable skeletal IS activity could be maintained at 280 °C, but when the feed was changed to an oxygenate containing Fischer–Tropsch-derived feed, the temperature had to be increased to 320 °C to achieve stable activity (Figure 5.24). The contribution of the undesirable alkene side reactions to catalyst deactivation is small compared with the contribution of oxygenates. This was supported by the analysis of spent catalysts, which indicated a much lower coke content for the oxygenate-free feed. Water, either from the feed or from the reaction of oxygenates over the catalyst, strongly adsorbed on the catalyst to inhibit conversion. The water could be readily desorbed from the catalyst at temperatures above 320 °C and did not cause permanent catalyst deactivation. The effect of water is reversible because water can be removed by increasing the temperature.[393] However, this precluded the use of the catalyst at operating temperatures below 320 °C with straight run feed from FTS. Catalyst deactivation by the formation and build-up of carbonaceous deposits occurs at a lower rate at lower temperatures and the inability to exploit the operating window at 280–320 °C decreased the

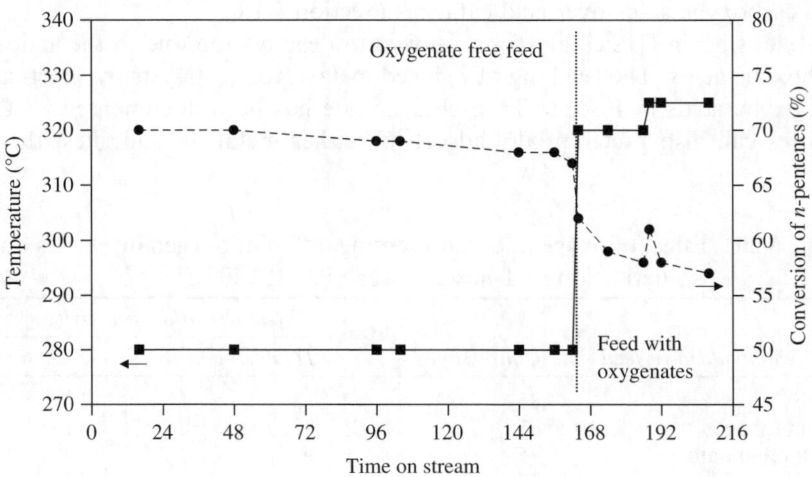

Figure 5.24 Deactivation behaviour during *n*-pentene skeletal IS over a metal-promoted non-zeolitic molecular sieve catalyst at 1.7 MPa with H_2 co-feed. After 162 h on-stream, the feed was changed from an oxygenate-free feed to an oxygenate-containing Fischer–Tropsch feed. The temperature (■) had to be increased to regain some of the lost conversion (●).

catalyst cycle length. With oxygenate-free feed materials, a cycle length of 10–12 months can be expected, but with oxygenates the cycle length is reduced to 1–2 months. Although the main cause of catalyst deactivation was the formation of carbonaceous deposits, this occurred due to the higher operating temperatures and more frequent temperature increases required by the presence of oxygenates.

Promoting the acidic support with a metal did not influence the inhibiting effect of the oxygenates. The inhibiting effect of water has also been specifically noted with other metal-promoted acid catalysts. For example, the inhibiting effect of water on metal-promoted acidic resin-catalysed reactions was reported by du Toit and Nicol.[394] Although the conversion in question was not HIS, it is a bifunctional catalyst and the work illustrates the point.

The way in which the oxygenates affect the isomerisation catalysis is dependent on both the nature of the catalyst and the nature of the oxygenates. The effects of different oxygenate classes on the isomerisation of 1-hexene over SPA is shown in Table 5.20.[66]

Of the oxygenates tested, the adverse effect of 2-pentanone and butanoic acid on conversion and skeletal IS was the least evident. The other oxygenates, namely 2-propanol, 1-butanol, propanal, ethyl ethanoate, 1,1-dimethoxyethane and ethoxyethane, significantly suppressed conversion and/or skeletal IS. This is partly related to the oxygenate functionality and formation of water, but also to the strong interaction of short-chain alkenes thus formed with SPA. Although the ketones and carboxylic acids had the least impact on the acid catalysis, the same is not true of bifunctional catalysts containing reduced metal sites. In this respect it is important to remember that ketones can be converted into carboxylic acids over acid catalysts (Section 5.1.6).

Metal sites in HIS catalysts are subject to deactivation due to the action of carboxylic acids. The leaching of reduced metals from catalysts by short-chain carboxylic acids in Fischer–Tropsch syncrude has been documented.[395] Oxygenates can also preferentially adsorb on either metal or acid sites, thereby

Table 5.20 Effect of oxygenates representing different oxygenate classes on the isomerisation of 1-hexene over SPA at 140 °C.

Oxygenate added to feed	*Conversion (%)*	*Isomerisation selectivity (%)* Double bond IS	*Isomerisation selectivity (%)* Skeletal IS[a]
None	85	83	17
2-Pentanone	81	88	12
Ethyl ethanoate	75	94	0
Butanoic acid	70	88	12
Ethoxyethane	40	96	4
1-Butanol	15	87	13
Propanal	7	88	12
2-Propanol	2	47	53
1,1-Dimethoxyethane	0	–	–

[a]Skeletal IS precedes OLI for hexenes over SPA; OLI products counted towards skeletal IS.

changing the metal-to-acid site ratio of bifunctional catalysts.[396] Although this does not result in deactivation, carboxylic acids can lower the effective metal-to-acid site ratio, which may cause an increase in the rate of deactivation by the formation of carbonaceous deposits.

Some oxygenate-induced deactivation can be beneficial in processes where IS is not desirable. For example, it has been reported that during the conversion of oxygenate-containing feed from FTS over Al_2O_3, the acid sites responsible for IS were selectivity deactivated, which permitted 1-alcohol dehydration to achieve high 1-alkene selectivity.[397]

5.2.4.2 Deactivation by Carbonaceous Deposits

To various extents, the formation of carbonaceous deposits during operation affects catalyst performance. This may be attributed to blocking active sites or restricting access to active sites due to such deposits. Moreover, the shape selectivity of the catalyst may be modified if coke deposits are formed in pores. However, in some instances such deposits have a beneficial effect, with the increase in butene skeletal IS selectivity over ferrierite being a case in point (Section 5.2.3.1).[398,399]

The most common industrial catalysts for HIS of naphtha range materials have catalyst lifetimes varying from 2–3 years for chlorinated Pt/Al_2O_3 to over 10 years for Pt/MOR.[248] In spite of fairly clean systems being used and the long catalyst lifetimes that can be achieved, catalyst deactivation by deposition of carbonaceous material during IS/HIS can ultimately not be avoided. This is especially true of skeletal IS processes employing unpromoted acidic catalysts or olefinic feeds. Although the formation of carbonaceous deposits is usually the main factor contributing to deactivation, other deactivation mechanisms may also contribute to deactivation. For example, recrystallisation of the active phase that may affect catalyst activity cannot be ruled out during operation at high temperatures.

Over bifunctional IS catalysts, it is expected that the structure of coke formed during the IS/HIS will differ markedly from that observed on the spent catalysts used in the hydroprocessing of heavy petroleum feeds. For example, Cowley reported that the carbonaceous deposits formed during the IS of pentenes over a bifunctional acidic non-zeolitic molecular sieve catalyst was not aromatic and did not comprise hydrogen-deficient polynuclear aromatics.[256] The deposits were not hard coke, but rather paraffinic, olefinic or polyolefinic structures with an H:C ratio exceeding unity. The structure of the carbonaceous deposits, and whether these deposits can be classified as coke, depend on the type of catalyst and operating conditions. Therefore, it is possible that the formation of an aromatic coke during the HIS of hydrocarbons also occurs, as was recorded during industrial high-temperature IS of pentenes from FTS over an Al_2O_3 catalyst.[400]

The catalyst type has a pronounced effect on catalyst deactivation during alkene IS (Table 5.21).[242] The reported deactivation was caused mainly by coke

Table 5.21 Deactivation behaviour of different catalysts during the IS of *n*-butene (5% butene in N_2).

Catalyst	*Isobutene (%)*	*By-products (%)*	*Temperature (°C)*	*Decrease in activity*
Ferrierite	41.2	9.7	350	Stable after 400 h
Chlorinated Al_2O_3	41.0	6.0	350	40% after 150 h
MnAPO-11	42.6	3.6	400	Stable after 16 h
SAPO-11	40.1	1.3	440	10% after 95 h
ZSM-22	33.6	9.8	420	25% after 20 h
H_3PO_4/SiO_2 (65% P_2O_5)	42.0	4.6	440	50% after 17 h

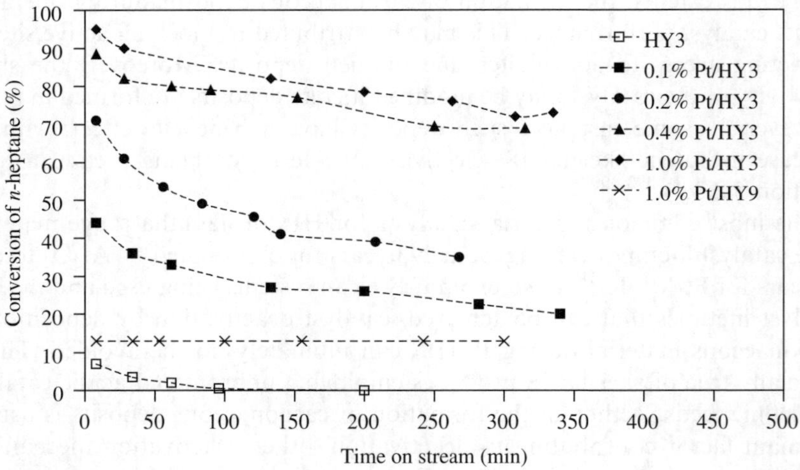

Figure 5.25 Effect of catalyst composition on catalyst stability during HIS of *n*-heptane over HY-zeolites at 250 °C, near atmospheric pressure and H_2:*n*-heptane ratio 9:1. The Pt loading and Si:Al ratios are indicated. The HY3 catalysts had an Si:Al ratio of 3:1, whereas the HY9 catalyst had an Si:Al ratio of 9:1.

deposition.[335] On oxidative regeneration, the catalyst activity could be restored to its original level. For the H_3PO_4/SiO_2 catalyst, the coke deposition increased with increasing concentration of *n*-butene in the feed mixture. The stability of this catalyst was increased when a small amount of water (3 kPa partial pressure) was added to the reaction mixture to maintain the hydration state of the catalyst. The catalyst stability may also be modified by ion exchange. For example, it was reported that catalyst deactivation was slowed for Li-exchanged ferrierite, whereas the same exchange with Cs had an adverse effect on catalyst stability.[267]

The stability of a catalyst can be improved by metal promotion and controlling the acid strength distribution. This is illustrated by Figure 5.25,[284] showing the effect of the Pt content and the Si:Al ratio of different HY zeolite catalysts on catalyst stability. An increase in Pt addition from 0 to 0.4% had a pronounced effect on the activity of the catalysts, but a further increase in Pt loading resulted in only incremental changes. As the Si:Al ratio increased, the acid site density decreased and for a constant Pt loading the catalyst stability increased. In general terms, it can therefore be stated that as the metal-to-acid site ratio increases, the catalyst stability increases. This makes sense, since the catalyst becomes more hydrogenating. Similar trends in heptane conversion with time on-stream as shown in Figure 5.25 were observed over a Pt catalyst supported on activated carbon.[391]

The results in Figure 5.25 can be considered as the initial activities, because of the short duration of the experiments. In a similar fashion, Corma *et al.* evaluated catalysts for the HIS of *n*-decane and observed a decline in activity within the first 3 h, after which the activity stabilised and exhibited little change until the end of the experiment lasting almost 300 h.[231]

The Si:Al ratio also had a pronounced effect on catalyst deactivation in Pt/MOR. In a study by Lenoi *et al.*, the HIS of *n*-pentane was evaluated over two different Pt/MOR catalysts, Pt/MOR(5) and Pt/MOR(18), having Si:Al ratios of about 5:1 and 18:1, respectively.[401] The Pt/MOR(5), which had a higher acid site density and a higher acid strength, deactivated more rapidly. A decrease in Pt dispersion during the experiment also contributed to catalyst deactivation. This was more evident for the Pt/MOR(18) catalyst than for Pt/MOR(5). On oxidative regeneration, it was easier to recover the activity of the Pt/MOR(5) catalyst than that of the Pt/MOR(18) catalyst. This was attributed to the diminished dispersion of Pt in the spent Pt/MOR(18) catalyst. XPS and NMR analyses revealed that the chemical structures of the coke on both catalysts were similar, but some differences in coke morphology were observed. These differences were caused by the different porosities of the catalysts.

Catalyst deactivation by carbonaceous deposits affects not only catalyst activity, but also catalyst selectivity. This can be seen from the results of 1-hexene HIS over Pd/MOR (Table 5.22), where the overall conversion and selectivity to 2,2-dimethylbutane for fresh and coke-deactivated catalysts differ.[402] Selectivity over the coked catalysts in the kinetically controlled regime leads to a difference in product selectivity, which is not apparent at higher

Table 5.22 Effect of catalyst deactivation by coking on the activity and selectivity of 1-hexene HIS over Pd/MOR at 260–280 °C and 2 MPa.

	Conversion (%)		*Selectivity to 2,2-dimethylbutane (%)*	
Catalyst	*260 °C*	*280 °C*	*260 °C*	*280 °C*
Fresh Pd/MOR (no coke)	78	80	23	24
Pd/MOR with 4.1 mass% coke	22	45	38	22
Pd/MOR with 6.1 mass% coke	18	28	42	23

temperatures in the thermodynamically controlled regime. It was postulated that coke deposited mainly in the pore openings. The H:C ratio of the coke was less than 1.0, indicating its aromatic nature. Aromatic compounds were also detected in the products. The amount and structure of the coke on the catalyst changed from top to bottom with position in the catalyst bed.

Catalyst morphology affects product selectivity and deactivation behaviour. This is shown by the IS of 20% *n*-butene in N_2 at 400 °C over SAPO-5 (7.3 × 7.3 Å channels), SAPO-11 (6.5 × 4.0 Å channels) and SAPO-34 (3.8 × 3.8 Å channels).[144] Significant skeletal IS was evident only over SAPO-11 and SAPO-34, although some isobutene also formed over SAPO-5, but disappeared after about 2 h on-stream. The high isobutene selectivity of SAPO-11 was attributed to the catalyst morphology, which allowed sufficient space for isobutene to be formed, but was restrictive enough to limit coke deposition. The large pore opening of SAPO-5 favoured coke deposition and the catalyst deactivated rather quickly. The SAPO-34 was the most acidic, but its small channels inhibited coke formation.

The influence of H_2 pressure from atmospheric to 0.5 MPa on catalyst stability during the HIS of *n*-octane over Pt/SAPO-5 and Pt/SAPO-11 at 375 °C was investigated by Campelo *et al.*[322] Deactivation with time on-stream was more pronounced for the Pt/SAPO-5 catalyst, due to its larger pore size, as indicated earlier. However, a gradual an increase in H_2 pressure to 0.3 and 0.5 MPa resulted in an improvement in catalyst activity with time on-stream. Although catalyst activity continued to decline with time on-stream, increased H_2 pressure reduced the rate of coke formation and also coincided with a decreased amount of coke deposited on the catalyst with increasing H_2 pressure. For Pt/SAPO-11, the same change in H_2 pressure resulted in a significant increase in the activity. Moreover, for Pt/SAPO-11, no catalyst deactivation during the entire run that lasted almost 16 h was observed. The use of metal promoters with a hydrogen co-feed is consequently an effective way to reduce the rate of coke formation and thereby limit catalyst deactivation during IS/HIS.

In terms of catalyst stability, the beneficial effect of adding metals to acidic catalysts from IS/HIS is clear, since it provides a hydrogenating function to limit coke formation and thereby reduce catalyst deactivation. In industrial practice, the use of noble metals is sometimes limited by the presence of sulfur in the feed, which may significantly suppress HIS activity.[306] For example, at 120 $\mu g\,g^{-1}$ S in the feed, the activity of a Pt/Beta-zeolite declined by about half. The sulfur sensitivity of noble metal-containing catalysts is not a concern when processing feed from FTS, because the primary hydrocarbon products from FTS are sulfur free.

5.2.4.3 *Deactivation of Sulfated Catalysts*

In the case of sulfated catalysts, the formation of carbonaceous deposits is the main source of catalyst deactivation (Section 5.2.4.2). It could be shown that during HIS of *n*-butane over SZ, removing alkenes from the feed significantly reduced catalyst deactivation.[403,404] Catalyst deactivation was observed even for a very small amount of deposited coke and catalyst activity can be restored by oxidative regeneration. However, this is not the only deactivation mechanism for sulfated catalysts.

Corma and co-workers reported that during the IS of 2-butene, the decline in catalyst activity over an SZ catalyst was much more pronounced than that over an H-Beta zeolite.[338,340] Although coke deposition was the main cause of deactivation, the elimination of sulfur as H_2S from the SZ catalyst contributed to the activity loss. The concentration of sulfur species is correlated with the number of protonic sites and loss of sulfur therefore contributed to activity loss.[405]

Li *et al.* reviewed and listed several causes of the loss of IS/HIS activity of SZ-based catalysts.[403] These included a reduction of the S^{6+} state to a lower oxidation state,[406,407] coke formation,[408] sulfur loss as H_2S,[409] surface phase changes[410] and the formation of organosulfur complexes *via* carbon–sulfur interactions during the deactivation process.[411] Sulfur may also be lost during calcination, which has an equally detrimental effect on SZ activity.[412]

As in the case of other IS/HIS catalysts, catalyst stability can be improved by promoting the sulfated catalyst with a metal. It has been shown that unpromoted SZ rapidly deactivates during HIS of *n*-butane at 300 °C and with an H_2:*n*-butane ratio of 6:1, but that deactivation is diminished for Ni/SZ and no deactivation was observed for Pt/SZ.[363] Similarly, the addition of Fe and Mn to SZ resulted in diminished initial deactivation during the HIS of *n*-pentane.[368]

5.3 Cracking and Hydrocracking

Cracking is one of the key technologies for the upgrading of FTS-derived waxes (atmospheric residue) to lower boiling products for the production of transportation fuels. The conversion of residual feed into lighter boiling fractions requires C–C bond scission. This can only be achieved at higher temperatures, even in the presence of a catalyst. Three main classes of commercial cracking technology can be differentiated:

1. hydrocracking (HCR), which requires operation in the presence of catalyst and H_2;
2. catalytic cracking, which requires operation in the presence of catalyst, but in the absence of H_2, such as fluid catalytic cracking (FCC);
3. thermal cracking, where operation is conducted in the absence of both catalyst and H_2.

Industrially, HCR has been adopted as standard FTS wax upgrading technology.[1] However, the study by Choi *et al.* indicated that fluid catalytic cracking (FCC) is more economical for transportation fuels production than HCR.[413] The industrial preference of HCR over FCC for upgrading of waxes from FTS is related to the production of distillate blending stock specifically.

Thermal cracking was found to be less efficient than HCR for the upgrading of FTS waxes to transportation fuels.[414] Thermal cracking of wax will not be discussed in this chapter (see Section 6.2.1). Most catalytic cracking applications are based on FCC technology,[415] hence further reference to catalytic cracking will focus on FCC.

5.3.1 Mechanism of Cracking

5.3.1.1 Mechanism of Catalytic Cracking

It is important that a clear distinction is made between catalytic cracking and HCR. The cracking studies conducted in the absence of a hydrogen co-feed are considered catalytic cracking, whereas HCR implies that hydrogen is a co-feed.

Catalytic cracking investigations over acidic catalysts in the absence of H_2 are generally conducted under FCC conditions. This brings the hydrocarbon feed in contact with the catalyst at a high temperature for a short period of time, typically only few seconds. Initial contact of the feed with the hot catalyst results in some thermal cracking ('thermal shock' conditions), but under typical FCC temperature conditions (480–550 °C) catalytic cracking dominates.

FCC catalysts are normally only monofunctional acidic catalysts and do not have metal sites that have the ability to dehydrogenate the feed. However, the involvement of hydrogen during catalytic cracking should not be discounted. Hydrogen that is transferred from the hydrocarbon feed to the catalyst surface is not desorbed as molecular hydrogen (H_2), but can be transferred between adsorbed species. This results in a comparative enrichment of the H:C ratio of some compounds (usually the lighter compounds), while reducing the H:C ratio of other compounds (usually the heavier compounds). In this way, carbon is rejected as coke on the catalyst surface, whereas the lighter reaction products from cracking are comparatively hydrogen enriched.[416]

Before significant concentrations of alkenes are created through cracking, direct protonation of the paraffinic feed takes place and cracking by protolysis is an important reaction pathway.[417] Protonation of an alkane will yield a pentacoordinated carbon structure that can crack by α-scission (protolysis) to yield products different from those from β-scission, including products that would otherwise require a primary carbocation intermediate to form *via* *β*-scission (Figure 5.26). Cracking by protolysis is also referred to as the Haag–Dessau mechanism of cracking.

The same basic cracking mechanism also applies to HCR. The main difference is that in hydrocracking the catalyst is bifunctional and the metal sites introduce additional catalytic pathways not present on a monofunctional acid catalyst.

5.3.1.2 Mechanism of Hydrocracking

The mechanism for HCR follows the same basic steps as that for HIS (Figure 5.15). The main difference is that during HCR the protonated isomerised

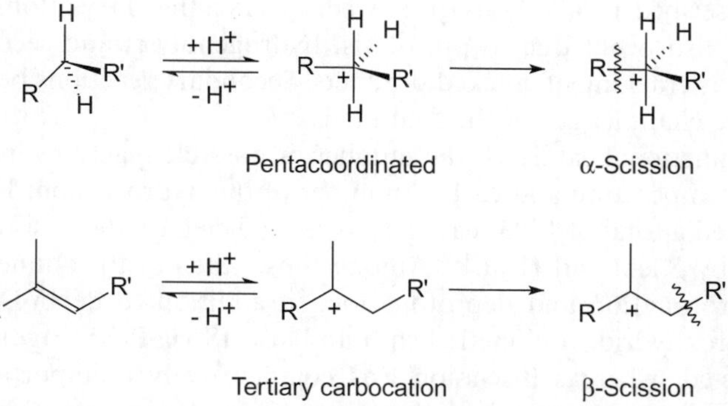

Figure 5.26 Catalytic cracking by protonation of an alkane to form a pentacoordinated carbocation resulting in α-scission (protolysis) and protonation of an alkene to form a carbocation resulting in β-scission.

intermediate undergoes β-scission before it is hydrogenated. The catalysts employed for both HIS and HCR are consequently similar in many respects and are bifunctional, with both metal sites and acid sites.

During the HCR of alkanes, the alkenes produced by dehydrogenation on the metal sites in the first step (Figure 5.15), are protonated on acidic sites. Subsequently, the carbocations undergo typical acid-catalysed reactions. Given sufficient reaction time and/or at high enough temperature, the carbocation intermediates are isomerised and then cracked. The IS/HIS reaction pathway is therefore always part of the overall HCR mechanism. The extent of HIS relative to HCR depends on temperature in relation to the acid strength of the catalyst and on the overall metal-to-acid site ratio. In Figure 5.21, the yield of cracking products increases as conversion increases and isomerisation selectivity decreases. Thus, at low temperature and/or low conversion, an HCR catalyst will typically behave like a HIS catalyst. With further increase in temperature and/or conversion, the contribution of HCR will gradually increase until it becomes the main reaction.

In the case of ideal HCR, two cracked products are formed from one parent reactant by cracking in a random fashion along the length the hydrocarbon chain at any position three or more carbons from the end. This type of ideal HCR may be approached when the deHYD/HYD function, which is provided by the metal sites, is properly balanced with the cracking function provided by the Brønsted acid sites. The conditions of ideal HCR are approached when using catalysts with a strong HYD function.[418] Ideal HCR is reflected by a symmetric distribution of the cracked products among the carbon numbers around the mean.[419,420]

The concept of an ideal HCR catalyst can be approached in practice with feed in the heavy naphtha range. For example, during the HCR of *n*-octane over an ideal HCR catalyst, a symmetrical distribution around C_4 was obtained up to 97% conversion.[421] However, on increasing the H_2 pressure from 1 to 5–10 MPa, a slight asymmetry in the amounts of C_3 and C_5 was noticed. This has been ascribed to alkylation followed by cracking. Deviations from ideal HCR may also occur due to protolysis, hydrogenolysis and secondary reactions, such as cracking of cracked products. Secondary reactions becomes more likely as the chain length of the feed increases.

Over bifunctional catalysts, the number of possible reactions increases with increasing temperature and carbon number of the hydrocarbon. For example, for a C_{16} feed a total of 1503 reactions were predicted on the basis of the model developed by Klein and Hou.[422] Among those, the most prominent reactions included protonation and deprotonation, in addition to deHYD and HYD. Furthermore, hydride and methyl shift and also IS *via* PCP (Figure 5.16) were also predicted, whereas β-scission was comparatively unimportant. Alkenes and carbocations were the most abundant among 465 species identified.

5.3.2 Commercial Processes for Cracking

The shrinking market for atmospheric residues (boiling point >360 °C) as heavy fuels, more stringent environmental regulations and high feedstock price forced refiners to convert residues into distillate boiling ranges. One way of accomplishing this is by cracking the heavy material into lighter boiling distillates. In conventional crude oil refineries, several residue processing technologies can be found. These may be hydrogen addition processes, such as HCR, and carbon rejection processes, such as fluid catalytic cracking, deasphalting, coking and visbreaking.[423] The quality of crude oil that can be processed in a refinery and the targeted product slate determine the type and extent of residue conversion.

Considering the significant difference in the composition between LTFT waxes and a typical crude oil residue fraction, one would not expect wax upgrading to follow the same refining strategy as that employed for heavy crude oil fractions. However, when it comes to cracking, the same basic conversion technologies can be considered for both, albeit with some modification.

Since LTFT waxes are already clean feed materials, a hydrogen addition strategy is not costly in terms of hydrogen use. HCR is consequently a preferred technology for upgrading waxes. Using the same argument, one would typically not consider a carbon rejection technology for products from FTS, because such products are already hydrogen rich. However, in refining practice, carbon rejection technologies are sources of alkenes that are essential for units such as aliphatic alkylation, etherification and oligomerisation. Although it seems wasteful, one cannot disregard catalytic cracking for the upgrading of Fischer–Tropsch products.

5.3.2.1 Commercial Hydrocracking Processes

The commercial processes employed for HCR can be classified according to reactor type as fixed bed, moving bed, ebullated bed and slurry bed reactor technologies. Because of the high severity of the operation (typically >350 °C and >5 MPa), primary products often require additional upgrading steps before being hydrocracked.

The commercial processes employed for HCR have been reviewed recently.[423–425] As the nature of the catalyst bed changes from fixed to moving, ebullated and slurry bed, the reactor technology becomes increasingly tolerant of metals and particulate matter in the feed. In advanced reactors, catalyst deactivation becomes less of a problem, since the catalyst can be replenished on-line. The selection of a commercial reactor technology is therefore dependent on the nature of the feed. Although FTS-derived feed materials are generally considered 'clean', some FTS residues and waxes may have a metal content that is sufficiently high to make fixed bed operation problematic.[426,427]

The conditions necessary for HCR are determined by the feed quality and the catalyst type, but in general, conventional hydrocrackers are operated in the range of 360–440 °C and 10–20 MPa.[428–430] Although mild HCR processes operate under less severe conditions, the hydrocracking of Fischer–Tropsch waxes requires even milder conditions, while achieving much higher conversions (Table 5.23).[429] The catalysts employed for mild HCR and HCR of Fischer–Tropsch waxes are typically less acidic.

Unsulfided base metal HCR catalysts would seem to be ideal for HCR of LTFT waxes, but many Ni- and Co-based HCR catalysts display high methane selectivity.[431–434] Unsulfided noble metal catalysts seem to work very well, not only on a small scale,[435–437] but also on a commercial scale, as used in the SMDS process in Bintulu, Malaysia.[438] It is likely that the proprietary catalyst employed by Shell in their SMDS process is a noble metal catalyst such as Pt on ASA support.

The HCR of LTFT waxes is much more facile than that of crude-derived residues.[437,438] It is consequently surprising that the Oryx GTL facility does not employ an unsulfided noble metal HCR catalyst. Chevron's Isocracking technology has been selected for this LTFT facility, which employs a sulfided base metal HCR catalyst operating at medium pressure. Hydrocracking has been adopted as the main upgrading technology for the conversion of waxes from Co-LTFT synthesis and will also be employed in facilities that are

Table 5.23 Typical processing conditions for conventional crude oil hydrocrackers, mild hydrocrackers and Fischer–Tropsch wax hydrocrackers.

	Hydrocracker type		
Description	*Conventional*	*Mild*	*FT wax*
Temperature (°C)	350–430	380–440	325–375
Pressure (MPa)	10–20	5–8	3.5–7
LHSV (h^{-1})	0.2–2	0.2–2	0.5–3
H_2:feed (normal m^3 m^{-3})	800–2000	400–800	500–1800
Reactor technology	Trickle bed	Trickle bed	Trickle bed
Conversion (%)	70–100	20–40	20–100

still under construction at the time of writing, namely Pearl GTL and Escravos GTL.[439,440]

When dealing with crude oil-derived feed that contains sulfur, the use of unsulfided catalysts can only be considered after the feed has been substantially hydrotreated. Conversely, FTS-derived feed is sulfur free and the use of sulfided HCR catalysts, as opposed to unsulfided catalysts, requires the co-feeding of a sulfiding agent, such as dimethyl disulfide (DMDS). It is consequently preferable to select unsulfided noble metal catalysts for HCR of FTS-derived waxes.[2,429] Nevertheless, fixed bed processes employing both sulfided and unsulfided catalysts have been used commercially for the HCR of FTS derived wax,[1] under lower severity operation than conventional hydrocrackers.[430]

The HCR of HTFT residue differs from that of the LTFT wax and resembles crude oil HCR. The HTFT residue fraction contains more than 25% aromatics, although its polynuclear aromatic content is low (<1%).[441] The same principles employed for the HCR of crude oil residues can be applied to HTFT residue, but less severe operating conditions are required on account of the low sulfur and nitrogen content of HTFT residue.

The heavy fraction from both HTFT and LTFT synthesis contains some metals as metal carboxylates. It has already been pointed out that HDM catalysts are ineffective for metal carboxylate removal.[426] There is consequently scope for the application of different HCR reactor technologies to overcome the problems associated with the deposition of metals from FT heavy fractions on catalyst surfaces, particularly in fixed bed reactors.

5.3.2.2 Commercial Fluid Catalytic Cracking Processes

There are various technology offerings for FCC.[415,442–444] The basic design principle for all technologies is the same. Normal FCC is performed at high temperature (480–550 °C), low pressure (0.1–0.3 MPa) and short contact time (<10 s). The hot catalyst is brought into contact with the feed before the feed and catalyst are separated again. The yield is influenced by the operating conditions, and also the nature of the feed. The deactivated catalyst is regenerated by controlled burn-off of the coke formed during the reaction. The heat generated during regeneration heats the catalyst again to supply the hot catalyst for the reaction.

Most commercial FCC catalysts are based on Y-zeolite (10–50%) mixed with a diluent, such as kaolin, to reduce the catalyst cost. Various catalyst additives may be added to the catalyst mixture to suit a specific feed or adjust the product slate. The catalyst may additionally contain additives such as pseudoboehmite to increase cracking activity and various other promoters. During FCC operation, various additives may either be added to the catalyst mixture or be co-fed with the catalyst. Some of these additives are combustion promoters (Pt or Pd salts), SO_x transfer agents (basic metallic oxides), metal traps and octane improvers (H-ZSM-5 zeolite).

HTFT residue can in principle be upgraded by standard FCC technology, but it constitutes less than 5% of the total syncrude and it is unlikely to be economically justifiable. On the other hand, about half of LTFT syncrude is wax. Although it has been shown that LTFT wax can easily be converted by FCC,[445–451] this technology has not yet been applied commercially with LTFT syncrude. At present, the only commercial application of FCC to FTS-derived feed is the conversion of olefinic HTFT naphtha in the high-temperature Superflex process.[444]

Superflex catalytic cracking (SCC) technology is used to convert an oxygenate-rich C_6–C_7 HTFT naphtha over an H-ZSM-5-based catalyst into ethene, propene and motor gasoline blending components. The SCC technology differs from standard FCC technology mainly in terms of operating temperature, which is 50–80 °C higher. This implies that there is a significant contribution from thermal cracking. The SCC technology has been designed to operate at end-of-riser temperatures above 600 °C, which is even higher than deep catalytic cracking (DCC) processes that are typically operated in the temperature range 525–595 °C.[452] The process consequently produces a combination of thermal cracking and catalytic cracking products, with propene being the main product.

5.3.3 Catalysts for Cracking

Efforts have been made to expand the list of catalysts used in commercial operations by developing novel catalysts that exhibit better activity, selectivity and stability. In this regard, novel catalyst formulations based on silica–alumina-based zeolites, silicoaluminophosphates, amorphous silica–aluminas and tungstated zirconia in combination with various metals have attracted attention.

Investigations of acid catalysts alone and bifunctional acidic supported metal catalysts have been reported with model compounds and realistic feed materials. In conjunction with FTS, noble metals received most attention, although conventional base metals, such as Ni and Mo, in combination with acidic supports have also been studied. The bifunctional nature of metal-promoted acid catalysts makes them suitable for simultaneous HIS and HCR and generally bifunctional catalysts require less severe operating conditions than acid catalysts employed for catalytic cracking.

5.3.3.1 Bifunctional Zeolitic Silica–Alumina HCR Catalysts

Table 5.24 shows the distribution of carbon numbers from the HCR of *n*-alkanes over a 1% Pt/USY catalyst.[453] It is evident that at low conversions, the *n*-alkane is cracked preferentially in the centre of the hydrocarbon chain. With increasing carbon chain length, propane formation decreased. Since cracking that involves a primary carbocation intermediate has a low probability during HCR, little methane formation was observed. Analogous results were observed over a 0.5% Pt/CaY catalyst with the hydrocarbons being cracked preferentially towards the centre of the hydrocarbon chain.[217,454] This

Table 5.24 Distribution by carbon number of cracked products from hydrocracking of different *n*-alkane feed materials over a 1% Pt/USY zeolite catalyst.

		Product yield per carbon number (mol per 100 mol cracked)						
Feed	*Conversion (%)*	C_3	C_4	C_5	C_6	C_7	C_8	C_9
Octane	7	41	118	41	–	–	–	–
Nonane	6	18	82	82	18	–	–	–
Decane	3	10	57	66	57	10	–	–
Undecane	30	7	44	49	49	44	7	–
Dodecane	8	5	33	41	42	41	33	5

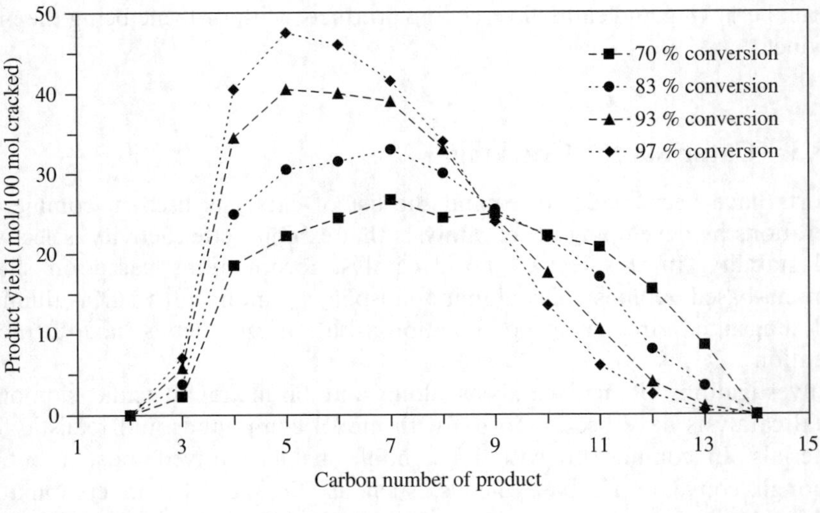

Figure 5.27 Hydrocracking of *n*-heptadecane over Pt/USY at increasing severity.

type of cracking selectivity was attributed to the absence of shape-selective constraints in these catalysts.

In catalysts with decreasing pore and/or cage diameter, a gradual decrease in the preference for cracking around the centre of the chain of *n*-decane was observed.[230] The cracking product distribution is affected by secondary cracking, when the conversion and/or the temperature is increased (Figure 5.27).[455] In this regard, the C_{12} fraction was the most affected. The susceptibility of cracked products for secondary cracking decreased with decreasing carbon number, with C_6 and lighter hydrocarbons being mechanistically more resistant to cracking, since it would involve the formation of a secondary or primary carbocation intermediate. Secondary cracking was also dominated by cracking at the central position. This suggests that the overall product distribution for such catalysts can be predicted provided that secondary IS does not proceed to a great extent.

A CaY zeolite that was prepared by ion exchange of an NaY zeolite was used as support for the preparation of a series of transition metal (Ni, Co, Fe, Mo, Ru, Rh, Pd, W, Re, Ir and Pt) catalysts by Welters *et al.*[418] The catalysts were prepared by pore volume impregnation. After sulfiding, the HCR activity of the catalysts was evaluated in a microreactor at 3 MPa using *n*-decane as feed. Most of the catalysts showed much higher cracking activity than the CeY zeolite support. However, ideal HCR behaviour was approached only on the Rh/CaY and Ir/CaY catalysts and with little secondary cracking being observed, which was in agreement with the work of Jacobs and Martens.[455] This was attributed to the efficient distribution of Rh and Ir on the support in a metallic form, rather than in a sulfided form.

In an attempt to improve on the HCR activity of monometallic HCR catalysts, Welters *et al.* prepared a bimetallic NiMo/CaY catalyst.[456] However, the activity of the bimetallic catalyst was similar to that of the monometallic catalysts when compared under similar conditions.

The HCR of *n*-decane could be further influenced by modification of zeolitic support.[455] Jacobs and Martens prepared several supports by varying the degree of dealumination of the Y-zeolite.[456] The supports were combined with Pt as the deHYD/HYD metal. Dealumination resulted in a gradual change in distribution of isomerised and cracked products as the metal-to-acid site ratios of the catalysts were changed.

The metal vapour deposition method used for preparation of the Ni/ZSM-5 and NiRe/ZSM-5 catalysts by Akhmedov *et al.* ensured a very efficient distribution of metals on the zeolite.[388] For the latter catalyst, almost complete conversion of C_8 and C_{16} hydrocarbons was achieved at 190–220 °C and near atmospheric H_2 pressure. The low H_2 pressure favoured dehydrogenation and implied that there was a high alkene concentration over the catalyst, which explained the high conversion at low temperature. As expected from the mechanism, the HCR reactivity increased with increasing carbon number in the order *n*-pentane < cyclopentane ≤ *n*-hexane < *n*-heptane < *n*-octane. The effect of temperature on HIS and HCR selectivities for *n*-heptane and *n*-octane exhibited the expected trend shown in Figure 5.21.

In a study of Heck and Chen conducted over a sulfided Ni/erionite catalyst, the product distribution could not be explained by simple primary and secondary cracking of *n*-butane and *n*-heptane used for the experiments.[457] There was some evidence for the involvement of reactants in a set of reactions such as OLI, deHYD/HYD, hydrogen transfer and cracking. The study by Heck and Chen highlighted the importance of alkene partial pressure over the catalyst. It should be noted that the temperature in this study of was more than 200 °C higher than that used in the study by Akhmedov *et al.*,[388] which increased the complexity of the reaction network.

The HCR of *n*-heptane in the temperature range 187–437 °C and 0.2 MPa of H_2 was investigated over a Co- and Ni-containing H-ZSM-5 catalyst by Lugstein *et al.*[458] At low conversion (less than 10%), the reaction selectivity over

Ni/H-ZSM-5 was dominated by HCR, giving propane and isobutane as the main products, whereas hydrogenolysis to small hydrocarbons prevailed over the Co/H-ZSM-5 catalyst, and also over the Ni/H-ZSM-5 catalyst at higher conversion. Maximum *n*-heptane conversion was observed for a catalyst with an Ni:Al ratio of 1.0. For the same Co:Al ratio, the overall conversion was very low. Compared with H-ZSM-5 alone, only saturated hydrocarbons were formed over the Ni- and Co-promoted H-ZSM-5 catalysts. At 0.2 MPa total pressure and with the H_2 partial pressure being varied from 0 to 0.2 MPa, the HCR activity increased with increasing H_2 partial pressure, whereas hydrogenolysis activity passed through a maximum. This was explained by the increasing HYD of olefinic products to promote desorption and thereby freeing up occupied acid sites to increase HCR.

Corma *et al.* prepared a HCR catalyst comprising ITQ-21 zeolite as the acidic component that was promoted with NiMo and used for the HCR of heavy gas oil (90% of feed boiling above 375 °C) with the aim of maximising the yield of middle distillates.[459] Before the impregnation with Ni and Mo metals, the ITQ-21 zeolite was mixed with γ-Al_2O_3 in a 1:1 ratio. Using the same procedure, catalysts were also prepared from USY and Beta zeolites and used for comparison with the ITQ-21-based catalyst. It was expected that the particular topology of the ITQ-21 zeolite would enhance the diffusion of bulky intermediate products through the six 12-membered ring openings while minimising undesired reactions. Indeed, the ITQ-21-based catalyst gave the largest conversion to the gas oil fraction (280–380 °C), whereas the USY-based catalyst was more selective towards the kerosene fraction (150–250 °C).

The effects of temperature and the Si:Al ratio of 0.27% Pd/SSZ-35 (STF) catalysts on the HIS and HCR of *n*-decane (H_2:*n*-decane ratio 100:1) were investigated by Tontisirin and Ernst.[460] They observed the usual trends with increasing temperature; HIS reaching a maximum at around 230 °C followed by a decline in HIS and an increase in HCR with further temperature increase. The catalytic activity for overall conversion increased with decreasing Si:Al ratio. Moreover, the product distribution was influenced by shape selectivity effects caused by the 10-membered ring sections in the one-dimensional pores.

Sulfided Ni-, Mo- and NiMo-loaded USY catalysts were tested for the HCR of *n*-decane at 400 °C and 3 MPa by Egia *et al.*[461] All metal-containing catalysts showed much higher HCR activity than USY zeolite alone, in spite of a strong imbalance between the deHYD/HYD and acidic functions. At 325 °C, a correlation between conversion, degree of sulfiding of metals and acidity could be established. The metals that were unavailable for sulfiding were not involved in the HCR catalysis. Little synergetic effect between Ni and Mo phases in the bimetallic NiMo catalyst was observed. All catalysts deactivated during very early stages of conversion, then reaching almost constant steady-state activity.

Increasing the concentration of H_2S in an *n*-heptane feed also resulted in a decrease in HCR conversion over NiMo/Y-zeolite at 380 °C and 5.7 MPa.[462] Inhibition of HCR was accompanied by an increased amount of coke on the catalyst. The coke comprised mainly of sulfur-containing polymers such as polysulfides.

Sulfided NiW zeolitic catalysts have been evaluated for the HCR of HTFT residue.[441] It was found that zeolite-based HCR catalysts were too active for Fischer–Tropsch feed, which resulted in a higher naphtha yield than obtained with catalysts based on ASA at comparable conversion and operating conditions.

The pore constraints imposed by zeolite catalysts make such catalysts less efficient for the HCR of LTFT waxes than larger pore amorphous materials. Nevertheless, a number of investigations have been reported that studied LTFT wax HCR over the metal-promoted zeolites USY, Beta and mordenite.[429,463,464]

There is also the possibility that in the future zeolite catalysts may be developed to exploit the 'window effect' for hydrocracking of Fischer–Tropsch waxes.[465] The 'window effect' was reported for HCR over ERI zeolite catalysts, which yielded a bimodal product distribution with maxima at C_3–C_4 and C_{10}–C_{12}, but few products in the C_5–C_8 range.[466] This is clearly not in line with the standard description of HCR. Although the phenomenon was initially explained in terms of diffusion, it was an incomplete explanation. In some zeolites, the pore or cage structure results in alkane adsorption where the heat of adsorption does not increase linearly with carbon number for all carbon numbers, but exhibits local adsorption minima. This is called the 'window effect'. Descriptions of non-linear phenomena such as these that have been published by Wei[467] and Maesen *et al.*[468] indicated that the 'window effect' is indeed theoretically possible.

5.3.3.2 Zeolitic Silica–Alumina FCC Catalysts

Zeolites have been widely used as catalysts for FCC and residue FCC. The two dominant zeolite catalysts in FCC are ultra-stable Y-zeolite (USY) and H-ZSM-5. In most FCC units USY is the main catalyst type and it is typically employed in conjunction with some catalyst additives. When propene production or motor gasoline octane number is important, H-ZSM-5 is generally added to the FCC unit.[443]

Although most FCC units are employed for residue upgrading, FCC of naphtha is employed for petrochemical applications to maximise the yields of ethene and propene. It was reported that among seven zeolites that were tested for petrochemical applications in order to produce light alkenes (C_2–C_3), the 10-membered ring zeolites, such as ferrierite, gave the highest yields of ethene and propene.[469] Since zeolite-catalysed FCC of naphtha is currently the only industrial application of catalytic cracking of Fischer–Tropsch syncrude, much of the subsequent discussion will focus on FCC of naphtha.

Significant contributions to the understanding of hydrocarbon cracking reactions over zeolites were made by Corma, Wojciechowski and coworkers.[470–477]

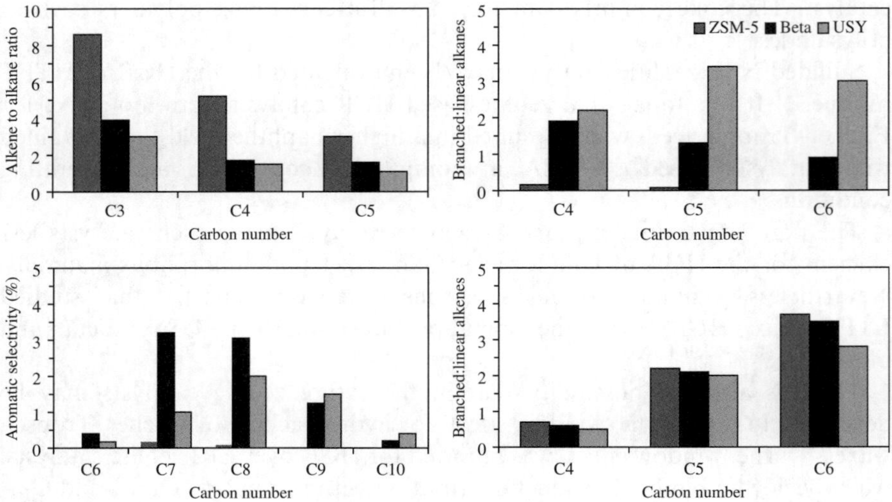

Figure 5.28 Product selectivity at 40% conversion of *n*-tetradecane over fresh ZSM-5, Beta and USY zeolite catalysts.

The effect of time on-stream on the cracking of C_7, C_{10}, C_{12} and C_{14} *n*-alkanes over ZSM-5, Beta and USY zeolites was investigated by Corma *et al.* employing a continuous flow reactor using an N_2:hydrocarbon molar ratio of 9:1.[470] One of the objectives of this study was to maximize the yields of C_3–C_5 alkenes and branched alkanes as potential blending stocks for reformulated gasoline. The reaction system designed by the authors was suitable for determining the instantaneous conversion and also the conversion at long reaction time.[471,472] In Figure 5.28, the effect of catalyst type on the product selectivity during cracking of *n*-tetradecane is shown.[470] For the alkene-to-alkane ratio of the C_3–C_5 fraction, the selectivity order is ZSM-5 > Beta > USY, whereas the order changes to USY > Beta > ZSM-5 for the selectivity towards C_4–C_6 branched alkanes. At the same time, the trend for branched alkenes was the opposite of that for the branched alkenes. The more stringent steric limitations imposed on bimolecular reactions in the channels of ZSM-5 zeolite compared with the other zeolites were responsible for the low selectivity of ZSM-5 towards aromatics. The changes with time on-stream did not have large effects on the selectivity, although these effects were more pronounced for *n*-alkanes with less than 10 carbon atoms. Increasing conversion and time on-stream decreased the alkene-to-alkane ratio. The alkene-to-alkane ratio increased with increasing chain length.

It has been observed that the activity of zeolites in cracking of *n*-hexane,[478] *n*-heptane[479] and *n*-octane[480] could be directly related to the presence of strong Brønsted acidity only. Apparently, Lewis acid sites can also be involved in cracking, although the role of Lewis acid sites has not yet been clearly defined.[481,482] The acid site distribution can be varied by changing the Si:Al ratio during synthesis or by dealumination through steaming.

Catalytic cracking of butenes was conducted with the aim of maximising the yield of propene as a feed for petrochemical industry.[483] The experiments were conducted in a continuous fixed bed system over H-ZSM-5, H-MOR and H-SAPO-34 at 550 °C and atmospheric pressure. It was found that the H-ZSM-5 catalyst had the highest stability and activity, and also good selectivity to propene. H-SAPO-34 was the least active and its selectivity for propene initially exceeded that of H-ZSM-5. Overall, H-ZSM-5 was a good catalyst for the cracking of butenes to propene. It was also reported that the H-ZSM-5 catalyst with a small crystal size exhibited higher stability than the catalyst with a larger crystal size.

The catalytic cracking of *n*-hexane over H-ZSM-48 zeolite between 420 and 500 °C produced C_2, C_3 and C_4 alkenes as the major products.[484] Various skeletal isomers of *n*-hexane such as 2,2-dimethylbutane, 2,3-dimethylbutane, 2-methylpentane and 3-methylpentane were also present, in addition to C_2 and C_3 alkanes and C_{6+} aliphatics. The lower yield of butanes compared with ethane and propane was attributed to the steric inhibition of the formation of a transition state complex between *n*-hexane and *tert*-butyl carbocation in the channel of the H-ZSM-48 zeolite.

The monomolecular cracking of *n*-hexane over H-ZSM-5, H-MOR, H-USY and dealuminated Y-zeolite was investigated Babitz *et al.*[485] The study indicated that the mechanisms of cracking on all of these catalysts were similar and that differences in activity was not due to acid site strength, but rather to strength of *n*-hexane adsorption on the different zeolites. The selectivity of the catalysts for the formation of propene varied from 40 to 60% when the conversion was below 30% and the selectivity for propane was around 10%. The levels of formation of ethane and ethene were both around 10%. Methane, butenes and isobutane were other important products. The highest selectivity for the formation of ethene and butenes was exhibited by H-ZSM-5 catalyst, whereas H-USY gave the largest yield of isobutane.

For the H-ZSM-5 zeolite, the activity during cracking of *n*-hexane increased with decreasing Si:Al ratio.[486,487] The selectivity for isobutene formation exhibited the same trend. However, for every Si:Al ratio in the range from 10:1 to 75:1, the cracking was dominated by the scission of the central C–C bond.

Antia and co-workers studied binderless H-ZSM-5 zeolite-coated monolithic reactors for the cracking of *n*-hexane.[488,489] A monolithic substrate, such as cordierite, was used to fabricate the reactor. The surface of the monolith substrate was coated with the zeolite before being mounted in a stainless-steel reactor. The *n*-hexane was introduced in a vapour phase either alone or in a mixture with N_2. Below 450 °C the *n*-hexane conversion was kinetically controlled, whereas above this temperature the involvement of mass transfer became evident. At 450 °C aliphatics (alkanes and alkenes) and aromatics accounted for about 84 and 16% of the products, respectively, whereas at 540 °C they accounted for about 61 and 39%, respectively.

The study of mechanistic modelling of *n*-heptane cracking over ZSM-5 zeolite between 450 and 550 °C (under N_2) revealed that the primary products included C_1–C_5 hydrocarbons, whereas isobutane and isopentane were formed in secondary reactions.[490] Propene was the major product. Other important products included H_2, ethane, ethene, propane, *n*-butane and butenes, whereas methane and C_5 hydrocarbons were minor products. Linear alkanes were formed predominantly *via* adsorption of the alkane followed by carbocation cracking. Alkene products resulted from β-scission and carbocation desorption. The high H_2 content in the product is of interest. Acid catalysts have a poor ability to desorb molecular hydrogen (H_2) and there are two likely explanations for the results. One possibility is that some thermal cracking occurred in parallel with catalytic cracking,[417] but this was ruled out by the investigators. The investigation included a blank run that showed only 1% conversion in the absence of the H-ZSM-5 catalyst. The second possibility is that the H-ZSM-5 catalyst and experimental conditions were conducive to protolytic cracking (Haag–Dessau mechanism), which is favoured by a low partial pressure of hydrocarbons and low acid site density (high Si:Al ratio).[491] This was indeed the case and illustrates the importance of both catalyst and operating conditions on the product spectrum that is obtained.

The mixing effect of USY and ZSM-5 zeolites was studied at 350 °C using heptane as a model feed and N_2 as carrier gas.[492] The formation of isomerised C_4 products over the USY–ZSM-5 mixtures (75:25 and 50:50) was higher than that of linear additive predictions. This enhancement was evident particularly with the ZSM-5 sample having a high acid strength. The catalysts prepared without template were more active and their activity exhibited little decline with time on-stream. The selectivity for C_3 and C_4 alkenes increased with time on-stream, when catalysts became deactivated. At the same time, the selectivity for C_3 and C_4 alkanes decreased. The decrease in alkane selectivity is indicative of a decrease in hydrogen transfer activity and aromatics formation.

Cubic and hexagonal faujasites with various Si:Al ratios were used for the catalytic cracking of *n*-heptane (10% heptane in N_2) at 450 °C.[493] Catalytic activity could be related to the concentration of acidic sites of the zeolite framework. The conversion of *n*-heptane was dominated by reactions leading to the formation of C_3 and C_4 products. Within the range of conversions from 1 to 68%, the combined selectivity to C_3 and C_4 products approached 92%. However, a rapid decline in catalyst activity was observed with time on-stream. The activity could be restored by oxidative regeneration. Other studies also showed that cracking of *n*-heptane over Y-zeolites yielded C_3 and C_4 hydrocarbons as the major products.[478,494]

Zeolites such as ZSM-5, Beta, Y, USY and their composites were used to study the cracking of *n*-octane, 2,2,4-trimethylpentane and 1-octene at 500 °C using He as carrier gas.[495] Under such conditions, the selectivity for alkenes was higher and for aromatics lower over Y-zeolite than over ZSM-5. For the composites of Y-zeolite with either ZSM-5 or Beta-zeolite, the C_3 and C_4 alkane selectivities approached weighted averages of the individual zeolites, whereas for the USY zeolite containing composites, the selectivities could be higher than those for the individual zeolites.

The cracking of *n*-octane over H-MOR occurred in mainly two positions, initially yielding C_3, C_4 and C_5 products.[496] The cracking of *n*-octane over

H-MOR at 400 °C ultimately yielded 14% isobutene, 25% C_3, 48% C_2 and 13% methane. The cracking reaction occurred on Brønsted acid sites. There was little evidence supporting the involvement of Lewis acid sites during cracking; however, Lewis acid sites may have participated in the reactions leading to the formation of aromatics and coke.

The residue fraction of HTFT syncrude contains very little wax and resembles a conventional crude oil residue, albeit with different heteroatom composition. The combined distillate and vacuum gas oil fraction from HTFT contains 27% aromatics, 3.3% oxygen (O content, not oxygenate content) and has a high olefinicity, 63 g Br per 100 g.[441] It can be expected that catalytic cracking of HTFT residue can be performed over zeolites employed for the cracking of petroleum feedstocks. The HTFT feed is more reactive and lower temperatures and/or shorter contact times may be advisable when processing such feed to produce light alkenes.

The catalytic cracking of LTFT waxes was investigated under typical FCC conditions by various researchers employing zeolites Y, Beta and H-ZSM-5.[445–451] It was found that waxes are readily cracked to lighter products and that waxes had a low tendency to form coke on the cracking catalysts. Converting LTFT waxes by FCC therefore requires an additional fuel source to maintain the heat balance over an FCC unit. This is understandable, since the H:C ratio of wax that consists mainly of *n*-alkanes is around 2, whereas that of aromatic coke is around 1. Significant hydrogen transfer is consequently required for wax to form coke.

The product selectivity depends on the choice of catalyst. H-ZSM-5 produced light products with a higher alkene-to-alkane ratio than H-Y and the alkanes from H-ZSM-5 cracking were mainly linear, rather than branched (Table 5.25).[446] The branched species formed by H-ZSM-5 cracking were exclusively monomethyl species, whereas H-Y cracking produced monomethyl and multi-branched species.

An extensive study on the catalytic cracking of LTFT wax was performed with support of the United States Department of Energy. In this study, a small-scale catalytic cracking system was employed and typical results comparing the performance of different zeolite catalysts are presented in Table 5.26.[449] The product selectivities changed with conversion. In general, it was found that the

Table 5.25 Initial selectivities during the catalytic cracking of LTFT wax over H-ZSM-5 and H-Y zeolite catalysts at 405 °C.

	Alkene:alkane ratio		*Linear:branched alkenes ratio*		*Linear:branched alkanes ratio*	
Product	*H-ZSM-5*	*H-Y*	*H-ZSM-5*	*H-Y*	*H-ZSM-5*	*H-Y*
C_3	2.38	1.14	–	–	–	–
C_4	3.37	1.93	0.64	0.85	3.4	0.32
C_5	4.05	2.06	0.38	0.39	18.1	1.1
C_6	6.05	1.55	0.61	0.58	4.6	0.07

Table 5.26 Product yield obtained during the catalytic cracking of LTFT wax over steamed H-Y, H-Beta and H-ZSM-5 zeolite catalysts at 470 °C and 83–84% conversion.

	FCC product yield (mass%)		
Product	*H-Y*	*H-Beta*	*H-ZSM-5*
C_2 and lighter	0.6	0.6	1.5
Propene	7.4	8.9	17.5
Propane	0.8	0.9	2.7
n-Butenes	7.4	8.3	15.1
Isobutene	5.8	9.4	12.3
Butanes	3.7	3.6	3.6
n-Pentenes	4.0	4.3	4.1
Isopentenes	7.7	9.2	9.8
Pentanes	3.6	2.2	2.0
C_6–220 °C	41.7	35.8	15.3
>220 °C	17.0	16.8	16.2
Coke	0.3	0.2	0.1

yield of short-chain alkenes increased with wax conversion over H-Y and H-Beta, but over H-ZSM-5 the alkene yield passed through a maximum and then decreased at high conversion.

5.3.3.3 *Bifunctional Silico-aluminophosphate HCR Catalysts*

The hydroconversion patterns of *n*-hexane over Pt/SAPO-5 catalysts differed from those of *n*-heptane.[320] For the latter, conversion was dominated by HIS, with isomerised C_7 material reaching a maximum at about 50% conversion, whereas for *n*-hexane, HCR became evident only above 80% conversion. The activity of the Pt/SAPO-5 could be influenced by the method of preparation, but the general trends in product distribution remained similar.

In comparison with a Pt/SAPO-5 catalyst, more cracked products were formed over a Pt/SAPO-11 catalyst. The HCR of *n*-octane over Pt/SAPO-5 differed from that over Pt/SAPO-11.[321] During HCR over a Pt/SAPO-5 catalyst in the temperature range 300–400 °C, the product distribution was symmetrical; C_4 hydrocarbons were the major products with an equivalent amount of C_3 and C_5 hydrocarbons also being formed. This 'ideal' HCR behaviour of the Pt/SAPO-5 catalyst was also observed over Pt/Y zeolite,[422,456] and was attributed to the similar pore diameters of the two catalysts. However, over a Pt/SAPO-11 catalyst, methane and C_2 and also C_3 in excess of C_5 were formed. Similar results were obtained with branched C_8 as feed material. In the case of *n*-octane, for both catalysts, HIS preceded HCR, as is generally observed (Figure 5.21).

A Pd/SAPO-11 catalyst was compared with a Pt/SAPO-11 catalyst during the HIS and HCR of heptane in the temperature range 400–500 °C using an H_2:heptane ratio of 15:1.[497] The latter catalyst was more resistant to deactivation and had a high aromatics selectivity. Much more aromatics were formed at 500 than at 400 °C. The formation of branched C_7 hydrocarbons was evident at 400 °C, but toluene and C_1–C_6 hydrocarbons were the main products at both temperatures. The product selectivity could be significantly changed by the addition of Na to the Pt/SAPO-11 catalyst. Of significance is the marked increase in ring closure activity on Na addition, with an alkylcycloalkane selectivity of 58% at 10% conversion reported for 1.5% Pt + 1.3% Na on SAPO-11 catalyst at 500 °C.[497] This type of behaviour has some significance for diesel fuel production from Fischer–Tropsch syncrude, since cycloalkanes have acceptable cetane numbers and density.[498]

Catalysts of the MeAPO-36 (Me = Mg, Zn and Co) type exhibited a good activity and selectivity during the HCR of gas oil at 400 °C and 5 MPa.[499] In this case, the objective was to maximize the yield of naphtha (C_5–177 °C) and middle distillate (177–343 °C) by converting the fractions boiling above 343 °C. The highest conversion of this fraction was achieved over CoAPO-36. The content of the heavy fraction was decreased from almost 70% in the feed to less than 20% in the products.

5.3.3.4 Bifunctional Amorphous Silica–Alumina HCR Catalysts

The mildly acidic Pt/SiO_2–Al_2O_3 catalysts have been identified as one of the key catalyst types for the HCR of FTS products in commercial applications.[500] It is consequently not surprising to find that some studies focused specifically on HCR of material from FTS over Pt/SiO_2–Al_2O_3.

An unsulfided Pt-promoted amorphous mesoporous silica–alumina (MSA) catalyst with a bulk silica-to-alumina ratio of 100:1 formed the basis of numerous HIS and HCR studies employing *n*-alkanes and waxes as feed materials.[231,435,501–504] The product distributions obtained from HCR of *n*-decane at high conversion over different Pt/MSA catalysts are shown in Figure 5.29.[231] A more symmetrical distribution around the C_5 fraction was observed for 0.6% Pt/MSA, suggesting a better balance between metal and acidic functions of the catalyst. The 1.2% Pt/MSA catalyst gave higher amounts of C_1 and C_2 hydrocarbons. This indicated that the metal function was dominant and that significant hydrogenolysis occurred over the metal function of the catalyst. During HCR of C_{10} and heavier *n*-alkanes with a carbon number distribution similar to that of FTS wax with an α-value of 0.87, 75–85% distillate selectivity was reported at 90% conversion.[503]

A series of Pt-promoted silicated amorphous silica–alumina catalysts were used for the HCR of *n*-hexadecane and FTS waxes.[437] Most of the experiments with *n*-hexadecane were performed in the temperature range 340–380 °C, H_2 pressure 5 MPa, LHSV 1.5 h^{-1} and an H_2:feed ratio of 1000 normal m^3 per m^3 feed in a trickle bed reactor. The properties of the Pt- and PtW-promoted silicated ASA catalysts are compared with that of Pt/MSA to indicate the similarities (Table 5.27).[231,437] Although the Siral40 and Siral75 catalyst supports have bulk SiO_2:Al_2O_3 ratios of 40:60 and 75:25, respectively, these

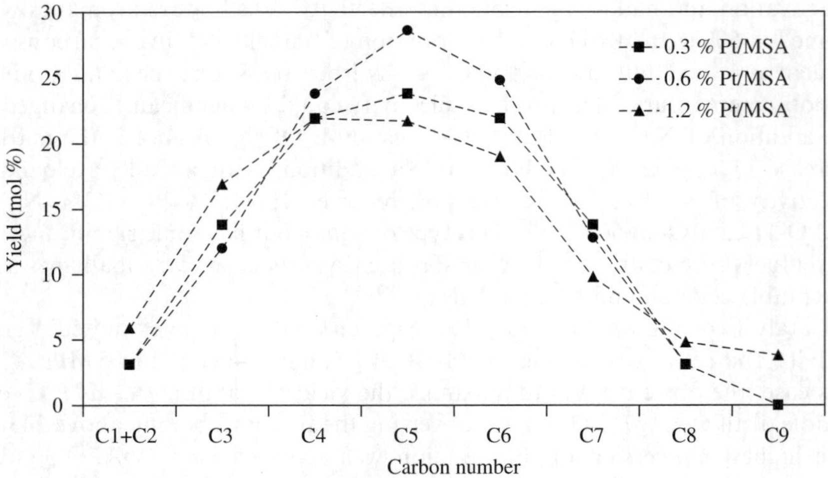

Figure 5.29 Product distribution obtained at 85% conversion during HCR of *n*-decane over Pt-promoted mesoporous silica–alumina catalysts at 3 MPa and H_2:*n*-decane ratio 4:1. The Pt content in the catalysts was varied: 0.3% (■), 0.6% (●) and 1.2% (▲).

Table 5.27 Catalyst properties of Pt-promoted mesoporous silica–alumina (MSA) and silicated amorphous silica–alumina catalysts that were employed in HCR studies.

Catalyst	*Surface area ($m^3\,g^{-1}$)*	*Bulk SiO_2:Al_2O_3 ratio*	*Pt dispersion (%)*	*Acidity (μmol g⁻¹)* Brønsted	*Acidity (μmol g⁻¹)* Lewis
Pt/MSA	750[a]	100:1	80	19.6	85.5
Pt/Siral40	332	40:60	–	2.1	94
PtW/Siral40	318	40:60	36	5.4	80
Pt/Siral75	407	75:25	–	10	30
PtW/Siral75	357	75:25	76	14	36

[a]Surface area of MSA support material; the surface areas for Siral40 and Siral75 support materials are 498 and 402 $m^3\,g^{-1}$, respectively.

materials are silicated and the surface concentration of alumina is much lower. Figure 5.30 compares the HCR activity of these Pt-promoted silicated ASA catalysts.[437]

A linear correlation was found between the activity, as expressed by the first-order rate constant for HCR, and the concentration of Brønsted acid sites.[437] The Siral40-based catalysts resulted in over-cracking of the reactant and C_1–C_2 products were obtained in high yields. The Siral75-based catalysts performed better than or equal to a commercial catalyst used for wax HCR.

The performance of PtMo/SiO_2–Al_2O_3 catalysts that were evaluated in parallel[436,437,505] indicated that PtMo did not perform as well as PtW-promoted

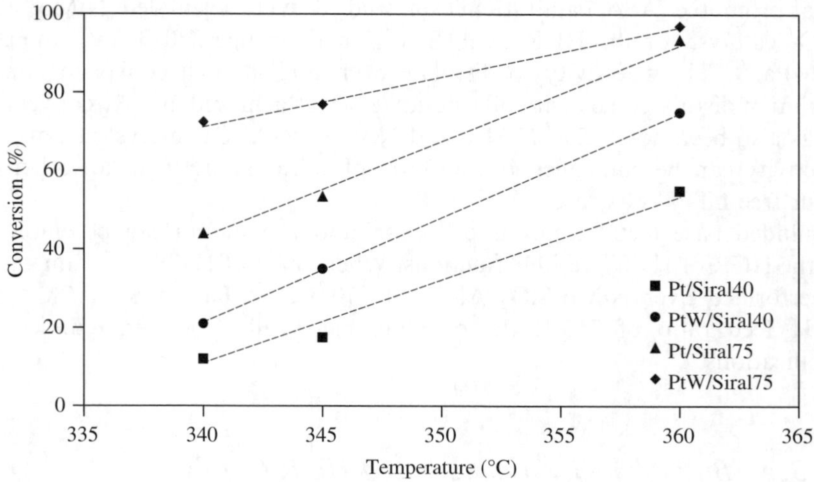

Figure 5.30 Conversion of *n*-hexadecane over different Pt-promoted silicated amorphous silica–alumina catalysts at 5 MPa, LHSV 1.5 h^{-1} and H_2:feed ratio 1000:1.

silicated ASA catalysts. A PtMo/Siral40 catalyst was nevertheless successfully used for the HCR of HTFT vacuum gas oil at 390 °C, 5 MPa and LHSV 0.5 h^{-1}, without and with hydrotreating pre- and post-treatment.[441] The distillate thus produced met all major final diesel fuel specifications.

Reports of *n*-alkane and FTS wax HCR using unsulfided and sulfided base metal-promoted ASA catalysts can also be found in the literature. A CoMo/SiO_2–Al_2O_3 catalyst was employed in its reduced form (not sulfided) for the HCR of *n*-tetradecane.[432,434] The testing was conducted in a trickle bed reactor at 330 °C, 4 MPa and an H_2:*n*-tetradecane ratio of 10:1. The catalyst exhibited good activity and produced liquid products with little branching and consequently had a high cetane number. However, a high yield of gaseous products, particularly methane, was a drawback of this catalyst. It has been pointed out that reduced Co and Ni catalysts are prone to hydrogenolysis.

In a related study, various Ni/SiO_2–Al_2O_3 catalysts were employed in their reduced form for the HCR of *n*-hexadecane.[433] Hydrogenolysis resulted in a high C_1–C_2 selectivity, which was in the range 1.8–11.5% at 38.7–42.6% conversion. The 4.5% Ni/SiO_2–Al_2O_3 catalyst was also tested with LTFT wax at 360 °C, 7 MPa, WHSV 2.8 h^{-1} and H_2:wax of 800 normal m^3 per m^3 wax, and compared with a sulfided commercial NiMo/SiO_2–Al_2O_3 HCR catalyst. Under these conditions, both catalysts had a conversion of around 52% and a distillate selectivity of around 73–75%. The main difference was in C_1–C_2 products, where the unsulfided reduced Ni/SiO_2–Al_2O_3 catalyst had a selectivity of 2.8% compared with 0.06% for the sulfided commercial NiMo/SiO_2–Al_2O_3 catalyst.

The Chevron Isocracking technology that is used commercially for the conversion of FTS wax in the Oryx GTL facility employs a sulfided base metal-promoted ASA catalyst. Leckel studied such a sulfided NiMo/SiO_2–Al_2O_3 catalyst for the HCR of FTS wax in the range 350–370 °C and 3.5–7.0 MPa.[506] The selectivity to distillate decreased at high conversion and a maximum distillate selectivity of around 78% was achieved at 80% conversion. It has also been noted that NiMo- and NiW/SiO_2–Al_2O_3 catalyst activity and selectivity can be controlled by the level of sulfur addition to an otherwise sulfur-free LTFT wax feed.[507]

Sulfided base metal catalysts on ASA support material can be employed for the HCR of HTFT residue fractions. When HCR of HTFT vacuum gas oil is performed over NiMo/SiO_2–Al_2O_3 at 370 °C, 5 MPa, LHSV 0.5 h^{-1} and an H_2:feed ratio of 715:1, the distillate meets all major final diesel fuel specifications.[441]

5.3.3.5 Bifunctional Zirconia-based HCR Catalysts

Superacidic Pt-promoted sulfated zirconia (SZ) catalysts were used to study the HCR of *n*-hexadecane in an autoclave at 150 °C and 3.5 MPa H_2.[359] It was shown that the addition of less than 0.5% of Pt has a dramatic effect on conversion. Further increase in Pt content had little effect on HCR and HIS selectivities. The usual trend in HCR and HIS selectivities with increasing conversion (Figure 5.21) was observed in this study. At lower conversions and low Pt concentrations (0.03 and 0.3 mass%), the maximum yield occurred at C_8. At higher conversions and higher Pt concentrations (3 and 5% Pt), the maximum yield occurred at C_7. The cracked products obtained with 0.6 mass% Pt were shifted to lower carbon numbers with the maximum yield at C_6–C_7.

Catalysts based on Pt-promoted SZ and TZ were also employed for HIS and HCR of LTFT waxes and model *n*-alkane feed materials.[463]

Another study, conducted by Grau *et al.*, focused on the HCR and HIS of *n*-octane with the aim of maximising the yield of branched C_4–C_7 products.[508] The testing was conducted at 300 °C and 0.1 MPa. A high yield of branched octanes was obtained over Pt/TZ. The incorporation of SO_4^{2-} into this catalyst increased the acidity and cracking activity. The most active Pt/TZ catalyst was obtained by calcination around 700 °C. The catalyst calcined at this temperature had the best liquid yield and selectivity to branched alkanes; the stability and the RON gain were relatively good. This maximum coincided with the maximum concentration of Brønsted acid sites and a Brønsted-to-Lewis acid ratio of 1.3–1.6. The performance of this catalyst was also evaluated at 400 °C and 1.5 MPa using *n*-decane as model compound. This test confirmed the high activity and stability of the bi-promoted (WO_3 and SO_4^{2-}) Pt on zirconia catalyst, producing an isomerate with the highest molar ratio of branched C_4–C_7 to total branched products.[509] The acidity of the bi-promoted catalyst could be regulated by the amount of WO_3 in the catalyst. In addition to the Brønsted-to-Lewis acid ratio, the calcination temperature also had a pronounced effect on the final metal-to-acid site balance of the catalysts. The highest liquid yield and yield of branched alkanes were again obtained by calcination at 700 °C. The catalyst calcined at 800 °C had the highest cracking activity and gave the highest yield of isobutane and propane.

5.3.3.6 *Other Cracking Catalysts*

Historically, amorphous silica–alumina materials and clays were extensively used for catalytic cracking,[510] but have since been replaced by zeolites for this purpose. Despite the tremendous variety of catalysts and catalyst additives available for FCC, the main catalytic component remains zeolitic.

5.3.4 Catalyst Deactivation During Cracking

The discussion on the deactivation of IS/HIS catalysts (Section 5.2.4) is equally applicable to HCR and FCC catalysts. The main difference is that the operating temperature of HCR and FCC units is typically higher.

During FCC, catalyst deactivation caused by thermal effects and hydrothermal dealumination during regeneration cannot be avoided. In addition to these deactivation mechanisms, there are also other deactivation mechanisms, such as the deposition of metals that cannot be removed during oxidative regeneration. For conventional FCC operations, considerable literature is available dealing with these and other phenomena.[442]

5.3.4.1 *Oxygenate-related Deactivation*

The potentially adverse effects of oxygenates on catalytic cracking are not well documented. Under typical FCC conditions, a high conversion of oxygenates is expected. In this regard, attention should be give to the action of water produced during oxygenate conversion, and also the action of the oxygenates during the conversion.

The effect of water adsorption can be reversed by increasing the temperature to desorb the adsorbed water. However, it is known that the structure of silica–alumina-based catalysts, such as zeolites, can be modified by prolonged steaming.[511] Hydrothermal dealumination of zeolites through the action of water results in activity and selectivity changes in the catalyst.[512] The way in which dealumination takes place also plays a role.[513,514] This process takes place by catalyst exposure to water generated during regeneration of FCC catalysts, and also steam being co-fed as diluent with the FCC feed. It has been reported that the addition of phosphorus improves hydrothermal stability.[515]

When processing feed from FTS that contains oxygenates, another source of water becomes available. The oxygenates can potentially produce water as a product from reaction. However, the water produced during oxygenate conversion is available as an adsorbed reaction intermediate in contact with an acidic aluminium site on the catalyst. This begs the question: would dealumination by oxygenates during the process of dehydration on the catalyst surface not result in more severe dealumination than steaming? This question has not yet been satisfactorily answered, but it is likely that this is indeed the case.

Oxygenates also affect HCR catalysts. Leckel used *n*-hexadecane as a model compound to study catalyst deactivation under conditions typically applied for the HCR of FT wax.[516] A sulfided NiMo/Al_2O_3–SiO_2 catalyst and an unsulfided Pt/Al_2O_3–SiO_2 catalyst were used to test the influence of various oxygenates on conversion. Inhibition of HCR due to competitive adsorption by the oxygenates occurred for all oxygenates. The unsulfided noble metal catalyst was more prone to inhibition than the sulfided base metal catalyst. The presence of 3-hexanone in the feed resulted in a rapid loss of HCR activity, whereas inhibition was less extensive in the presence of carboxylic acids and esters. Alcohols also caused a decrease in the yield of cracked products over the Pt/Al_2O_3–SiO_2 catalyst. The HIS activity decreased, which was attributed to the formation of water that changed the equilibrium between Lewis and Brønsted acid sites. Inhibition of HCR by oxygenates has also been reported in wax HCR studies,[506] and also HCR of HTFT residue.[441] It has been reported that HCR of a straight run FTS wax over a sulfided NiMo/SiO_2–Al_2O_3 catalyst required a 15 °C higher operating temperature to achieve the same conversion as when a hydrotreated FTS wax was used.[506]

Further work by Leckel showed that different oxygenates affected the balance of acid and metal sites on the catalyst.[396,516] Carboxylic acids preferentially adsorbed on the metal sites, whereas alcohols preferentially adsorbed on the acid sites.

5.3.4.2 Deactivation by Carbonaceous Deposits

The combination of the acid site distribution and shape selectivity makes ZSM-5 zeolite suitable for the selective cracking and HCR of long-chain alkanes without excessive coke formation. The situation is different for larger pore zeolites. For example, the activity loss during the cracking of *n*-heptane over HY was 80% afer 30 min on-stream, but only 50% after 70 h on-stream for HZSM-5.[109,517] In the commercial Mobil catalytic dewaxing process, a steady performance of ZSM-5 zeolite can be maintained for several months,[518] which can partly be ascribed to H_2 recirculation that slows coke formation. The stability of ZSM-5 zeolites is further increased by adding a HYD component, such as Zn, Ni and Pd.[519,520]

Catalyst deactivation by coking can be reduced by lowering the temperature and increasing the hydrogen pressure. One would therefore expect catalysts with a strong HYD function to be less susceptible to coking than similar catalysts with a less hydrogenating metal. The nature and number of acid sites in relation to the HYD function are equally important. In a series of the HY zeolites evaluated by Moljord *et al.*, the resistance to coke formation was observed to decrease with increasing number of protonic acid sites.[521]

The zeolites ZSM-20 and USY were compared during the cracking of *n*-heptane at 450 °C in a mixture with N_2.[522] The ZSM-20 catalyst exhibited higher cracking and coking activity, as evidenced by a greater amount of coke than that formed on USY zeolite. The structure and molecular weight of the extractable portion of coke from both zeolites were similar.

Coking is not necessarily a detrimental attribute. In FCC, where coking is required for carbon rejection and energy, it can be beneficial to employ a catalyst with a high coking propensity. The conversion *versus* time on-stream correlations for USY, Beta and ZSM-5 zeolites reported by Corma and *et al.* indicated that the activity decay decreased in the order USY > Beta ≫ ZSM-5.[471] For ZSM-5, very fast, small decay within the first few seconds was followed by practically constant activity. At the same time, for USY and Beta zeolites, the decays were more gradual but much more extensive. However, a decrease in activity is not necessarily directly correlated with degree of coking. Coke on H-ZSM-5 has less of an effect on catalyst activity than coke on USY zeolite. For example, little deactivation was observed during *n*-hexane conversion over H-ZSM-5 until the amount of coke exceeded 4 mass%.[523] When the amount of coke exceeded 4 mass% on the H-ZSM-5 catalyst, the activity loss was greater than could be accounted for by the loss in total or in strong acid sites, and the rapid significant activity loss was attributed to pore blockage and reduced reactant diffusivity.

5.4 Hydrotreating

Hydrotreating is the mainstay of refining. It fulfils two functions in the refinery, both related to the removal of specific functional groups. First, it is useful as a feed pretreatment step for refinery operations that are sensitive to impurities, *e.g.* the HYD of dienes to monoenes as feed pretreatment before an acid-catalysed conversion step in order to prevent the formation of heavy polymers. Second, it is used to meet final product specifications in terms of composition.

Hydrotreating can be classified in terms of its function, which is also a convenient way of indicating the fields that are most relevant to the refining of primary products from FTS:

1. Hydrodesulfurisation (HDS).[524–526] There is essentially no sulfur in Fischer–Tropsch syncrude. This type of hydroprocessing is relevant only when material from FTS is co-refined with sulfur-containing materials, for example co-refining with crude oil, oil shale liquids, direct coal liquefaction, low-temperature coal gasification or coal pyrolysis liquids.
2. Hydrodenitrogenation (HDN).[527,528] The same comments as for HDS apply; Fischer–Tropsch syncrude is essentially free from nitrogen-containing compounds.
3. Hydrodeoxygenation (HDO).[529–531] This is one of the most important hydrotreating reactions for the refining of Fischer–Tropsch syncrude. Material from FTS invariably contains oxygenates. Depending on the application or subsequent refining steps, it may be necessary to hydrotreat the material from FTS as a feed pretreatment step. In the case of

conversion processes such as catalytic reforming or HIS that employ a chlorinated Pt/Al_2O_3 catalyst, it is critical that the feed be oxygenate free. The degree of HDO differs from application to application and complete HDO is not always necessary and may even be undesirable.[465,532] Some applications include the selective conversion of carbonyl compounds to alcohols,[533–536] deep deoxygenation of waxes for food applications[537] and oxygenate conversion for the production of transportation fuels.[538]

4. Hydrodearomatisation (HDAr).[539] The polynuclear aromatic content of Fischer–Tropsch syncrude is low. This class of hydroprocessing is nonetheless relevant in specific applications, such as the upgrading of the atmospheric residue from HTFT synthesis. Highly aromatic tar from low-temperature coal pyrolysis, which may be a by-product of coal-to-liquids applications of FTS, requires extensive HDAr before it can be employed as a fuel.
5. Hydrogenation of alkenes (HYD).[540] This is an important class of hydrotreating on account of the high alkene content of Fischer–Tropsch syncrude. Its two main applications are the HYD of the products from OLI and HYD of straight run syncrude (in conjunction with partial HDO) to produce transportation fuels.[532,538,541,542] Without HYD, specifications such as bromine number, acid number and oxidation stability of the final fuel products cannot not be met.
6. Hydrodemetallisation (HDM).[543–545] The main metals present in Fe-HTFT and Fe-LTFT syncrudes are iron and sodium. These metals are present as metal carboxylates that are produced during corrosion and catalyst loss by leaching. Likewise, one would expect some of the metals present in Co-LTFT syncrude to be related to the LTFT catalyst composition. (Catalyst attrition also contributes to metal containing suspended particulate matter in the syncrude.) Unfortunately, conventional HDM catalysts are ineffective in the removal of these metals.[426] These metal carboxylate species can be stable under hydroprocessing conditions. When hydroprocessing is performed with a sulfided base metal catalyst, a sulfiding agent must be added to the syncrude to keep the catalyst in a sulfided state, which may cause stable metal sulfides to be formed. The decomposition of iron carboxylates to yield stable iron sulfides is especially troublesome in FT refineries.[426,546]

Despite the prominent place of hydrotreating in Fischer–Tropsch product refining, there is surprisingly little literature dealing specifically with this subject. Hydrotreating of material from FTS relies on the same basic technologies and commercial catalysts as those encountered in a conventional crude oil refinery. However, there are two important differences between hydrotreating crude oil and Fischer–Tropsch syncrude, namely the refining focus and total heat release during hydrotreating (Table 5.28).[547]

Due to the scope of hydrotreating and its ubiquitous use in refining, the subsequent discussion of hydrotreating will not follow the pattern set by the previous topics. The focus will be on hydrotreating in the context of FTS.

Table 5.28 Differentiating features between hydrotreating conventional crude oil and Fischer–Tropsch syncrude.

Differentiating feature	*Conventional crude oil*	*Fischer–Tropsch syncrude*
Feed material	Alkanes, aromatics, S, N	Alkenes, O
Hydrotreating focus	HDS (also HDN)	HDO and alkene saturation
Heat of hydrogenation	–2 to –8 $kJ\,g^{-1}$ S	–6 to –16 $kJ\,g^{-1}$ O
Total heat release	$<450\,kJ\,kg^{-1}$	$>950\,kJ\,kg^{-1}$

5.4.1 Commercial Hydrotreating Processes and Catalysts

Most commercial refinery hydrotreating catalysts are bi- or tri-metallic, with NiMo, NiW, CoMo and NiCoMo on γ-Al_2O_3 being the main types encountered in practice.[548] On account of the sulfur content of conventional crude oil, these catalysts are all designed to be operated as sulfided metal catalysts and are called sulfided catalysts for short.[549] Although Fischer–Tropsch syncrude is sulfur free, in commercial refining practice there are a surprising number of hydroprocessing units associated with FTS that operate with sulfided catalysts.

The frequent use of sulfided catalysts for hydrotreating products from FTS is related to the oxygenate content of the syncrude and specifically the carboxylic acid content. Reduced (unsulfided) metal-promoted catalysts can be deactivated by carboxylic acid leaching of the active metal.[395] Leaching is a problem because the oil products from FTS may contain corrosive short-chain carboxylic acids. Although the short-chain carboxylic acids preferentially dissolve in the Fischer–Tropsch aqueous product, C_3–C_4 carboxylic acids are amphiphilic and may dissolve in the oil product from FTS. The distribution of carboxylic acids in the oil product is dependent on the separation efficiency after FTS (Section 4.2.1). The oil phase from FTS cannot be described as an apolar hydrocarbon phase; it contains percentage levels of dissolved oxygenates, which gives it some polar character. The short-chain carboxylic acids boil in the naphtha range and the acid content of Fischer–Tropsch-derived naphtha can be fairly high, especially in the case of HTFT naphtha (Table 4.6). It has therefore been pointed out that that stainless-steel units or stainless-steel linings are required when processing the acid-containing naphtha from FTS.[550]

A smaller group of hydroprocessing catalysts are used for selective HYD and are used in the absence of sulfur. Generally, these catalysts are based on Ni, Pd or Pt on γ-Al_2O_3. Such catalysts are ideal for hydrotreating heavier fractions from FTS that contain less corrosive longer chain carboxylic acids or little oxygenates, such as LTFT waxes.

The selection of hydroprocessing catalysts is very application specific.[551] In practice, hydroprocessing reactors are not loaded with a single type of catalyst, but with different layers, each performing a specific function. However, it is not only the catalyst activity that is important, but also its deactivation behaviour with the intended feed.[552] Special catalyst types are often loaded on top of the main catalyst beds to help with feed distribution and to remove feed impurities that can lead to deposit formation. Catalyst grading with an HDM catalyst on top to trap metals and avoid pressure drop problems is therefore common practice.

Fixed bed reactors are most commonly used for hydrotreating. Specific applications may benefit from catalytic distillation, where the fixed bed catalyst is contained within a distillation column. In cases where metals in the Fischer–Tropsch syncrude are a problem, moving bed, ebullated bed or even slurry phase reactors can be considered. However, due to the higher cost and complexity associated with these reactor types, fixed bed hydrotreating is generally preferred.

5.4.2 Hydrotreating Fischer–Tropsch Syncrude

5.4.2.1 Hydrotreating Fischer–Tropsch Oil

In a hydrotreating study by Lamprecht, the feed materials were HTFT stabilised light oil (SLO), and also a 60:40 blend of HTFT SLO with LTFT Arge distillate (Arge distillate is straight run distillate derived from fixed bed Fe-LTFT synthesis).[538] In this study, both HDO and HYD were important. The properties of the catalysts that were evaluated are shown in Table 5.29. The $NiMo/Al_2O_3$ and $CoMo/Al_2O_3$ catalysts were sulfided, whereas the Ni/Al_2O_3 catalyst was unsulfided.

Commercial sulfided base metal catalysts were used and typical distillate hydrotreater conditions for processing material from FTS were stated as 288 °C, 5.8 MPa, LHSV $1.2\,h^{-1}$ and H_2:feed ratio 247:1.[538] The operating temperature given is a bed average and under commercial operation the adiabatic temperature rise during distillate hydrotreating is 30 °C. The performance of the catalysts listed in Table 5.29 with the 60:40 blend of HTFT SLO with LTFT Arge distillate is given in Table 5.30.[538] The aim was to produce a diesel fuel and the desired product specifications were an alkene content of less than 7 g Br per 100 g, an acid content of less than 0.25 mg KOH g^{-1} and an oxidation stability of better than $2\,mg\,l^{-1}$.

Acceptable oxidation stability could be achieved over all catalysts employed. However, a higher than specified bromine number was observed over the

Table 5.29 Properties of the commercial catalysts that were evaluated for the hydrotreating of HTFT and mixed HTFT–LTFT distillate range materials to produce fuels.

Property	*$NiMo/Al_2O_3$*	*$CoMo/Al_2O_3$*	*Ni/Al_2O_3*
Nominal diameter (mm)	1.1	1.3	3.5
Catalyst shape	Quadrulobe	Cylindrical	Spherical
Surface area ($m^2\,g^{-1}$)	138	265	58
Metal promoters (mass%)			
NiO/Ni[a]	4.0	–	10
CoO	–	5.0	–
MoO_3	19.5	16.0	–

[a]NiO for sulfided $NiMo/Al_2O_3$ catalyst and Ni for unsulfided Ni/Al_2O_3 catalyst.

Table 5.30 Hydrogenation of a 60:40 blend of HTFT SLO and LTFT Arge distillate over different catalysts at 2.5 MPa, LHSV $0.5\,h^{-1}$ and H_2:feed ratio 540:1.

		Composite product properties[a]		
Hydrogenation catalyst	*Temperature (°C)*	*Alkenes (g Br per 100 g)*	*Acids (mg KOH g^{-1})*	*Oxidation stability (mg l^{-1})*
None – feed	–	53	3.76	3.1
$NiMo/Al_2O_3$ (sulfided)	239	11.2	0.08	1.5
	272	2.0	0.4	0.9
$CoMo/Al_2O_3$ (sulfided)	270	14.2	0.3	0.9
Ni/Al_2O_3 (reduced)	240	22.2	2.5	0.9
	271	5.3	2	1

[a]Diesel fuel obtained by distillation from composite has lower values than the composite.

$CoMo/Al_2O_3$ catalyst, unless the temperature was increased to at least 300 °C. Further, acceptable alkene hydrogenation was achieved at 270 °C over both reduced Ni/Al_2O_3 and sulfided $NiMo/Al_2O_3$ catalysts. This is attributed to the higher HYD activity of the Ni-based catalysts than the Co-based catalyst. The reduced Ni/Al_2O_3 performed poorly in the HDO of carboxylic acids, whereas both sulfided catalysts were able to remove carboxylic acids to an acceptable level.

The reduced Ni/Al_2O_3 catalyst deactivated measurably with time on-stream. Over a period of 6 days at 240 °C, the alkene conversion in the HTFT SLO/LTFT Arge feed decreased from 80% to less than 40%.[538] It was speculated that this may have been caused by acid leaching of the Ni, which was confirmed by a later study.[395] Leaching of reduced Ni/Al_2O_3 catalysts by carboxylic acids resulted in the formation of nickel carboxylates. Carboxylic acid leaching can in principle be prevented by operating at a temperature above the nickel carboxylate decomposition temperature, which was found to be in the range 280–305 °C for the C_2–C_5 nickel carboxylates. Unfortunately, this is not industrially practical for hydrotreating over reduced Ni/Al_2O_3 catalysts, because of the hydrogenolysis propensity of reduced nickel catalysts under these conditions. It was also found that nickel leaching did not increase monotonically with temperature, but was inhibited at >200 °C, probably due to polymerisation of the nickel carboxylates. However, this inhibition is insufficient to make Ni/Al_2O_3 catalysts suitable for hydrotreating Fischer–Tropsch materials containing short-chain carboxylic acids.

During hydrotreating of the Fischer–Tropsch syncrude over sulfided base metal catalysts, it is necessary to co-feed sulfur-containing compounds with the sulfur-free syncrude to keep the hydrotreating catalyst sulfided. With insufficient sulfur in the feed, the stability of catalysts may be affected. This was illustrated by the deactivation of the sulfided $CoMo/Al_2O_3$ hydrotreating catalyst when the H_2S content in tail gas was decreased during hydroprocessing of

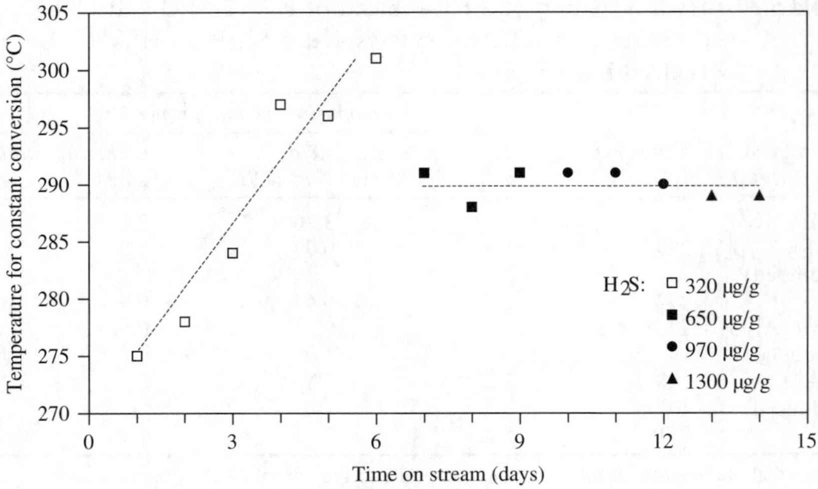

Figure 5.31 Temperature required to maintain a constant level of alkene hydrogenation (7 g Br per 100 g) in an HTFT straight run distillate over a sulfided $CoMo/Al_2O_3$ catalyst at 5.8 MPa, LHSV $1\ h^{-1}$, H_2:feed ratio 270:1 and different levels of sulfur co-feeding. The H_2S content in the tail gas was 320 (□), 650 (■), 970 (●) and 1300 (▲) $\mu g\ g^{-1}$.

Table 5.31 Hydrotreating of straight run HTFT distillate over low- and high-activity sulfided $NiMo/Al_2O_3$ catalysts at 5 MPa.

		Hydrogenated over $NiMo/Al_2O_3$	
Description	*HTFT feed*	*Low activity*	*High activity*
Density ($kg\ m^{-3}$)	822	813.1	804.1
Cetane number	55	57	63
Lubricity HFRR wear scar (µm)	<460	506	546
Alkene content (g Br per 100 g)	63	5.07	1.18
Acid content (mg KOH g^{-1})	12.8	0.02	0.004
Aromatic content (mass%)			
Mononuclear	26.3	24.4	22.2
Dinuclear	0.6	0.51	0.24
Polynuclear	0.1	0.09	Not detected

oxygenate-containing syncrude (Figure 5.31).[538] Deactivation may be due to the replacement of the catalytic sulfur with oxygen.[553] It was postulated that in the absence of a sufficient amount of S-donating species, the replacement of S by O may occur on the catalyst surface with the OH^- anion being a less efficient donor than the SH^- anion.[529]

The hydrotreating of straight run HTFT distillate over different sulfided $NiMo/Al_2O_3$ catalysts has been investigated by Leckel (Table 5.31).[441] The lower activity $NiMo/Al_2O_3$ catalyst produced a better quality distillate than the higher activity supertype-II active reaction sites (STARS) catalyst. The feed from FTS is too reactive and high-activity catalysts that are beneficial for crude oil hydrotreating are not always beneficial for Fischer–Tropsch syncrude hydrotreating.

It is clear that conventional sulfided base metal catalysts can be used to hydrotreat material from FTS. However, the addition of sulfur to the sulfur-free FTS-derived oil is undesirable, since it causes the hydroprocessed products from FTS synthesis to have a similar sulfur content as severely hydroprocessed crude oil-derived products. It is believed that there is still considerable opportunity to develop catalysts suitable for hydrotreating FTS-derived oil. The objective would be a stable catalyst in the absence of a sulfiding agent. This may require novel catalytic phases combined either with the conventional γ-Al_2O_3 support or different supports. The opportunity to develop catalysts employing a more apolar support to improve discrimination between HYD and HDO has also been suggested.[532]

5.4.2.2 Hydrogenation of Fischer–Tropsch Alkenes

Alkene hydrogenation (HYD) can be performed with sulfided base metal, reduced base metal and reduced noble metal catalysts. The catalyst selection depends on the refining objective (partial or complete alkene HYD), the feed matrix (presence of oxygenates and aromatics) and engineering considerations (heat management). In fuel applications, HYD is far more prominent, especially to ensure fuel stability in order to meet fuel specifications. Trends observed generally indicate that fuel stability decreases in the order alkanes > cycloalkanes > branched alkanes > aromatics > alkenes, with monofunctional alkenes being more stable than dienes.

Noble metal hydrotreating is typically considered when alkyne or diene saturation is required in an alkene-containing feed. This is a partial HYD process, where HYD of the monofunctional alkene is undesirable. In a refining context, the catalyst may be selected to allow concomitant double bond IS (typically over Pd-based catalysts), as employed in the CDHydro units at the Sasol Synfuels HTFT refinery.[554] When complete alkene HYD is necessary, noble metal catalysts tend to be too active. One exception is application of noble metal catalysts in situations that require both HYD and HDAr. In such instances, proper heat management is critical.

It has been reported that the commercial use of sulfided base metal catalysts for alkene HYD associated with FTS leads to a deterioration in product quality (octane number of the motor gasoline) with time on-stream.[555] This deterioration was not due to operating temperature and the deactivation behaviour was not explained. Therefore, the selection of a suitable catalyst may be an issue, although the HYD of alkenes requires fairly mild conditions.

The HYD of the alkene-rich product from OLI typically requires that only part of the product should by hydrotreated. Some opportunities for efficiency improvement in these situations have been pointed out. The main recommendations were:[532]

1. Select the most appropriate configuration of units (partial HYD *versus* complete HYD with some by-pass).
2. Use an isomerisation catalyst during partial HYD in order to isomerise the double bond of 1-alkenes to higher octane internal alkenes.
3. Consider operating conventional base metal hydrotreating catalysts in dual mode, first as reduced (unsulfided) catalysts at low temperature and then later as sulfided catalysts to extend the operating temperature range and lifetime of the catalysts.

When sulfided base metal catalysts are employed for HYD of alkenes, control of the H_2S concentration in relation to the temperature and H_2 partial pressure is very important.[556] At lower temperatures H_2S may react with the alkenes to produce thiols. This is an equilibrium-limited conversion. Although the thiols are readily hydrogenated, they may undergo side-reactions to form more stable sulfur-containing products that are more difficult to hydrogenate. In this way, sulfur may be incorporated into the product.

5.4.2.3 Hydrotreating Fischer–Tropsch Waxes

Wax hydrogenation is mainly employed to improve the wax properties, such as odour, colour and stability. Little detail has been provided about the catalyst selection for wax HYD in the Shell Bintulu facility, apart from the fact that alkenes and oxygenates are saturated over a non-isomerising catalyst.[557] The absence of any sulfur in the products indicates that it is likely to be a reduced (unsulfided) base metal or noble metal hydrotreating catalyst.

The performance of a sulfided $NiMo/Al_2O_3$ catalyst for LTFT wax hydrogenation has been reported by Bolder.[537] The operating conditions required to meet the desired product quality was 290–330 °C, 6 MPa hydrogen pressure and LHSV 1 h^{-1}. Industrial hard wax hydrogenation is performed at around 260 °C, 5 MPa and LHSV 0.3–0.5 h^{-1} over reduced Ni-based catalysts.

Additional information on the hydrotreating of LTFT waxes can be found in the next chapter (Section 6.3.1).

5.4.2.4 Hydrotreating Fischer–Tropsch Aqueous Products

The oxygenates that can be recovered from the Fischer–Tropsch aqueous product have value as chemicals. Nevertheless, it is beneficial to reduce the complexity of the aqueous product refinery.[534,535] Hydrotreating the carbonyl compounds to alcohols simplifies the product slate and in the case of ethanal, partial hydrogenation to ethanol converts a normally gaseous product into a liquid product. Industrially reduced $Ni/SiO_2–Al_2O_3$ performs well for the partial hydrogenation of Fischer–Tropsch carbonyls to alcohols.[534]

5.4.2.5 Hydrotreating Coal Liquids Associated with FTS

In a study by Leckel, four conventional catalysts were evaluated for hydroprocessing of the liquids produced as by-products of coal gasification.[558] It should be noted that for such liquids, porosity of the catalyst is a more important parameter than that of catalysts used for hydroprocessing of the Fischer–Tropsch syncrude. In a subsequent study, it was pointed out that the best HDO performance over NiW/Al_2O_3 catalysts was obtained for catalysts with a peak pore diameter in the range 6.8–16 nm.[559] Hydrotreating of coal pyrolysis liquids typically requires severe conditions, such as those employed in the aforementioned study, namely 377–480 °C and 12.5–17.5 MPa of H_2.

There is a significant body of literature dealing specifically with the catalysis of coal conversion and the hydroprocessing of coal liquids.[560] Reference to coal liquids in a Fischer–Tropsch context is included due to the possible need to co-hydrotreat coal pyrolysis liquids in an FTS-based coal-to-liquids facility. It is important to realise that there is a significant difference in operating parameters and that coal liquids cannot just be co-hydrotreated with Fischer–Tropsch syncrude. The severity of hydrotreating required to produce fuels typically increases in the order Fischer–Tropsch syncrude < conventional crude oil < coal liquids.

References

1. A. de Klerk, *Prepr. Am. Chem. Soc. Div. Pet. Chem.*, 2008, **53**(2), 105.
2. A. de Klerk, *Green Chem.*, 2008, **10**, 1249.
3. A. G. Oblad, G. A. Mills and H. Heinemann, in *Catalysis. Volume VI. Alkylation, Isomerization, Polymerization, Cracking and Hydroreforming*, ed. P. H. Emmett, Reinhold, New York, 1958, p. 341.
4. J. F. McMahon, C. Bednars and E. Solomon, in *Advances in Petroleum Chemistry and Refining*, ed. K. A. Kobe and J. J. McKetta, Vol. **7**, Wiley, New York, 1963, p. 285.
5. F. Asinger, *Mono-olefins Chemistry and Technology*, Pergamon Press, Oxford, 1968.
6. J. Skupinska, *Chem. Rev.*, 1991, **91**, 613.
7. M. Sanati, C. Hörnell and S. G. Järås, *Catalysis*, 1999, **14**, 236.
8. P. Leprince, in *Petroleum Refining. Volume 3. Conversion Processes*, ed. P. Leprince, Technip, Paris, 2001, p. 321.
9. P. R. Pujadó and D. J. Ward, in *Handbook of Petroleum Processing*, ed. D. S. J. Jones and P. R. Pujadó, Springer, Dordrecht, 2006, p. 372.
10. R. J. Quann, L. A. Green, S. A. Tabak and F. J. Krambeck, *Ind. Eng. Chem. Res.*, 1988, **27**, 565.
11. C. T. O'Connor, in *Handbook of Heterogeneous Catalysis*, Vol. **5**, ed. G. Ertl, H. Knözinger and J. Weitkamp, Wiley-VCH, Weinheim, 1997, p. 2380.
12. L. Schmerling and V. N. Ipatieff, *Adv. Catal.*, 1950, **2**, 21.
13. A. de Klerk, *Ind. Eng. Chem. Res.*, 2005, **44**, 3887.

14. A. Corma, *Chem. Rev.*, 1995, **95**, 559.
15. T. Mole, J. R. Anderson and G. Creer, *Appl. Catal.*, 1985, **17**, 141.
16. J. Kanai and N. Kawata, *J. Catal.*, 1988, **114**, 284.
17. J. S. Buchanan, J. G. Santiesteban and W. O. Haag, *J. Catal.*, 1996, **158**, 279.
18. A. de Klerk, *Ind. Eng. Chem. Res.*, 2006, **45**, 578.
19. R. A. Keogh and B. H. Davis, *in Conference Procedings*, AIChE Spring Meeting, New Orleans, LA, 2003, p. 686.
20. J. P. G. Pater, P. A. Jacobs and J. A. Martens, *J. Catal.*, 1999, **184**, 262.
21. A. de Klerk, *Energy Fuels*, 2007, **21**, 3084.
22. V. N. Ipatieff, B. B. Corson and G. Egloff, *Ind. Eng. Chem.*, 1935, **27**, 1077.
23. V. N. Ipatieff and G. Egloff, *Oil Gas J.*, 1935, 16 May, 31.
24. S. Raseev, *Thermal and Catalytic Processes in Petroleum Refining*, Marcel Dekker, New York, 2002.
25. P. C. Weinert and G. Egloff, *Pet. Process.*, 1948, June, 585.
26. J. M. Meister, S. M. Black, B. S. Muldoon, D. H. Wei and C. M. Roeseler, *Hydrocarbon Process.*, 2000, **79**(5), 63.
27. M. Marchionna, M. di Girolamo and R. Patrini, *Catal. Today*, 2001, **65**, 397.
28. R. Birkhoff and M. Nurminen, in *Handbook of Petroleum Refining Processes*, ed. R. A. Meyers, McGraw-Hill, New York, 2004, p. 1.3.
29. A. Convers, D. Commerenc and B. Torec, *Rev. Inst. Fr. Pet.*, 1994, **49**, 437.
30. F. Favre, A. Forestière, F. Hughes, H. Olivier-Bourbigou and J. A. Chodorge, *Oil Gas*, 2005, **31**(2), 83.
31. F. Nierlich, *Hydrocarbon Process.*, 1992, **71**(2), 45.
32. E. Köhler, F. Schmidt, H. Wernicke, M. de Pontes and H. L. Roberts, *Hydrocarbon Technol. Int.*, 1995, Summer, 37.
33. G. K. Chitnis, A. B. Dandekar, B. S. Umansky, G. B. Brignac and J. Stokes, presented at the NPRA 103rd National Meeting, San Francisco, 2005, AM-05-77.
34. G. K. Chitnis, A. B. Dandekar, B. S. Umansky, G. B. Brignac, J. Stokes and W. A. Leet, *Hydrocarbon Eng.*, 2005, **10**(6), 21.
35. L. P. Dancuart, R. de Haan and A. de Klerk, *A. Stud. Surf. Sci. Catal.*, 2004, **152**, 482.
36. G. Egloff and P. C. Weinert, in *Proceedings of the 3rd World Petroleum Congress*, 1951, Section IV, p. 201.
37. V. N. Ipatieff and B. B. Corson, *Ind. Eng. Chem.*, 1936, **28**, 860.
38. W. F. Deeter, *Oil Gas J.*, 1950, 23 March, 252.
39. V. N. Ipatieff and R. E. Schaad, *Ind. Eng. Chem.*, 1938, **30**, 596.
40. J. H. Coetzee, T. N. Mashapa, N. M. Prinsloo and J. D. Rademan, *Appl. Catal. A*, 2006, **308**, 204.
41. A. de Klerk and N. M. Prinsloo, *Prepr. Pap. Am. Chem. Soc. Div. Fuel Chem.*, 2009, **54**(1), 211.

42. J. S. Swart, G. J. Czajkowski and R. E. Conser, *Oil Gas J.*, 1981, **79**(35), 62.
43. A. de Klerk, D. J. Engelbrecht and H. Boikanyo, *Ind. Eng. Chem. Res.*, 2004, **43**, 7449.
44. A. de Klerk and P. L. de Vaal, *Ind. Eng. Chem. Res.*, 2008, **47**, 6870.
45. A. de Klerk, *Energy Fuels*, 2006, **20**, 439.
46. T. N. Mashapa and A. de Klerk, *Appl. Catal. A*, 2007, **332**, 200.
47. G. Egloff, J. C. Morrell and E. F. Nelson, *Pet. Refiner*, 1937, **16**, 497.
48. N. M. Prinsloo, *Fuel Process. Technol.*, 2006, **87**, 437.
49. V. N. Ipatieff, *Ind. Eng. Chem.*, 1935, **27**, 1067.
50. A. Farkas and L. Farkas, *Ind. Eng. Chem.*, 1942, **34**, 716.
51. A. de Klerk, *Ind. Eng. Chem. Res.*, 2004, **43**, 6325.
52. V. N. Ipatieff and H. Pines, *Ind. Eng. Chem.*, 1935, **27**, 1364.
53. T. R. Krawietz, P. Lin, K. E. Lotterhos, P. D. Torres, D. H. Barich, A. Clearfield and J. F. Haw, *J. Am. Chem. Soc.*, 1998, **120**, 8502.
54. V. N. Ipatieff, H. Pines and R. E. Schaad, *J. Am. Chem. Soc.*, 1934, **56**, 2696.
55. R. J. J. Nel and A. de Klerk, *Ind. Eng. Chem. Res.*, 2007, **46**, 2902.
56. R. Schwarzer, E. du Toit and W. Nicol, *Appl. Catal. A*, 2008, **340**, 119.
57. J. March, *Advanced Organic Chemistry*, 3rd edn, Wiley, New York, 1985.
58. J. Abbot and B. W. Wojciechowski, *J. Catal.*, 1987, **108**, 346.
59. G. E. Langlois and J. E. Walkey, *Pet. Refiner*, 1952, **21**, 79.
60. R. Bekker and N. M. Prinsloo, *Ind. Eng. Chem. Res.*, 2009, **48**, 10156.
61. R. A. Munson, *J. Phys. Chem.*, 1964, **68**, 3374.
62. R. A. Munson, *J. Phys. Chem.*, 1965, **69**, 1761.
63. S. R. Bethea and J. H. Karchmer, *Ind. Eng. Chem.*, 1956, **48**, 370.
64. A. de Klerk, D. O. Leckel and N. M. Prinsloo, *Ind. Eng. Chem. Res.*, 2006, **45**, 6127.
65. J. Villadsen and H. Livbjerg, *Catal. Rev. Sci. Eng.*, 1973, **17**, 203.
66. A. de Klerk, R. J. J. Nel and R. Schwarzer, *Ind. Eng. Chem. Res.*, 2007, **46**, 2377.
67. R. F. Jameson, *J. Chem. Soc.*, 1959, **752**.
68. E. H. Brown and C. D. Whitt, *Ind. Eng. Chem.*, 1952, **44**, 615.
69. D. I. MacDonald and J. R. Boyack, *J. Chem. Eng. Data*, 1969, **14**, 380.
70. F. Cavani, G. Girotti and G. Terzoni, *Appl. Catal. A*, 1993, **97**, 177.
71. Z. Zhu, Z. Xie, Y. Chen, R. Wang and Y. Yao, *React. Kinet. Catal. Lett.*, 2000, **70**, 379.
72. J. P. Smith, W. E. Brown and J. R. Lehr, *J. Am. Chem. Soc.*, 1955, **77**, 2728.
73. R. N. Bell, *Ind. Eng. Chem.*, 1948, **40**, 1464.
74. A. L. Huthi and P. A. Gartaganis, *Can. J. Chem.*, 1956, **34**, 785.
75. S. Ohashi and H. Sugatani, *Bull. Chem. Soc. Jpn.*, 1957, **30**, 864.
76. B. J. Fontana, *J. Am. Chem. Soc.*, 1951, **73**, 3348.
77. W. H. Ross and R. M. Jones, *J. Am. Chem. Soc.*, 1925, **47**, 2165.
78. E. Thilo, *Adv. Inorg. Chem. Radiochem.*, 1962, **4**, 1.

79. R. F. Jameson, in *Phosphoric Acid*, ed. A. V. Slack, Marcel Dekker, New York, 1968, p. 983.
80. J. R. van Wazer, *J. Am. Chem. Soc.*, 1956, **78**, 5709.
81. H. J. de Jager and A. M. Heyns, *J. Phys. Chem. A*, 1998, **102**, 2838.
82. R. N. Bell, *Ind. Eng. Chem.*, 1947, **39**, 136.
83. H. Ohtsuka and K. Aomura, *Bull. Jpn. Pet. Inst.*, 1962, **4**, 3.
84. N. M. Prinsloo, *Ind. Eng. Chem. Res.*, 2007, **46**, 7838.
85. R. B. Anderson, A. Krieg, B. Seligman and W. Tarn, *Ind. Eng. Chem.*, 1948, **40**, 2347.
86. F. Papmahl and H. F. Rase, *Ind. Eng. Chem. Prod. Res. Dev.*, 1969, **8**, 352.
87. C. M. Fougret, M. P. Atkins and W. F. Hölderich, *Appl. Catal. A*, 1999, **181**, 145.
88. N. S. Kotsarenko, V. P. Shmachkova, I. L. Mudrakovskii and V. M. Mastikhin, *Kinet. Catal.*, 1989, **30**, 974.
89. C. Lin-na, E. S. Boichinova and O. N. Setkina, *Russ. J. Inorg. Chem.*, 1964, **9**, 798.
90. Y. Maki, K. Sato, A. Isobe, N. Iwasa, S. Fujita, M. Shimokawabe and N. Takezawa, *Appl. Catal. A*, 1998, **170**, 269.
91. R. Calvert, *Diatomaceous Earth*, ACS Monograph Series 52, Chemical Catalog Company, New York, 1930.
92. J. B. Moffatt, *Catal. Rev. Sci. Eng.*, 1978, **18**(2), 199.
93. S. Göbölös, E. Tálas, M. Hegedüs, I. Bertóti and J. L. Margitfalvi, *Appl. Catal. A*, 1997, **152**, 63.
94. V. N. Ipatieff and H. J. Pines, *J. Am. Chem. Soc.*, 1936, **58**, 1056.
95. M. N. Timofeeva, *Appl. Catal. A*, 2003, **256**, 19.
96. J. S. Vaughan, C. T. O'Connor and J. C. Q. Fletcher, *J. Catal.*, 1994, **147**, 441.
97. J. D. Burrington, J. R. Johnson and J. K. Pudelski, *Top. Catal.*, 2003, **23**, 175.
98. G. Chen, J. Li, X. Yang and Y. Wu, *Appl. Catal. A*, 2006, **310**, 16.
99. J. Zhang, R. Ohnishi, T. Okuhara and Y. Kamiya, *Appl. Catal. A*, 2009, **353**, 68.
100. C. T. O'Connor and M. Kojima, *Catal. Today*, 1989, **6**, 329.
101. S. Schwartz, M. Kojima and C. T. O'Connor, *Appl. Catal.*, 1989, **56**, 263.
102. H. H. Mooiweer, K. P. de Jong, B. Kraushar-Czarnetzki, W. H. J. Stork and B. C. H. Krutzen, *Stud. Surf. Sci. Catal.*, 1994, **84**, 2327.
103. W. E. Garwood, *ACS Symp. Ser.*, 1983, **218**, 383.
104. S. A. Tabak, F. J. Krambeck and W. E. Garwood, *AIChE J.*, 1986, **32**, 1526.
105. C. T. O'Connor, S. T. Langford and J. C. Q. Fletcher, in *Proceedings of the 9th International Zeolite Conference*, Montreal, 1992, p. 467.
106. O. R. Minnie, The effect of 1-hexene extraction on the COD process, MTech dissertation, University of South Africa, Pretoria, 2006.
107. S. A. Tabak and F. J. Krambeck, *Hydrocarbon Process.*, 1985, **64**(9), 72.
108. C. Knottenbelt, *Catal. Today*, 2002, **71**, 437.
109. S. J. Miller, *Stud. Surf. Sci. Catal.*, 1987, **38**, 187.

110. I. P Dzikh, J. M. Lopes, F. Lemos and F. R. Ribeiro, *Appl. Catal. A*, 1999, **177**, 245.
111. J. Bandiera and Y. Ben Taarit, *Appl. Catal. A*, 1995, **132**, 157.
112. M. Kojima, M. W. Rautenbach and C. T. O'Connor, *Ind. Eng. Chem. Res.*, 1988, **27**, 248.
113. J. F. Knifton, J. R. Sanderson and P. E. Dai, *Catal. Lett.*, 1994, **28**, 223.
114. L. M. Tiako Ngandjui and F. C. Thyrion, *Ind. Eng. Chem. Res.*, 1996, **35**, 1269.
115. A. L. Lapidus, Y. I. Isakov, T. F. Tishkina, V. G. Lipovich and K. M. Minachev, *Pet. Chem. USSR*, 1976, **16**(3), 121.
116. M. L. Occelli, J. T. Hsu and L. G. Galya, *J. Mol. Catal.*, 1985, **32**, 377.
117. M. Bjørgen, K.-P. Lillerud, U. Olsbye, S. Bordiga and A. Zecchina, *J. Phys. Chem. B*, 2004, **108**, 7862.
118. L. R. Martens, J. Verduyn and G. M. Mathijs, *Catal. Today*, 1997, **36**, 451.
119. M. Tiitta, E. Harlin, J. Makkonen, A. Root, F. Sandelin and H. Osterholm, *Stud. Surf. Sci. Catal.*, 2004, **154**, 2323.
120. M. Golombok and J. de Bruijn, *Ind. Eng. Chem. Res.*, 2000, **39**, 267.
121. A. de Klerk, *Energy Fuels*, 2006, **20**, 1799.
122. A. de Klerk, *Energy Fuels*, 2007, **21**, 625.
123. G. Bellussi, C. Perego, A. Carati, S. Peratello, E. Previde-Massara and G. Perego, *Stud. Surf. Sci. Catal.*, 1994, **84**, 85.
124. B. Chiche, E. Sauvage, F. di Renzo, I. I. Ivanova and F. Fajula, *J. Mol. Catal. A*, 1998, **134**, 145.
125. S. Peratello, M. Molinari, G. Bellussi and C. Perego, *Catal. Today*, 1999, **52**, 271.
126. R. Catani, M. Mandreoli, S. Rossini and A. Vaccari, *Catal. Today*, 2002, **75**, 125.
127. C. Flego, S. Peratello, C. Perego, L. M. F. Sabatino, G. Bellussi and U. Romano, *J. Mol. Catal. A*, 2003, **204–205**, 581.
128. J. M. Escola, R. van Grieken, J. Moreno and R. Rodríguez, *Ind. Eng. Chem. Res.*, 2006, **45**, 7409.
129. P. S. E. Dai, J. R. Sanderson and J. F. Knifton, *Stud. Surf. Sci. Catal.*, 1994, **84**, 1701.
130. S. M. Pillai and M. Ravindranathan, *J. Chem. Soc., Chem. Commun.*, 1994, 1813.
131. J. P. Hogan, R. L. Banks, W. C. Lanning and A. Clark, *Ind. Eng. Chem.*, 1955, **47**, 752.
132. V. C. F. Holm, G. C. Bailey and A. Clark, *Ind. Eng. Chem.*, 1957, **49**, 250.
133. H. Imai and H. Uchida, *Bull. Chem. Soc. Jpn.*, 1965, **38**, 925.
134. R. L. Espinoza, R. Snel, C. J. Korf and C. P. Nicolaides, *Appl. Catal.*, 1987, **29**, 295.
135. M. D. Heydenrych, C. P. Nicolaides and M. S. Scurrell, *J. Catal.*, 2001, **197**, 49.
136. H. Imai, T. Hasegawa and H. Uchida, *Bull. Chem. Soc. Jpn.*, 1968, **41**, 45.
137. R. Spinicci and A. Tofanari, *Mater. Chem. Phys.*, 1990, **25**, 375.

138. D. Kiessling, G. Wendt, K. Hagenau and R. Schoellner, *Appl. Catal.*, 1991, **71**, 69.
139. J. Heveling, C. P. Nicolaides and M. S. Scurrell, *Appl. Catal. A*, 2003, **248**, 239.
140. S. G. Hindin, G. A. Mills and A. G. Oblad, *J. Am. Chem. Soc.*, 1951, **73**, 278.
141. R. G. Haldeman and P. H. Emmett, *J. Am. Chem. Soc.*, 1956, **78**, 2917.
142. M. R. Basila, T. R. Kantner and K. H. Rhee, *J. Phys. Chem.*, 1964, **68**, 3197.
143. J. S. Vaughan, C. T. O'Connor and J. C. Q. Fletcher, *Stud. Surf. Sci. Catal.*, 1994, **84**, 1709.
144. S. M. Yang, D. H. Guo, J. S. Lin and G. T. Wang, *Stud. Surf. Sci. Catal.*, 1994, **84**, 1677.
145. A. S. Chellappa, R. C. Miller and W. J. Thomson, *Appl. Catal. A*, 2001, **209**, 359.
146. S. Soled, N. Dispenziere and R. Saleh, *Stud. Surf. Sci. Catal.*, 1992, **73**, 77.
147. R. Sarin, D. K. Tuli, S. Sinharay, M. M. Rai, S. Ghosh and A. K. Bhatnagar, *Prepr. Am. Chem. Soc. Div. Pet. Chem.*, 1996, **41**(3), 625.
148. R. A. Keogh and B. H. Davis, in *AIChE Spring National Meeting*, New Orleans, 2003, p. 686.
149. A. K. Kolah, Q. Zhiwen and S. M. Mahajani, *Chem. Innov.*, 2001, **31**(3), 15.
150. W. O. Haag, *Chem. Eng. Prog. Symp. Ser.*, 1967, **63**, 140.
151. M. L. Honkela and A. O. I. Krause, *Catal. Lett.*, 2003, **87**, 113.
152. I. Bucsi and G. A. Olah, *J. Catal.*, 1992, **137**, 12.
153. M. di Girolamo, M. Lami, M. Marchionna, E. Pescarollo, L. Tagliabue and F. Ancillotti, *Ind. Eng. Chem. Res.*, 1997, **36**, 4452.
154. G. Scharfe, *Hydrocarbon Process.*, 1973, **52**(4), 171.
155. V. J. Cruz, J. F. Izquierdo, F. Cunill, J. Tejero, M. Iborra and C. Fité, *React. Funct. Polym.*, 2005, **65**, 149.
156. V. J. Cruz, R. Bringué, F. Cunill, J. F. Izquierdo, J. Tejero, M. Iborra and C. Fité, *J. Catal.*, 2006, **238**, 330.
157. S. Talwalkar, M. Chauhan, P. Aghalayam, Z. Qi, K. Sundmacher and S. Mahajani, *Ind. Eng. Chem. Res.*, 2006, **45**, 1312.
158. R. Thornton and B. C. Gates, *J. Catal.*, 1974, **34**, 275.
159. D. Smook and A. de Klerk, *Ind. Eng. Chem. Res.*, 2006, **45**, 467.
160. Y. Chauvin, J. F. Gaillard, J. Léonard, P. Bonnifay and J. W. Andrews, *Hydrocarbon Process.*, 1982, **61**(5), 110.
161. Y. Chauvin, J. F. Gaillard, D. V. Quang and J. W. Andrews, *Pet. Petrochem. Int.*, 1973, **13**(10), 108.
162. P. M. Kohn, *Chem. Eng.*, 1977, 23 May, 114.
163. J. Léonard and J. F. Gaillard, *Hydrocarbon Process.*, 1981, **60**(3), 99.
164. J. F. Boucher, G. Follain, D. Fulop and J. F. Gaillard, *Oil Gas J.*, 1982, 29 March, 84.
165. J. A. Chodorge, J. Cosyns, F. Hughes and H. Olivier-Bourbigou, in *DGMK Conference*, Aachen, 1997, p. 227.

166. J. T. Dixon, M. J. Green, F. M. Hess and D. H. Morgan, *J. Organomet. Chem.*, 2004, **689**, 3641.
167. A. Bollmann, K. Blann, J. T. Dixon, F. M. Hess, E. Killian, H. Maumela, D. S. McGuinness, D. H. Morgan, A. Neveling, S. Otto, M. J. Overett, A. M. Z. Slawin, P. Wasserscheid and S. Kuhlmann, *J. Am. Chem. Soc.*, 2004, **126**, 14712.
168. V. N. Ipatieff and H. Pines, *J. Org. Chem.*, 1936, **1**, 464.
169. V. N. Ipatieff and B. B. Corson, *Ind. Eng. Chem.*, 1935, **27**, 1069.
170. A. Ranwell, The oligomerisation of Sasol alpha-olefins (English translation), MSc dissertation, Randse Afrikaanse Universiteit, Johannesburg, 1994.
171. Y. Chauvin, D. Commereuc, F. Hughes and J. Thivolle-Cazat, *Appl. Catal.*, 1988, **42**, 205.
172. J.-K. Jeon, S.-K. Park and Y.-K. Park, *Catal. Today*, 2004, **93–95**, 467.
173. Y. I. Pae, S. H. Lee and J. R. Sohn, *Catal. Lett.*, 2005, **99**, 241.
174. Q. Zhang, M. Kantcheva and I. G. Dalla Lana, *Ind. Eng. Chem. Res.*, 1997, **36**, 3433.
175. T. Cai, L. Zhang, A. Qi, D. Wang, D. Cao and L. Li, *Appl. Catal.*, 1991, **69**, 1.
176. T. Cai, D. Cao, Z. Song and L. Li, *Appl. Catal. A*, 1993, **95**, L1.
177. T. Cai, *Catal. Today*, 1999, **51**, 153.
178. W. A. Horne, *Ind. Eng. Chem.*, 1950, **42**, 2428.
179. S. Yang, Z. Liu, X. Meng and C. Xu, *Energy Fuels*, 2009, **23**, 70.
180. A. de Klerk, *Energy Fuels*, 2005, **19**, 1462.
181. M. Cowley, *Org. Proc. Res. Dev.*, 2007, **11**, 286.
182. S. H. McAllister, W. A. Bailey Jr. and C. M. Bouton, *J. Am. Chem. Soc.*, 1940, **62**, 3210.
183. M. Demorest, D. Mooberry and J. D. Danforth, *Ind. Eng. Chem.*, 1951, **43**, 2569.
184. P. M. Loggenberg, A study of acid formation over a zeolite catalyst during the isomerisation of 1-hexene (English translation), MSc dissertation, Randse Afrikaanse Universiteit, Johannesburg, 1986.
185. G. S. Salvapati, K. V. Ramanamurthy and M. Janardanarao, *J. Mol. Catal.*, 1989, **54**, 9.
186. T. Xu, E. J. Munson and J. F. Haw, *J. Am. Chem. Soc.*, 1994, **116**, 1962.
187. C. D. Chang, N.-Y. Chen, L. R. Koenig and D. E. Walsh, *Prepr. Pap. Am. Chem. Soc. Div. Fuel Chem.*, 1983, **28**(2), 146.
188. C. H. Bartholomew, *Appl. Catal. A*, 2001, **212**, 17.
189. D. M. Bibby, R. F. Howe and G. D. McLellan, *Appl. Catal. A*, 1992, **93**, 1.
190. J. P. G. Pater, P. A. Jacobs and J. A. Martens, *J. Catal.*, 1998, **179**, 477.
191. B. Subramaniam and B. J. McCoy, *Ind. Eng. Chem. Res.*, 1994, **33**, 504.
192. C. D. Chang, *Catal. Rev. Sci. Eng.*, 1983, **25**, 1.
193. F. Bauer, H. Ernst, E. Giedel and R. Schodel, *J. Catal.*, 1996, **164**, 146.
194. R. M. Lago, W. O. Haag, R. J. Mikovsky, D. H. Olson, S. D. Hellring, K. D. Schmitt and G. T. Kerr, in *Proceedings of the 7th International Zeolite Conference*, Tokyo, 1986, p. 677.

195. N.-Y. Topsøe, A. Joensen and E. G. Derouane, *J. Catal.*, 1988, **110**, 404.
196. S. K. Sahoo, N. Viswanadham, N. Ray, J. K. Gupta and I. D. Singh, *Appl. Catal. A*, 2001, **205**, 1.
197. J. N. Finch and A. Clark, *J. Phys. Chem.*, 1969, **73**, 2234.
198. M. Guisnet and P. Magnoux, *Appl. Catal.*, 1989, **54**, 1.
199. M. Guisnet, P. Magnoux and C. Canaff, *Stud. Surf. Sci. Catal.*, 1986, **28**, 701.
200. E. G. Derouane, *Stud. Surf. Sci. Catal.*, 1985, **20**, 221.
201. J. R. Anderson, Y. F. Chang and R. J. Western, *J. Catal.*, 1989, **118**, 466.
202. B. Paweewan, P. J. Barrie and L. F. Gladden, *Appl. Catal. A*, 1998, **167**, 353.
203. N. S. Gnep, F. Bouchet and M. R. Guisnet, *Prepr. Am. Chem. Soc. Div. Pet. Chem.*, 1991, **36**(4), 620.
204. J. R. Anderson, Y. F. Chang and R. J. Western, *Stud. Surf. Sci. Catal.*, 1991, **68**, 745.
205. J. P. G. Pater, P. A. Jacobs and J. A. Martens, *J. Catal.*, 1999, **184**, 262.
206. Z.-X. Cheng and V. Ponec, *Catal. Lett.*, 1994, **27**, 113.
207. M. Guisnet, P. Andy, N. S. Gnep, C. Travers and E. Benazzi, *Stud. Surf. Sci. Catal.*, 1997, **105**, 1365.
208. J. Houžvièka and V. Ponec, *Ind. Eng. Chem. Res.*, 1997, **36**, 1424.
209. M. Guisnet, P. Andy, N. S. Gnep, C. Travers and E. Benazzi, *Ind. Eng. Chem. Res.*, 1998, **37**, 300.
210. J. Houžvièka and V. Ponec, *Ind. Eng. Chem. Res.*, 1998, **37**, 303.
211. M. A. Asensi and A. Martinez, *Appl. Catal. A*, 1999, **183**, 155.
212. D. Rutenbeck, H. Papp, D. Freude and W. Schwieger, *Appl. Catal. A*, 2001, **206**, 57.
213. D. Rutenbeck, H. Papp, H. Ernst and W. Schwieger, *Appl. Catal. A*, 2001, **208**, 153.
214. D. M. Brouwer and H. Hogeveen, in *Progress in Physical Organic Chemistry*, Vol. **9**, ed. A. Streitwieser Jr. and R. W. Taft, Wiley, New York, 1972, p. 197.
215. F. G. Ciapetta and J. B. Hunter, *Ind. Eng. Chem.*, 1954, **45**, 147.
216. H. N. Dunning, *Ind. Eng. Chem.*, 1953, **45**, 551.
217. F. E. Condon, in *Catalysis. Volume VI. Alkylation, Isomerization, Polymerization, Cracking and Hydroreforming*, ed. P. H. Emmett, Reinhold, New York, 1958, p. 121.
218. H. F. Schultz and J. Weitkamp, *Ind. Eng. Chem. Prod. Res. Dev.*, 1972, **11**, 46.
219. J. Weitkamp, *Erdoel Kohle Erdgas Petrochem.*, 1978, **31**, 13.
220. H. Pines, *Chemistry of Catalytic Hydrocarbon Conversion*, Academic Press, New York, 1981.
221. K. J. Chao, H. C. Wu and L. J. Leu, *Appl. Catal. A*, 1996, **143**, 223.
222. H. Deldari, *Appl. Catal. A*, 2005, **293**, 1.
223. J. D. Roberts, C. C. Lee and W. H. Saunders, *J. Am. Chem. Soc.*, 1954, **76**, 4501.
224. F. Fajula, *Stud. Surf. Sci. Catal.*, 1985, **20**, 361.

225. A. Ozaki and K. Kimura, *J. Catal.*, 1964, **3**, 395.
226. M. Misono, Y. Saito and Y. Yoneda, *J. Catal.*, 1978, **55**, 314.
227. J. Hightower and W. K. Hall, *J. Am. Chem. Soc.*, 1967, **89**, 778.
228. S. T. Sie, *Ind. Eng. Chem. Res.*, 1992, **31**, 1881.
229. S. T. Sie, in *Handbook of Heterogeneous Catalysis*, Vol. **5**, ed. G. Ertl, H. Knözinger and J. Weitkamp, Wiley-VCH, Weinheim, 1997, p. 1998.
230. J. A. Martens, M. Tielen, P. A. Jacobs and J. Weitkamp, *Zeolites*, 1984, **4**, 98.
231. A. Corma, A. Martinez, S. Pergher, S. Peratello, C. Perego and G. Bellusi, *Appl. Catal. A*, 1997, **152**, 107.
232. D. M. Brouwer and J. M. Oelderik, *Recl. Trav. Chim. Pays-Bas*, 1968, **87**, 721.
233. W. E. Alvarez, H. Liu, E. A. Garcia, E. H. Rueda, A. J. Rouco and D. E. Resasco, *Stud. Surf. Sci. Catal.*, 1996, **101**, 553.
234. J. E. Tabora and R. J. Davis, *J. Catal.*, 1996, **162**, 125.
235. V. Adeeva, G. D. Lei and W. M. H. Sachtler, *Appl. Catal. A*, 1994, **118**, L11.
236. J. Houžvička and V. Ponec, *Catal. Rev. Sci. Eng.*, 1999, **39**, 319.
237. F. Bauer, E. Bilz, W. H. Chen, A. Freyer, V. Sauerland and S. B. Liu, *Stud. Surf. Sci. Catal.*, 2007, **170**, 1096.
238. P. Meriaudeau, R. Bacaud, L. N. Hung and A. T. Vu, *J. Mol. Catal. A*, 1996, **110**, L177.
239. M. Guisnet, P. Andy, N. S. Gnep, E. Benazzi and C. Travers, *J. Catal.*, 1996, **158**, 551.
240. J. Čejka, B. Wichterlová and P. Sarv, *Appl. Catal. A*, 1999, **179**, 217.
241. M. Trombetta, G. Busca, S. A. Rossini, V. Piccoli and U. Cornaro, *J. Catal.*, 1997, **168**, 334.
242. J. Houžvička and V. Ponec, *Appl. Catal. A*, 1996, **145**, 95.
243. P. Andy, N. S. Gnep, M. Guisnet, E. Benazzi and C. Travers, *J. Catal.*, 1998, **173**, 322.
244. C. Travers, in *Petroleum Refining. Volume 3. Conversion Processes*, ed. P. Leprince, Technip, Paris, 2001, p. 229.
245. F. M. Floyd, M. F. Gilbert, M. Pérez and E. Köhler, *Hydrocarbon Eng.*, 1998, September, 42.
246. P. J. Kuchar, R. D. Gillespie, C. D. Gosling, W. C. Martin, M. J. Cleveland and P. J. Bullen, *Hydrocarbon Eng.*, 1999, March, 50.
247. M. J. Hunter, *Oil Gas Eur. Mag.*, 2003, **29**, 97.
248. H. Weyda and E. Köhler, *Catal. Today*, 2003, **81**, 51.
249. C. Slade and S. Zuijdendorp, *Catal. Courier*, 2006, **64**, 10.
250. N. A. Cusher, in *Handbook of Petroleum Refining Processes*, ed. R. A. Meyers, McGraw-Hill, New York, 2004, pp. 9.15, 9.29 and 9.41.
251. T. Kimura, *Catal. Today*, 2003, **81**, 57.
252. K. P. de Jong, H. H. Mooiweer, J. G. Buglass and P. K. Maarsen, *Stud. Surf. Sci. Catal.*, 1997, **111**, 127.
253. J. B. Wise and D. Powers, *ACS Symp. Ser.*, 1994, **552**, 273.
254. S. M. Ozmen, H. Abrevaya, P. Barger, M. Bentham and M. Kojima, *Fuel Reformulation*, 1993, **3**(5), 54.

255. J. L. Duplan, P. Amigues, J. Verstraete and C. Travers, in *Proceedings of Ethylene Production Conference*, 1996, Vol. **5**, p. 429.
256. M. Cowley, *Energy Fuels*, 2006, **20**, 1771.
257. S. Rossini, R. Catani, U. Cornaro, A. Guercio, R. Miglio and V. Piccoli, *Prepr. Am. Chem. Soc. Div. Pet. Chem.*, 1997, **42**(3), 588.
258. M. J. Girgis and Y. P. Tsao, *Ind. Eng. Chem. Res.*, 1996, **35**, 386.
259. C.-L. O'Young, W.-Q. Xu, M. Simon and S. L. Suib, *Stud. Surf. Sci. Catal.*, 1994, **84**, 1671.
260. J. Houžvička, J. G. Nienhuis, S. Hansildaar and V. Ponec, *Appl. Catal. A*, 1997, **165**, 443.
261. J. Houžvička, J. G. Nienhuis and V. Ponec, *Appl. Catal. A*, 1998 **174**, 207.
262. M. A. Asensi, A. Corma, A. Martinez, M. Derewinski, J. Krysciak and S. S. Tamhankar, *Appl. Catal. A*, 1998, **174**, 163.
263. S. H. Baeck and W. Y. Lee, *Appl. Catal. A*, 1997, **164**, 291.
264. S. H. Baeck and W. Y. Lee, *Appl. Catal. A*, 1998, **168**, 171.
265. H. C. Woo, K. H. Lee and J. S. Lee, *Appl. Catal. A*, 1996, **134**, 147.
266. C. T. O'Connor, E. van Steen and M. E. Dry, *Stud. Surf. Sci. Catal.*, 1996, **102**, 323.
267. G. Onyestyak, J. Valyon, G. Pal-Borbely and L. V. C. Rees, *Stud. Surf. Sci. Catal.*, 2004, **154**, 2316.
268. J. Abbot and B. W. Wojciechowski, *Can. J. Chem. Eng.*, 1985, **63**, 451.
269. J. Abbot, A. Corma and B. W. Wojciechowski, *J. Catal.*, 1985, **92**, 398.
270. S. J. Chong and J. B. Butt, *Korean J. Chem. Eng.*, 1993, **7**, 175.
271. D. Li, M. Li, Y. Chu, H. Nie and Y. Shi, *Catal. Today*, 2003, **81**, 65.
272. F. Sandelin, I. Eilos, E. Harlin, J. Hiltunen, J. Jakkula, J. Makkonen and M. Tiitta, *Stud. Surf. Sci. Catal.*, 2004, **154**, 2130.
273. C. Yin and C. Liu, *Prepr. Am. Chem. Soc. Div. Pet. Chem.*, 2004, **49**(2), 242.
274. X. Liu, Y. Li and B. Teng, *Prepr. Am. Chem. Soc. Div. Pet. Chem.*, 2002, **47**(2), 134.
275. A. Lugstein, A. Jentys and H. Vinek, *Appl. Catal. A*, 1998, **166**, 29.
276. A. Lugstein, A. Jentys and H. Vinek, *Appl. Catal. A*, 1999, **176**, 119.
277. P. Canizares, A. de Lucas, J. L. Valverde and F. Dorado, *Ind. Eng. Chem. Res.*, 1998, **37**, 2592.
278. F. Dorado, R. Romero and P. Canizares, *Ind. Eng. Chem. Res.*, 2001, **40**, 3428.
279. A. Zhang, I. Nakamura, K. Aimoto and K. Fujimoto, *Ind. Eng. Chem. Res.*, 1995, **34**, 1074.
280. K. Fujimoto, K. Maeda and K. Aimoto, *Appl. Catal. A*, 1992, **91**, 81.
281. S. Sivasanker, A. V. Ramaswamy and P. Ratnasamy, *Appl. Catal. A*, 1996, **138**, 369.
282. D. Karthikeyan, N. Lingappan, B. Sivasankar and N. J. Jabarathinam, *Ind. Eng. Chem. Res.*, 2008, **47**, 6538.
283. M. J. Ramos, A. de Lucas, V. Jiménez, P. Sánchez and J. L. Valverde, *Fuel Process. Technol.*, 2008, **89**, 721.

284. G. E. Giannetto, G. R. Perot and M. R. Guisnet, *Ind. Eng. Chem. Prod. Res. Dev.*, 1986, **25**, 481.
285. Z. B. Wang, A. Kamo, T. Yoneda, T. Komatsu and T. Yashima, *Appl. Catal. A*, 1997, **159**, 119.
286. L. Oswaldo Almanza, T. Narbershuber, P. d'Araujo, C. Naccache and Y. Ben Taarit, *Appl. Catal. A*, 1999, **178**, 39.
287. D. Lamprecht and A. de Klerk, *Chem. Eng. Commun.*, 2009, **196**, 1206.
288. P. A. Jacobs, J. A. Martens and H. K. Beyer, *Stud. Surf. Sci. Catal.*, 1985, **20**, 399.
289. G. Giannetto, G. Perot and M. Guisnet, *Stud. Surf. Sci. Catal.*, 1985, **20**, 265.
290. J. M. Grau and J. M. Perera, *Appl. Catal. A*, 1993, **106**, 27.
291. M. T. Tran, N. S. Gnep, G. Szabo and M. Guisnet, *Appl. Catal. A*, 1998, **170**, 49.
292. J. F. Allain, P. Magnoux, P. Schulz and M. Guisnet, *Appl. Catal. A*, 1997, **152**, 221.
293. A. Corma, C. Corell, F. Llopis, A. Martinez and J. Perez-Pariente, *Appl. Catal. A*, 1994, **115**, 121.
294. R. Ravishankar and S. Sivasanker, *Appl. Catal. A*, 1996, **142**, 47.
295. S. G. Hindin, A. G. Oblad and G. A. Mills, *J. Am. Chem. Soc.*, 1955, **77**, 535.
296. Y. Rezgui and M. Guemini, *Energy Fuels*, 2007, **21**, 602.
297. G. Shi and J. Shen, *Energy Fuels*, 2009, **23**, 320.
298. A. Melchor, E. Garbowski, M.-V. Mathieu and M. Primet, *J. Chem. Soc., Faraday Trans. 1*, 1986, **82**, 3667.
299. B. Ducourty, G. Szabo, J. P. Dath, J. P. Gilson, J. M. Goupil and D. Cornet, *Appl. Catal. A*, 2004, **269**, 203.
300. A. S. Kinyakin, E. A. Glushachenkova, P. N. Borutskii, A. S. Shuvalov and Y. A. Pisarenko, *Theor. Found. Chem. Eng. (USSR)*, 2008, **42**, 815.
301. B. A. Orkin, *Ind. Eng. Chem. Prod. Res. Dev.*, 1969, **8**, 154.
302. Z. X. Cheng and V. Ponec, *Appl. Catal. A*, 1994, **118**, 127.
303. E. C. Decario, J. W. Bruno, V. P. Nero and J. C. Edwards, *J. Catal.*, 1993, **140**, 84.
304. V. Adeeva and W. M. H. Sachtler, *Appl. Catal. A*, 1997, **163**, 237.
305. M. Skotak and Z. Karpiński, *Chem. Eng. J.*, 2002, **90**, 89.
306. E. A. Blekkan, C. Pham-Huu, M. J. Ledoux and J. Guille, *Ind. Eng. Chem. Res.*, 1994, **33**, 1657.
307. O. V. Kikhtyanin, A. V. Toktarev and G. V. Echevsky, *Stud. Surf. Sci. Catal.*, 2006, **163**, 897.
308. S. M. Yang, D. H. Guo, J. S. Lin and G. T. Wang, *Stud. Surf. Sci. Catal.*, 1994, **84**, 1677.
309. S. M. Yang, J. Y. Lin, D. H. Guo and S. G. Liaw, *Appl. Catal. A*, 1999, **181**, 112.
310. L. H. Gielgens, I. H. E. Veenstra, V. Ponec, M. J. J. Haanepen and J. H. C. van Hooff, *Catal. Lett.*, 1995, **32**, 195.

311. Y. Wei, G. Wang, Z. Liu, L. Xu and P. Xie, *Stud. Surf. Sci. Catal.*, 2007, **154**, 2359.
312. J. M. Campelo, F. Lafont and J. M. Marinas, *Appl. Catal. A*, 1997, **152**, 53.
313. P. Meriaudeau, V. A. Tuan, V. T. Nghiem, S. Y. Lai, L. N. Huang and C. Nacache, *J. Catal.*, 1997, **69**, 55.
314. P. Meriaudeau, V. A. Tuan, G. Sapaly, V. T. Nghiem, S. Y. Lai and C. Nacache, *Catal. Today*, 1999, **49**, 258.
315. J. M. Bennet, J. W. Richardson, J. J. Pluth and J. N. Smith, *Zeolites*, 1987, **7**, 160.
316. E. Jahn, D. Mueller and K. Becker, *Zeolites*, 1990, **10**, 151.
317. S. Zhang, S. L. Chen, P. Dong, G. Yuan and K. Xu, *Appl. Catal. A*, 2007, **332**, 46.
318. J. M. Campelo, F. Lafont and J. M. Marinas, *J. Catal.*, 1995, **156**, 11.
319. J. M. Campelo, F. Lafont and J. M. Marinas, *Zeolites*, 1995, **15**, 97.
320. J. M. Campelo, F. Lafont and J. M. Marinas, *J. Chem. Soc., Faraday Trans.*, 1995, **91**, 1551.
321. J. M. Campelo, F. Lafont and J. M. Marinas, *J. Chem. Soc., Faraday Trans.*, 1995, **91**, 4171.
322. J. M. Campelo, F. Lafont and J. M. Marinas, *Appl. Catal. A*, 1998, **170**, 139.
323. F. Zhang, C.-H. Geng, Z.-X. Gao and J.-L. Zhou, *Prepr. Am. Chem. Soc. Div. Pet. Chem.*, 2004, **49**(1), 88.
324. S. Zhang, S. L. Chen, P. Dong, Z. Ji, J. Zhao and K. Xu, *Catal. Lett.*, 2007, **118**, 109.
325. H. L. Zubkowa, G. Lischke, B. Parlitz, E. Schreier, R. Eckelt, G. Schulz and R. Fricke, *Appl. Catal. A*, 1994, **110**, 27.
326. A. K. Sinha, S. Sivasanker and P. Ratnasami, *Ind. Eng. Chem. Res.*, 1998, **37**, 2208.
327. Y. Liu, Z. Tian, Z. Xu, C. Liu and L. Lin, *Prepr. Am. Chem. Soc. Div. Pet. Chem.*, 2001, **46**(4), 402.
328. Z. Yu, A. Yan, Y. Liu and Q. Xu, *J. Inorg. Chem.*, 1989, **5**, 6.
329. W. Huang, D. Li, X. Kang, Y. Shi and H. Nie, *Stud. Surf. Sci. Catal.*, 2007, **154**, 2353.
330. R. J. Taylor and R. H. Petty, *Appl. Catal. A*, 1994, **119**, 121.
331. S. J. Miller, *Prepr. Am. Chem. Soc. Div. Pet. Chem.*, 1993, **38**(4), 788.
332. J. Hancsok, M. Krar, S. Magyar, L. Boda, A. Hollo and D. Kallo, *Stud. Surf. Sci. Catal.*, 2007, **170**, 1605.
333. D. McNeil and P. W. Reynolds, US Pat., 2554202, 1951.
334. B. P. Nilsen, M. Stoecker and T. Riis, *Stud. Surf. Sci. Catal.*, 1985, **20**, 197.
335. J. Houžvička, S. Hansildaar, J. G. Nienhuis and V. Ponec, *Appl. Catal. A*, 1999, **176**, 83.
336. H. Hino and K. Arata, *J. Chem. Soc., Chem. Commun.*, 1980, 851.
337. R. L. Marcus and R. D. Gonzales, *Prepr. Pap. Am. Chem. Soc. Div. Pet. Chem.*, 2001, **46**(3), 238.

338. A. Corma, A. Martinez and C. Martinez, *Appl. Catal. A*, 1996, **144**, 249.
339. D. E. Sparks, R. A. Keogh and B. H. Davis, *Appl. Catal. A*, 1996, **144**, 205.
340. A. Corma, V. Fornes, M. I. Juan-Rajadell and J. M. Lopez-Nieto, *Appl. Catal. A*, 1994, **116**, 151.
341. B. Li and R. D. Gonzales, *Catal. Lett.*, 1998, **54**, 5.
342. B. Li and R. D. Gonzales, *Ind. Eng. Chem. Res.*, 1996, **35**, 314.
343. M. T. Tran, N. S. Gnep, G. Szabo and M. Guisnet, *Appl. Catal. A*, 1998, **171**, 207.
344. R. A. Comelli, S. A. Canavese, S. R. Vaudugna and N. S. Figoli, *Appl. Catal. A*, 1996, **135**, 287.
345. V. Parvulescu, S. Coman, P. Grange and V. I. Parvulescu, *Appl. Catal. A*, 1999, **176**, 27.
346. S. Coman, V. Parvulescu, P. Grange and V. I. Parvulescu, *Appl. Catal. A*, 1999, **176**, 45.
347. J. C. Yori and J. M. Parera, *Appl. Catal. A*, 1995, **129**, L151.
348. A. Corma, M. I. Juan-Rajadell, J. M. Lopez-Nieto, A. Martinez and C. Martinez, *Appl. Catal. A*, 1994, **111**, 175.
349. A. Hess and E. Kemmitz, *Appl. Catal. A*, 1997, **149**, 373.
350. R. B. Gore and W. J. Thomson, *Appl. Catal. A*, 1998, **168**, 23.
351. C. Guo, S. Yao, J. Cao and Z. Qian, *Appl. Catal. A*, 1994, **107**, 229.
352. C. Guo, S. Liao, Z. Qian and K. Tanabe, *Appl. Catal. A*, 1994, **107**, 239.
353. B. Li and R. D. Gonzales, *Appl. Catal. A*, 1997, **165**, 291.
354. B. Li and R. D. Gonzales, *Appl. Catal. A*, 1998, **174**, 109.
355. F. Garin, D. Andriamasinoro, D. Abdulsamad and J. Sommer, *J. Catal.*, 1996, **131**, 199.
356. K. Ebitani, J. Tsuji, H. Hattori and H. Kita, *J. Catal.*, 1992, **135**, 609.
357. J. Zhao, B. P. Huffman and B. H. Davis, *Catal. Lett.*, 1994, **24**, 385.
358. A. Sayari and A. Dicko, *J. Catal.*, 1994, **145**, 561.
359. R. A. Keogh, R. Srinivansan and B. H. Davis, *Appl. Catal. A*, 1996, **140**, 47.
360. R. A. Keogh, D. Sparks, J. Hu, I. Wender, J. W. Tierney, W. Wang and B. H. Davis, *Energy Fuels*, 1994, **8**, 755.
361. K. R. Venkatesch, J. Hu, W. Wang, G. D. Holder, J. W. Tierney and I. Wender, *Energy Fuels*, 1996, **10**, 1163.
362. J. M. Grau, C. R. Vera and J. M. Parera, *Appl. Catal. A*, 1998, **172**, 311.
363. J. C. Yori and J. M. Parera, *Appl. Catal. A*, 1995, **129**, 83.
364. G. Larsen, E. Lotero, R. D. Parra, L. M. Petkovic, H. S. Silva and S. Raghavan, *Appl. Catal. A*, 1995, **130**, 213.
365. W. E. Alvarez, H. Liu and D. E. Resasco, *Appl. Catal. A*, 1997, **162**, 103.
366. H. Liu, G. D. Lei and W. M. H. Sachtler, *Appl. Catal. A*, 1996, **146**, 165.
367. H. Liu, G. D. Lei and W. M. H. Sachtler, *Appl. Catal. A*, 1996, **137**, 167.
368. J. M. Grau and J. M. Parera, *Appl. Catal. A*, 1997, **162**, 17.
369. J. M. Grau, C. R. Vera and J. M. Parera, *Appl. Catal. A*, 1998, **198**, 311.
370. J. M. Grau, J. C. Yori and J. M. Parera, *Appl. Catal. A*, 2001, **213**, 213.
371. J. M. Grau, C. R. Vera and J. M. Parera, *Appl. Catal. A*, 2002, **227**, 217.

372. J. M. Grau, J. C. Yori, C. R. Vera, F. C. Lovey, A. M. Condóó and J. M. Parera, *Appl. Catal. A*, 2004, **265**, 141.
373. C. Y. Hsu, C. R. Heimbuch, C. T. Armes and B. C. Gates, *J. Chem. Soc., Chem. Commun.*, 1992, 1645.
374. K. Saito, R. A. Keogh and B. H. Davis, *Prepr. Am. Chem. Soc. Div. Pet. Chem.*, 1999, **44**(2), 150.
375. G. Larsen, E. Lotero, S. Raghavan, R. D. Parra and C. A. Querini, *Appl. Catal. A*, 1996, **139**, 201.
376. E. Iglesia, D. G. Barton, S. L. Soled, S. Miseo, J. E. Baumgartner, W. E. Gates, G. A. Fuentes and G. D. Meitzner, *Stud. Surf. Sci. Catal.*, 1996, **101**, 533.
377. J. C. Yori, M. A. d'Amato, G. Costa and J. M. Parera, *J. Catal.*, 1995, **153**, 218.
378. G. Larsen and L. M. Petkovic, *Appl. Catal. A*, 1996, **148**, 155.
379. J. C. Yori, C. L. Pieck and J. M. Parera, *Appl. Catal. A*, 1999, **181**, 5.
380. J. C. Yori, C. R. Vera and J. M. Parera, *Appl. Catal. A*, 1997, **163**, 165.
381. M. Busto, V. M. Benítez, C. R. Vera, J. M. Grau and J. C. Yori, *Appl. Catal. A*, 2008, **347**, 117.
382. Y. Liu, G. Koyano, K. Na and M. Misono, *Appl. Catal. A*, 1998, **166**, L263.
383. M. Hino and K. Arata, *Appl. Catal. A*, 1998, **173**, 121.
384. B. G. Baker and N. J. Clark, *Stud. Surf. Sci. Catal.*, 1987, **30**, 483.
385. E. Lopez-Salinas, J. G. Hernandez-Cortez, M. A. Cortes-Jacome, J. Navarrete, M. A. Llanos, A. Vazquez, H. Armendariz and T. Lopez, *Appl. Catal. A*, 1998, **175**, 43.
386. V. M. Benitez, C. A. Querini, N. S. Figoli and R. A. Comelli, *Appl. Catal. A*, 1999, **178**, 205.
387. P. Patrono, A. la Ginestra, G. Ramis and G. Busca, *Appl. Catal. A*, 1994, **107**, 249.
388. V. M. Akhmedov, A. H. Al-Khowaiter, E. Akhmedov and A. Sadikhov, *Appl. Catal. A*, 1999, **181**, 51.
389. A. P. E. York, C. Pham-Huu, P. del Gallo, E. A. Blekkam and M. J. Ledoux, *Ind. Eng. Chem. Res.*, 1996, **35**, 672.
390. A. Guerrero-Ruiz, B. Bachiller-Baeza and I. Rodriguez-Ramos, *Appl. Catal. A*, 1998, **173**, 231.
391. I. Rodriguez-Ramos and A. Guerrero-Ruiz, *Appl. Catal. A*, 1994, **119**, 271.
392. A. Campos, B. C. Gagea, S. Moreno, P. Jacobs and R. Molina, *Stud. Surf. Sci. Catal.*, 2007, **170**, 1405.
393. A. Corma, O. Marie and F. J. Ortega, *J. Catal.*, 2004, **222**, 338.
394. E. du Toit and W. Nicol, *Appl. Catal. A*, 2007, **277**, 219.
395. A. D. Pienaar and A. de Klerk, *Ind. Eng. Chem. Res.*, 2008, **47**, 4962.
396. D. O. Leckel, *Energy Fuels*, 2007, **21**, 662.
397. F. H. A. Bolder and H. Mulder, *Appl. Catal. A*, 2006, **300**, 36.
398. G. Seo, H. S. Jeong, D.-L. Jang, D. L. Cho and S. B. Hong, *Catal. Lett.*, 1996, **41**, 189.

399. S. van Donk, J. H. Bitter and K. P. de Jong, *Appl. Catal. A*, 2001, **212**, 97.
400. D. Smook and A. de Klerk, in *Proceedings of South African Chemical Engineering Congress*, Sun City, 2003, cdP09.
401. C. Lenoi, F. Rohr, A. Allahverdiev, M. Stocker and P. Ruiz, *Stud. Surf. Sci. Catal.*, 2004, **154**, 2371.
402. C. L. Li, O. Novaro, E. Munoz, J. L. Boldu, X. Bokhimi, J. A. Wang, T. Lopez and R. Gomez, *Appl. Catal. A*, 2000, **199**, 211.
403. B. Li, R. Marcus and R. D. Gonzales, *Prepr. Am. Chem. Soc. Div. Pet. Chem.*, 1999, **44**(4), 430.
404. B. Li and R. D. Gonzales, *Appl. Catal. A*, 1998, **174**, 109.
405. M. T. Tran, N. S. Gnep. G. Szabo and M. Guisnet, *Appl. Catal. A*, 1998, **171**, 207.
406. J. R. Sohn and H. W. Kim, *J. Mol. Catal.*, 1989, **52**, 361.
407. J. C. Yori, J. C. Luy and J. M. Parera, *Appl. Catal.*, 1989, **46**, 103.
408. F. R. Che, G. Condurier, J. F. Joly and J. C. Nendrine, *J. Catal.*, 1989, **143**, 616.
409. K. B. Fogash, Z. Hong, J. M. Kobe and J. A. Dumesic, *Appl. Catal. A*, 1998, **172**, 107.
410. F. T. T. Ng and H. Norbert, *Appl. Catal. A*, 1995, **123**, L197.
411. C. Li and P. C. Stair, *Catal. Lett.*, 1996, **36**, 119.
412. B. Li and R. D. Gonzales, *Appl. Catal. A*, 1997, **165**, 291.
413. G. N. Choi, S. J. Kramer, S. S. Tam, J. M. Fox and W. J. Reagan, *Prepr. Pap. Am. Chem. Soc. Div. Fuel Chem.*, 1996, **41**(3), 1079.
414. A. de Klerk, *Ind. Eng. Chem. Res.*, 2007, **46**, 5516.
415. R. Bonifay and C. Marcilly, in *Petroleum Refining. Volume 3. Conversion Processes*, ed. P. Leprince, Technip, Paris, 2001, p. 169.
416. K. A. Cumming and B. W. Wojciechowski, *Catal. Rev. Sci. Eng.*, 1996, **38**, 101.
417. A. Corma, J. Planelles, J. Sánchez-Marín and F. Tomás, *J. Catal.*, 1985, **93**, 30.
418. W. J. J. Welters, O. H. van der Waerden, H. W. Zandbergen, V. H. J. de Beer and R. A. van Santen, *Ind. Eng. Chem. Res.*, 1995, **34**, 1156.
419. P. A. Jacobs, J. A. Martens, J. Weitkamp and H. K. Beyer, *Faraday Discuss. Chem. Soc.*, 1981, **72**, 353.
420. J. Weitkamp, P. A. Jacobs and J. A. Martens, *Appl. Catal.*, 1983, **8**, 123.
421. H. Vasina, M. A. Baltanas and G. F. Froment, *Ind. Eng. Chem. Prod. Res. Des. Dev.*, 1983, **22**, 526.
422. M. T. Klein and G. Hou, *Prepr. Pap. Am. Chem. Soc., Div. Fuel Chem.*, 2001, **46**(2), 372.
423. E. Furimsky, *Catalysts for Upgrading Heavy Petroleum Feeds*, Elsevier, Amsterdam, 2007.
424. M. S. Rana, V. Sámano, J. Ancheyta and J. A. I. Diaz, *Fuel*, 2007, **86**, 1216.
425. E. Furimsky, *Ind. Eng. Chem. Res.*, 2009, **48**, 2752.
426. A. de Klerk, *Catal. Today*, 2008, **130**, 439.
427. Anon, *Pet. Economist*, 2008, **75**(6), 36.

428. A. Billon and P.-H. Bigeard, in *Petroleum Refining. Volume 3. Conversion Processes*, ed. P. Leprince, Technip, Paris, 2001, p. 333.
429. C. Bouchy, G. Hastoy, E. Guillon and J. A. Martens, *Oil Gas Sci. Technol. Rev. IFP*, 2009, **64**, 91.
430. G. Valavarasu, M. Bhaskar and K. S. Balaraman, *Pet. Sci. Technol.*, 2003, **21**, 1185.
431. G. E. Langlois, R. F. Sullivan and C. J. Egan, *J. Phys. Chem.*, 1966, **70**, 3666.
432. W. Böhringer, A. Kotsiopoulos, M. de Boer, C. Knottenbelt and J. C. Q. Fletcher, in *Proceedings of South African Chemical Engineering Congress*, Durban, 2006, Oral PDO27.
433. R. de Haan, G. Joorst, E. Mokoena and C. P. Nicolaides, *Appl. Catal. A*, 2007, **327**, 247.
434. W. Böhringer, A. Kotsiopoulos, M. de Boer, C. Knottenbelt and J. C. Q. Fletcher, *Stud. Surf. Sci. Catal.*, 2007, **163**, 345.
435. V. Calemma, S. Peratello, S. Pavoni, G. Clerici and C. Perego, *Stud. Surf. Sci. Catal.*, 2001, **136**, 307.
436. D. O. Leckel and M. Liwanga-Ehumbu, *Energy Fuels*, 2006, **20**, 2330.
437. D. O. Leckel, *Ind. Eng. Chem. Res.*, 2007, **46**, 3505.
438. S. Zhou, Y. Zhang, J. W. Tierney and I. Wender, *Fuel Proc. Technol.*, 2001, **69**, 59.
439. N. Fabricius, in *Fundamentals of Gas to Liquids*, 2nd edn, Petroleum Economist, London, 2005, p. 12.
440. K. Fraser, in *Fundamentals of Gas to Liquids*, Petroleum Economist, London, 2005, p. 15.
441. D. O. Leckel, *Energy Fuels*, 2009, **23**, 38.
442. P. B. Venuto and E. T. Habib Jr., *Fluid Catalytic Cracking with Zeolite Catalysts*, Marcel Dekker, New York, 1979.
443. R. H. Harding, A. W. Peters and J. R. D. Nee, *Appl. Catal. A*, 2001, **221**, 389.
444. K. Eng, S. Heidenreich, S. Swart and F. Möller, in *Proceedings of the 18th World Petroleum Congress*, Johannesburg, 2005, cd122.
445. I. Kobolakis and B. W. Wojciechowski, *Can. J. Chem. Eng.*, 1985, **63**, 269.
446. J. Abbot and B. W. Wojciechowski, *Ind. Eng. Chem. Prod. Res. Dev.*, 1985, **24**, 501.
447. T. M. Leib, J. C. W. Kuo, W. E. Garwood, D. M. Nace, W. R. Derr and S. A. Tabak, in *Proceedings of AIChE Annual Meeting*, Washington DC, 1988, paper 61d.
448. W. J. Reagan, *Prepr. Pap. Am. Chem. Soc. Div. Fuel Chem.*, 1994, **39**(2), 337.
449. M. M. Schwartz, DOE Contract No. DE-AC22-91PC90057 – Final Report, Department of Energy, Washington, DC, 1995.
450. X. Dupain, R. A. Krul, M. Makkee and J. A. Moulijn, *Catal. Today*, 2005, **106**, 288.
451. X. Dupain, R. A. Krul, C. J. Schaverien, M. Makkee and J. A. Moulijn, *Appl. Catal. B*, 2006, **63**, 277.

452. W. Letzsch, in *Handbook of Petroleum Processing*, ed. D. S. J. Jones and P. R. Pujadó, Springer, Dordrecht, 2006, p. 239.
453. J. A. Martens, J. Weitkamp and P. A. Jacobs, *Stud. Surf. Sci. Catal.*, 1985, **20**, 427.
454. H. E. Schultz and J. Weitkamp, *Ind. Eng. Chem. Prod. Res. Dev.*, 1972, **11**, 46.
455. P. A. Jacobs and J. A. Martens, *Stud. Surf. Sci. Catal.*, 1991, **58**, 445.
456. W. J. J. Welters, O. H. van der Waerden, V. H. J. de Beer and R. A. van Santen, *Ind. Eng. Chem. Res.*, 1995, **34**, 1166.
457. R. H. Heck and N. Y. Chen, *Appl. Catal. A*, 1992, **86**, 83.
458. A. Lugstein, A. Jentys and H. Vinek, *Appl. Catal. A*, 1997, **152**, 93.
459. A. Corma, M. J. Diaz-Cabanas, C. Lopez and A. Martinez, *Stud. Surf. Sci. Catal.*, 2004, **154C**, 2380.
460. S. Tontisirin and S. Ernst, *Stud. Surf. Sci. Catal.*, 2007, **170**, 1351.
461. B. Egia, J. F. Cambra, P. L. Arias, M. B. Guemez, J. A. Legarreta, B. Pawelec and J. L. G. Fierro, *Appl. Catal. A*, 1998, **169**, 37.
462. C. Thomazeau, C. Canaff, J. L. Lemberton and M. Guisnet, *Appl. Catal. A*, 1993, **103**, 163.
463. Z. Zhou, Y. Zhang, J. W. Tierney and I. Wender, *Fuel Process. Technol.*, 2003, **83**, 67.
464. M. Kobayashi, M. Saitoh, S. Togawa and K. Ishida, *Energy Fuels*, 2009, **23**, 513.
465. A de Klerk, in *Proceedings of the 9th European Congress on Catalysis*, Salamanca, 2009, KN8.
466. D. Dubbeldam, S. Calero, T. L. M. Maesen and B. Smit, *Angew. Chem. Int. Ed.*, 2003, **42**, 3624.
467. J. Wei, *Ind. Eng. Chem. Res.*, 1994, **33**, 2467.
468. T. L. M. Maesen, E. Beerdsen, S. Calero, D. Dubbeldam and B. Smit, *J. Catal.*, 2006, **237**, 278.
469. H. Abrevaya, *Stud. Surf. Sci. Catal.*, 2007, **170**, 1244.
470. A. Corma, P. J. Miguel and A. V. Orchillés, *Appl. Catal. A*, 1996, **138**, 57.
471. A. Corma, P. J. Miguel and A. V. Orchillés, *J. Catal.*, 1994, **145**, 171.
472. A. Corma, P. J. Miguel and A. V. Orchillés, *J. Catal.*, 1994, **145**, 58.
473. A. Corma, P. J. Miguel and A. V. Orchillés, *Appl. Catal. A*, 1994, **117**, 29.
474. A. Corma and B. W. Wojciechowski, *Catal. Rev. Sci. Eng.*, 1982, **24**, 1.
475. A. Corma, V. Gonzales-Alfaro and A. V. Orchillés, *Appl. Catal. A*, 1995, **129**, 203.
476. J. Abbott and B. W. Wojciechowski, *J. Catal.*, 1987, **104**, 80.
477. D. A. Best and B. W. Wojciechowski, *J. Catal.*, 1977, **47**, 11.
478. J. T. Miller, P. D. Hopkins, B. L. Meyers, G. J. Ray, R. T. Roginski, G. W. Zajac and N. H. Rorenbaum, *J. Catal.*, 1992, **138**, 115.
479. F. Lemos, F. Ramos, M. Kern, G. Gianetto and M. Guisnet, *Appl. Catal.*, 1987, **29**, 43.
480. N. L. Spiridonov, S. E. Spiridonov and S. N. Khadshiev, *Kinet. Katal.*, 1989, **30**, 243.

481. G. Ohlmann, H. G. Jerschkewitz, G. Lischke, R. Eckelt, T. Gross, B. Parlitz, I. Schulz, K. Wehner and D. Trimm, *Stud. Surf. Sci. Catal.*, 1991, **65**, 415.
482. M. Richter, *Angew. Chem. Int. Ed.*, 1991, **30**, 607.
483. G. L. Zhao, J. W. Teng, Z. K. Xie, W. M. Yang, Q. L. Chen and Y. Tang, *Stud. Surf. Sci. Catal.*, 2007, **170**, 1307.
484. D. Bhattacharya, S. S. Tambe and S. Sivasanker, *Appl. Catal. A*, 1997, **154**, 139.
485. S. M. Babitz, B. A. Williams, J. T. Miller, R. Q. Snurr, W. O. Haag and H. H. Kung, *Appl. Catal. A*, 1999, **179**, 71.
486. S. Jolly, J. Saussey and J. C. Lavalley, *J. Mol. Catal.*, 1994, **86**, 401.
487. S. Jolly, J. Saussey, M. M. Bettahar, J. C. Lavalley and E. Benazzi, *Appl. Catal. A*, 1997, **156**, 71.
488. J. E. Antia, K. Israni and R. Govind, *Appl. Catal. A*, 1997, **159**, 89.
489. J. E. Antia and R. Govind, *Appl. Catal. A*, 1995, **131**, 107.
490. B. A. Watson, M. T. Klein and R. H. Harding, *Ind. Eng. Chem. Res.*, 1996, **35**, 1506.
491. A. Corma and A. V. Orchillés, *Microporous Mesoporous Mater.*, 2000, **35–36**, 21.
492. I. P. Dzikh, J. M. Lopes, F. Lemos and F. Ramoa Ribeiro, *Appl. Catal. A*, 1999, **176**, 239.
493. G. Lischke, E. Schreier, B. Parlitz, I. Pitsch, U. Lohse and M. Woettke, *Appl. Catal. A*, 1995, **129**, 57.
494. F. Dougnier, J. Patarin, J. L. Guth and D. Anglerot, *Zeolites*, 1992, **12**, 160.
495. P. G. Smirniotis and E. Ruckenstein, *Ind. Eng. Chem. Res.*, 1994, **33**, 800.
496. J. Abbot and F. N. Guerzoni, *Appl. Catal. A*, 1992, **85**, 173.
497. M. A. Chaar and J. B. Butt, *Appl. Catal. A*, 1994, **114**, 287.
498. A. de Klerk, *Energy Fuels*, 2009, **23**, 4593.
499. C. W. Ingram and K. Ghebreyessus, *Prepr. Am. Chem. Soc. Div. Pet. Chem.*, 2004, **49**(1), 92.
500. A. de Klerk, *Prepr. Pap. Am. Chem. Soc. Div. Fuel Chem.*, 2009, **54**(1), 116.
501. V. Calemma, S. Peratello and C. Perego, *Appl. Catal. A*, 2000, **190**, 207.
502. V. Calemma, S. Peratello, F. Stroppa, R. Giardino and C. Perego, *Ind. Eng. Chem. Res.*, 2004, **43**, 934.
503. V. Calemma, S. Correra, C. Perego, P. Pollesel and L. Pellegrini, *Catal. Today*, 2005, **106**, 282.
504. V. Calemma, C. Gambaro, W. O. Parker Jr., R. Carbone, R. Giardino and P. Scorletti, *Catal. Today*, 2010, **149**, 40.
505. D. O. Leckel, *Energy Fuels*, 2007, **21**, 1425.
506. D. O. Leckel, *Energy Fuels*, 2005, **19**, 1795.
507. D. O. Leckel, *Energy Fuels*, 2009, **23**, 2370.
508. M. Grau, V. M. Benitez, J. C. Yori, C. R. Vera, J. F. Padilha, L. A. Magalhaes Pontes and A. O. S. Silva, *Energy Fuels*, 2007, **21**, 1390.

509. M. Grau, C. R. Vera, V. M. Benitez and J. C. Yori, *Energy Fuels*, 2008, **22**, 1680.
510. L. B. Ryland, M. W. Tamele and J. N. Wilson, in *Catalysis. Volume VII. Oxidation, Hydration, Dehydration and Cracking Catalysts*, ed. P. H. Emmett, Reinhold, New York, 1960, p. 1.
511. A. Elo Jr. and P. Clements, *J. Phys. Chem.*, 1967, **71**, 1078.
512. A. De Lucas, P. Canizares, A. Durán and A. Carrero, *Appl. Catal. A*, 1997, **154**, 221.
513. B. A. Williams, J. T. Miller, R. Q. Snurr and H. H. Kung, *Microporous Mesoporous Mater.*, 2000, **35–36**, 61.
514. C. S. Triantafillidis, A. G. Vlessidis, L. Nalbandian and N. P. Evmiridis, *Microporous Mesoporous Mater.*, 2001, **47**, 369.
515. T. Blasco, A. Corma and J. Martínez-Triguero, *J. Catal.*, 2006, **237**, 267.
516. D. O. Leckel, in *Proceedings of the 7th European Congress on Catalysis*, Sofia, 2005, O5-05.
517. P. Magnoux, P. Cartraud, S. Mignard and M. Guisnet, *J. Catal.*, 1987, **106**, 242.
518. H. R. Ireland, C. Redini, A. S. Raff and L. Fava, *Hydrocarbon Process*, 1979, **58**(5), 119.
519. S. M. Csicseri, *Pure Appl. Chem.*, 1986, **58**, 841.
520. W. Holderich and R. Gallei, *Chem. Eng.*, 1985, **8**, 337.
521. K. Moljord, P. Magnoux and M. Guisnet, *Appl. Catal. A*, 1995, **122**, 21.
522. J. M. Lopes, F. Lemos, F. Ramoa Ribeiro, E. G. Derouane, P. Magnoux and M. Guisnet, *Appl. Catal. A*, 1994, **114**, 161.
523. P. D. Hopkins, J. T. Miller, B. L. Meyers, G. J. Ray, R. T. Roginski, M. A. Kuehne and H. H. Kung, *Appl. Catal. A*, 1996, **136**, 29.
524. R. R. Chianelli, *Catal. Rev. Sci. Eng.*, 1984, **26**, 361.
525. H. Topsoe, B. C. Clausen and F. E. Massoth, in *Catalysis, Science and Technology*, ed. J. Anderson and M. Boudart, Vol. **11**, Springer, Berlin, 1999, p. 1.
526. T. C. Ho, *Catal. Today*, 2004, **98**, 3.
527. T. C. Ho, *Catal. Rev. Sci. Eng.*, 1988, **30**, 117.
528. E. Furimsky and F. E. Massoth, *Catal. Rev. Sci. Eng.*, 2005, **47**, 297.
529. E. Furimsky, *Catal. Rev. Sci. Eng.*, 1983, **25**, 421.
530. E. Furimsky, *Appl. Catal. A*, 2000, **199**, 147.
531. D. C. Elliott, *Energy Fuels*, 2007, **21**, 1792.
532. A. de Klerk and M. J. Strauss, *Prepr. Pap. Am. Chem. Soc. Div. Fuel Chem.*, 2008, **53**(1), 313.
533. K. McGurk, in *Proceedings of South African Chemical Engineering Congress*, Sun City, 2003, P082.
534. R. J. J. Nel and A. de Klerk, *Ind. Eng. Chem. Res.*, 2007, **46**, 3558.
535. R. J. J. Nel and A. de Klerk, *Prepr. Pap. Am. Chem. Soc. Div. Fuel Chem.*, 2009, **54**(1), 118.
536. F. H. A. Bolder, A. de Klerk and J. L. Visagie, *Ind. Eng. Chem. Res.*, 2009, **48**, 3755.

537. F. H. A. Bolder, *Energy Fuels*, 2007, **21**, 1396.
538. D. Lamprecht, *Energy Fuels*, 2007, **21**, 2509.
539. A. Stanislaus and B. H. Cooper, *Catal. Rev. Sci. Eng.*, 1994, **36**, 75.
540. M. Bremaud, L. Vivier, G. Perot, V. Harle and C. Bouchy, *Appl. Catal. A*, 2005, **289**, 44.
541. M. S. Lylykangas, P. A. Rautanen and A. O. I. Krause, *Ind. Eng. Chem. Res.*, 2004, **43**, 1641.
542. M. S. Lylykangas, P. A. Rautanen and A. O. I. Krause, *Appl. Catal. A*, 2004, **259**, 73.
543. F. M. Dautzenberg and J. C. de Deken, *ACS Symp. Ser.*, 1987, **344**, 233.
544. H. Toulhoat, R. Szymanski and J. C. Plumail, *Catal. Today*, 1990, **7**, 531.
545. F. X. Long, B. S. Gevert and P. Abrahamsson, *J. Catal.*, 2004, **222**, 6.
546. D. Smook and T. Mashapa, in *Proceedings of South African Chemical Engineering Congress*, Durban, 2006, Poster PDO16.
547. A. de Klerk, presented at the 2nd Sub-Saharan Africa Catal. Symposium, Swakopmund, Namibia, 2001, paper 10.
548. Anon, *Oil Gas J.*, 2003, 6 October, 1.
549. O. Weisser, *Int. Chem. Eng.*, 1963, **3**, 388.
550. J. C. Hoogendoorn and J. M. Salomon, *Br. Chem. Eng.*, 1957, 238–244, 308–312, 368–373, 418–419.
551. E. Furimsky, *Appl. Catal. A*, 1998, **171**, 177.
552. E. Furimsky and F. E. Massoth, *Catal. Today*, 1999, **52**, 381.
553. T. R. Viljava, R. S. Komulainen and A. O. I. Krause, *Catal. Today*, 2000, **60**, 83.
554. A. M. van Wyk, M. A. Moola, C. J. de Bruyn, E. Venter and L. Collier, *Chem. Technol. (S. Afr.)*, 2007, April, 4.
555. T. Lourens, in *Proceedings of South African Chemical Engineering Congress*, Sun City, 2003, cd009.
556. M. Brémaud, L. Vivier, G. Pérot, V. Harlé and C. Bouchy, *Appl. Catal. A*, 2005, **289**, 44.
557. J. Ansorge, *Prepr. Pap. Am. Chem. Soc. Div. Fuel Chem.*, 1997, **42**(2), 654.
558. D. O. Leckel, *Energy Fuels*, 2006, **20**, 1761.
559. D. O. Leckel, *Energy Fuels*, 2008, **22**, 231.
560. J. A. Cusumano, R. A. Dalla Betta and R. B. Levy, *Catalysis in Coal Conversion*, Academic Press, New York, 1978.

CHAPTER 6

Upgrading of Fischer–Tropsch Waxes

The distributions of hydrocarbon fractions from low-temperature Fischer–Tropsch (LTFT) synthesis and high-temperature Fischer–Tropsch (HTFT) synthesis are shown in Table 1.1. The significantly higher yield of >360 °C boiling (C_{22} and heavier) products from LTFT synthesis is evident. In LTFT syncrude, this fraction is called the wax product and it contains mainly alkanes (95%), smaller amounts of alkenes and oxygenates and neither sulfur nor much aromatics.[1] The equivalent >360 °C boiling fraction found in HTFT syncrude is termed a residue. The HTFT residue is in fact very aromatic (>25%) and cannot be classified as a paraffin wax, although it is sometimes referred to as a waxy oil.[2] Wax upgrading therefore deals only with primary hydrocarbons from LTFT synthesis.

The average ratio of condensates to wax from the iron-based slurry bed LTFT synthesis is 38:62. The *n*-alkanes in the condensates and the *n*-alkane-rich waxes can be used as feed for the production of fuels, lubricants and chemicals.[3] In each instance, the objective and upgrading methodology are determined by the specifications of the commercial products being produced.

Figure 6.1 shows the carbon number distribution of the condensate and wax from iron-based slurry LTFT synthesis.[4] The wax fraction includes alkanes with carbon numbers exceeding C_{100}, peaking around C_{30}. It is evident from Figure 6.1 that for LTFT condensates, the carbon number distribution peaked at about C_{20} and that there is considerable overlap of the C_{15}–C_{35} fraction between condensate and wax. This is a consequence of the separation strategy after FTS (Section 4.2.1) and better separation can be achieved by appropriate design.

Generally, iron-based tubular fixed bed reactor products contain less alkenes than iron-based slurry bubble column reactor products. The products from fixed bed conversion are also more linear and contain less oxygenates, specifically alcohols and carbonyls. This can be understood from reactor engineering

RSC Catalysis Series No. 4
Catalysis in the Refining of Fischer–Tropsch Syncrude
By Arno de Klerk and Edward Furimsky

Published by the Royal Society of Chemistry, www.rsc.org

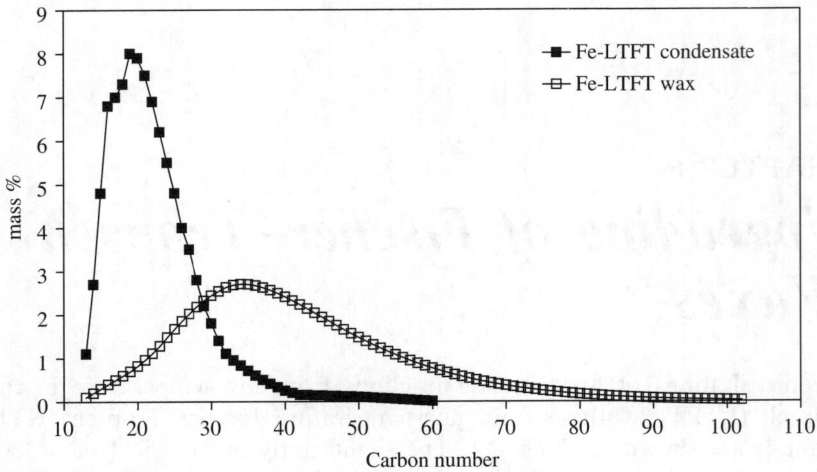

Figure 6.1 Carbon number distribution of iron-based low-temperature Fischer–Tropsch (Fe-LTFT) condensate and wax fractions that are produced commercially during FTS in a slurry bubble column reactor.

Table 6.1 Metal contaminants from industrial fixed bed Fe-LTFT synthesis.

Metal	*Concentration* ($\mu g\,g^{-1}$)
Na	0.5
K	0.2
Fe	4.3
Cu	<0.1

principles, with a fixed bed reactor approximating plug flow behaviour better than a slurry bed reactor. It is therefore to be expected that the products from fixed bed conversion would be more hydrogenated (Section 3.2.2). Analogous differences can be found in Co-LTFT synthesis. Cobalt is more hydrogenating than iron, and for the same reactor technology, the cobalt-based FTS products contain less alkenes and oxygenates. Nevertheless, the products from cobalt-based LTFT synthesis are very similar to those from iron-based LTFT synthesis.

Traces of metals are usually present in the iron-based LTFT wax (Table 6.1).[5] The metal content of LTFT wax from FTS in a slurry bubble column reactor is generally higher due to the added contribution of catalyst attrition. Catalyst attrition under slurry bed operating conditions is unavoidable and has been studied by various researchers.[6–8] A similar situation exists for cobalt-based LTFT wax and the commissioning problems and subsequent operational problems of the Oryx GTL facility have been mainly attributed to the formation of fine sediment as a result of Fischer–Tropsch catalyst attrition.[9] Various methods for removing metals from LTFT wax that cause problems with blockage of downstream refinery units have been suggested in both the journal[10,11] and patent literature (Section 10.1.2).

6.1 Commercial Upgrading of Fischer–Tropsch Waxes

Historically, the first commercial production of LTFT waxes took place in Germany during the 1930–40s. The waxes were produced by cobalt-based LTFT synthesis and were separated from the lighter products by fractionation. The waxy products were further separated by steam stripping to produce different wax grades. The soft wax and atmospheric bottoms (Gatsch) were either oxidised or thermally cracked. Air oxidation of the waxes produced oxygenated waxes and carboxylic acids and the yield depended on the severity of oxidation. The carboxylic acids were mainly used to manufacture soap. The products from thermal cracking were employed to produce lubricating oil by $AlCl_3$-catalysed OLI. The medium, hard and oxidised waxes all ended up as chemical products. Good overviews of early upgrading efforts have been compiled by Asinger[12] and Freerks.[13]

A similar approach was followed in the 1950s in South Africa, where the waxes produced by iron-based LTFT synthesis were first separated into different wax grades and then further refined depending on the wax grade. In addition to the different paraffin waxes, various grades of oxidised waxes were produced.

The main wax grades that were marketed are described in detail by Le Roux and Oranje (Table 6.2).[14] In later years, with the introduction of iron-based LTFT synthesis in a slurry bed,[15] the configuration of the Sasol 1 plant was changed. Two additional hard wax grades were produced, namely C80 and C105, with congealing points around 80 and 105 °C, respectively.

The Shell Co-LTFT facility in Bintulu, Malaysia, has been designed to produce mostly transportation fuels by HCR, although part of the production is also directed towards the wax market.[16] The wax grades produced are listed in Table 6.3.[17]

A similar refinery design to the Shell Bintulu facility has been used for the Oryx GTL facility in Ras Laffan, Qatar, with the waxes being hydrocracked to

Table 6.2 Selected properties of the main Fe-LTFT wax grades originally produced at Sasol 1.

Property	*Sasolwaks L1*	*Sasolwaks M*	*Sasolwaks H1*
Carbon range	C_{13}–C_{36}	C_{19}–C_{38}	$>C_{33}$
Average molecular formula	$C_{23}H_{48}$	$C_{28}H_{58}$	$C_{50}H_{102}$
Linear paraffin content (%)	84	96	90
Congealing point (°C)	37	58	98
Oil content (mass%)[a]	15	1.4	0.8

[a]ASTM D721, 2-butanone (MEK) solubility at −32 °C.

Table 6.3 Selected properties of the main Co-LTFT wax grades produced by Shell at their SMDS Bintulu facility.

Property	*SX30*	*SX50*	*SX70*	*SX100*
Congealing point (°C)	31	50	70	98
Oil content (mass%)[a]	5	2.5	0.4	0.1

[a]ASTM D721, 2-butanone (MEK) solubility at −32 °C.

produce mainly cracker naphtha and distillates.[18] However, the Oryx GTL facility does not make provision for the HYD of wax and separation to produce LTFT waxes as final products.[19]

Hydrocracking has been adopted as the main upgrading technology for the conversion of waxes from Co-LTFT synthesis and will also be employed in facilities that are still under construction at the time of writing, namely Escravos GTL[20] and Pearl GTL.[21]

Compared with conventional crude oil-derived feeds of a similar boiling range, a high HCR conversion of the FT wax can be achieved under milder conditions. Also, because of a lower H_2 pressure and high alkane content of the wax feed, hydrogen consumption during the HCR of the LTFT wax is much lower. The HIS/HCR of LTFT wax using commercial bifunctional catalysts has been investigated for several decades.[22] In most cases, diesel fuel and lube base oil have been the targeted products. However, the formation of naphtha and light gases as by-products cannot be avoided.

The schemes used for upgrading the FT wax differ from those used for upgrading residues obtained from conventional crude. For the latter, several residue processing technologies have been used commercially.[23] Residue conversion is important in crude oil refining, because it provides a way to improve refinery economics and it is a prerequisite for the production of good quality transportation fuels.[24] Wax may be a by-product of the production of lubricants in conventional petroleum refineries when solvent extraction methods are employed. Hence the wax upgrading processes that will be discussed are relevant to both petroleum and Fischer–Tropsch refineries.

6.2 Non-catalytic Upgrading of Waxes

It has already been mentioned that thermal cracking and oxidation have been employed on a commercial scale for the upgrading of FT waxes. In addition to these two processes, it has been pointed out that lighter Fischer–Tropsch *n*-alkanes (paraffins) and waxes make good feed materials for sulfochlorination and nitration,[12] but little else has been reported on these subjects. The products of direct chlorination of FT waxes have been studied and such products were reportedly more stable than crude oil- and coal liquid-derived products. However, this topic has not been extensively studied in the context of Fischer–Tropsch wax upgrading.

6.2.1 Thermal Cracking of Waxes

The thermal cracking of Fischer–Tropsch waxes has been described in a number of studies.[12,25–28]

More recently, thermal cracking was re-evaluated as a possible route for upgrading LTFT waxes.[5] In this work, a hard wax, which had a congealing point around 100 °C and consisted of mostly *n*-alkanes in the C_{20}–C_{120} range, was used. The study evaluated the applicability of the kinetic descriptions postulated by Bragg,[29] Voge and Good,[30] and Kossiakoff and Rice,[31] to describe thermal cracking of hard wax for alkanes up to C_{120}, although these descriptions were developed with wax no heavier than C_{40}.

Thermal cracking of LTFT wax resulted in a shift of carbon numbers of heavier fractions to lighter fractions as cracking progressed, with a bimodal distribution developing at intermediate conversion (Figure 6.2).[5] The cracking rate increased with increasing carbon chain length, at least until C_{90} and likely to C_{120}. This suggested that the Voge and Good description of increased cracking rate with increasing carbon chain length, and also the Kossiakoff–Rice description of the product distribution, held true for *n*-alkane waxes over the C_{20}–C_{120} range.

The product distribution from thermal cracking shown in Table 6.4 indicates that less than 50% conversion of the vacuum residue (> 500 °C boiling fraction) was achieved under the cracking conditions studied.[5] As expected, the olefinicity of the distillates approached 50%. The yield of C_1–C_4 hydrocarbons was low.

6.2.2 Autoxidation of Waxes

An overview of the early efforts to oxidise Co-LTFT waxes was given by Asinger.[12] In these studies, the oxidation was fairly severe and the main aim

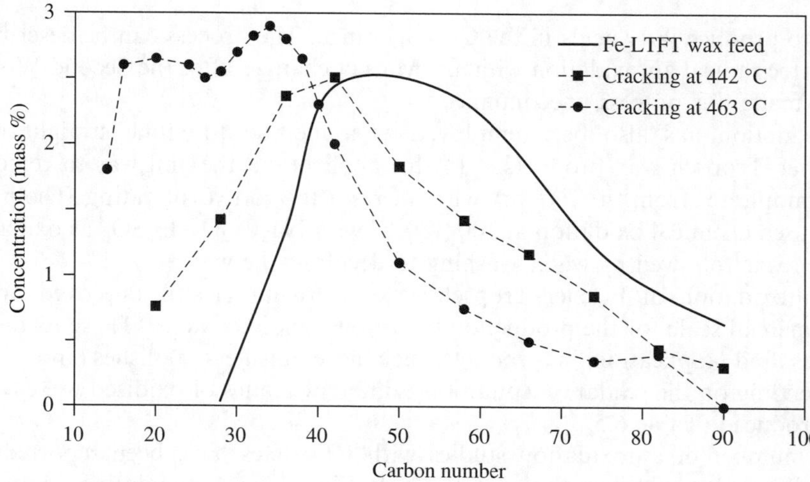

Figure 6.2 Thermal cracking of Fe-LTFT hard wax at 442 °C, 2 MPa and 1 h residence time (■) and at 463 °C, 6 MPa and 0.1 h residence time (●).

Table 6.4 Product distribution obtained during the thermal cracking of LTFT wax at different temperatures, pressure 2 MPa, residence time 1 h and using H_2 at 250 $m^3\ m^{-3}$ wax as stripping gas.

Description	*Product distribution (mass%)* 434 °C	438 °C	442 °C
Gas (C_1–C_4)	2	2	2
Naphtha and distillate (C_5–370 °C)	13	16	20
Vacuum distillate (370–500 °C)	24	24	26
Vacuum residue (>500 °C)	61	58	52
Alkenes in C_5–370 °C fraction	–	48	42
1-Alkenes in C_5–370 °C fraction	–	41	39

Table 6.5 Operating conditions and selected properties of some oxidised waxes produced commercially by the batch-mode autoxidation of Fe-LTFT waxes with air.

Property	*Oxidised wax grade* A1	A6	A28.1
Fe-LTFT feed material	H2-wax	H2-wax	C105-wax
Oxidation temperature (°C)			
First phase	175	180	175
Second phase	140	180	140
Acid number (mg KOH g^{-1})	27	37	28
Ester number (mg KOH g^{-1})	28	65	27
Penetration at 25 °C, ASTM D1321 (mm)	0.6	2.5	–
Congealing point, ASTM D 938 (°C)	87	79	94

was to produce fatty acids in the C_{12}–C_{18} range. The process can be described as a free radical autoxidation with air. Market changes after the Second World War made this process uneconomic.[32]

Oxidation has also been employed as a method to refine straight run Fischer–Tropsch wax products.[33] In this application, the aim was to remove chromophores from the FT wax without resorting to hydrotreating. The process used chemical oxidation at 100–130 °C with $Na_2Cr_2O_7$–H_2SO_4 as oxidant, which was followed by water washing to decolour the wax.

Autoxidation of Fischer–Tropsch waxes with air is still employed on a commercial scale for the production of various oxidised waxes. These oxidised waxes find application in products such as emulsifiers, polishes and inks. Depending on the oxidation conditions, different grades of oxidised waxes can be produced (Table 6.5).[34]

A number of autoxidation studies with FT waxes have been reported to describe and manipulate the oxidation selectivity.[35–37] Autoxidation temperature, oxygen availability and autoxidation time have been highlighted as the main factors determining oxidation selectivity. Primary oxidation products (alcohols, ketones and hydroperoxides) dominated the product spectrum at low temperatures (<165 °C), with secondary oxidation products (esters and carboxylic acids) becoming more prevalent at higher temperatures. Ketone selectivity was increased by high oxygen availability.

In order to overcome some shortcomings of commercial batch-mode operation, continuous-mode wax oxidation was investigated to explore ways to improve wax oxidation selectivity.[37] It was shown that continuous-mode wax oxidation was more efficient and differentiated itself from the batch-mode wax oxidation by the ability to achieve high selectivity (>85%) to alcohols and ketones. It was also possible to suppress acid formation completely.

The effect of stainless steel on wax autoxidation was investigated to determine its effect on the transportation and storage of such materials, and also to determine the possible influence that metals may have on autoxidation process.[38] It was concluded that steel materials in contact with wax and air had little effect on the wax oxidation. The enhancement of oxidation by metals reported in the literature is mainly due to the action of metal ions (mainly Fe, Mn, Co and Cu compounds) in decomposing hydroperoxide species.[39]

6.3 Catalytic Upgrading of Waxes

6.3.1 Hydrogenation of Waxes

Hydrogenated waxes have various applications. Medium wax is especially well suited for use in candles, whereas hard wax finds application in, among others, cosmetics, coatings, lubricants, adhesives and plasticisers. For food applications, the non-paraffinic compounds have to be below the limits specified by regulating bodies, for example, the United States Food and Drug Administration (FDA). Properties such as odour, colour and high-temperature stability provide information on the purity of the final product, but are not necessarily regulated. The HYD of wax may be performed with unsulfided base or noble metal catalysts, such as employed by Shell,[17] or with sulfided base metal catalysts.[40]

The wax hydrogenation study of Bolder was undertaken with the objective of obtaining a Saybolt value of +24.[40] Fischer–Tropsch wax fractions having a congealing point of 98 °C were investigated in a flow reactor over a conventional sulfided $NiMo/Al_2O_3$ catalyst at 255–330 °C, 3–6 MPa, LHSV 0.5–2.0 h^{-1} and different H_2:wax ratios in the range 100:1–600:1. The darker wax (Saybolt colour–42) required a 40 °C higher operating temperature than the lightly coloured wax (Saybolt colour –7) to obtain a similar product colour (Table 6.6).[40] In all cases, a pressure of 6 MPa (the highest pressure investigated) produced the best results and Saybolt colours of better than +24 could be obtained at 330 °C from a feedstock with a colour of −41 Saybolt units. Under these conditions, minimal HIS and hydrogenolysis of the wax was observed.

Prior to upgrading, LTFT wax may also be mildly hydrogenated as feed pretreatment to remove small amounts of alkenes and oxygenates without

Table 6.6 Effect of operating conditions on the properties of the products obtained during the hydrotreating of different LTFT waxes over a sulfided $NiMo/Al_2O_3$ catalyst.

	Product hydrogenated at 3 MPa			*Product hydrogenated at 6 MPa*		
Description	*290 °C*	*290 °C*	*330 °C*	*255 °C*	*290 °C*	*330 °C*
Saybolt colour of wax feed	−7	−18	−42	−7	−18	−42
Saybolt colour of product	+22	+20	+1	+14	+23	+27
Alkene content (g Br per 100 g)	0.3	<0.1	<0.1	<0.1	<0.1	<0.1
Aromatic content (absorption)[a]	<0.001	0.001	0.003	<0.001	<0.001	0.001
Penetration at 65 °C (mm)	18	23	24	20	20	24
C_1–C_6 in off-gas (mass%)	0.24	0.1	0.3	0.18	0.26	0.41
Product <280 °C (mass%)	0.9	1.8	2.4	0.6	1.4	1.6
Viscosity at 135 °C (mPa · s)	8	8	9	9	9	10

[a]Ultraviolet absorption at 290 nm.

substantial HIS and HCR. In the study by Leckel,[4] the HCR performances of unhydrogenated and hydrogenated LTFT waxes were compared with the aim of identifying the effect of the feed pretreatment on the product selectivity and yield. The wax feed from commercial slurry bubble column Fe-LTFT operation that was identified as Sasol C80 wax consisted of hydrocarbons between C_{20} and C_{60} with a peak at C_{38}. The conversion of the unhydrogenated and hydrogenated waxes at different operating temperatures is shown in Figure 6.3.[4] Compared with the unhydrogenated wax, HCR of the hydrogenated wax required 10–15 °C lower operating temperatures for the same conversion.

The need for higher operating temperatures in HCR of unhydrogenated wax compared with hydrogenated wax was a surprising result. One would expect that the higher alkene content of the unhydrogenated waxes would aid HCR, because mechanistically, deHYD precedes IS and cracking. It turned out that the harsher conditions required for the unhydrogenated wax could be attributed to oxygenates. Oxygenates adsorb strongly on the catalyst and inhibit HCR, with different oxygenate classes preferentially adsorbing on different active sites.[41,42] Interestingly, at 70% conversion the distillate selectivity was better during HCR of the unhydrogenated wax (74%) than during HCR of the hydrogenated wax (68%), despite the lower HCR temperature of the latter.

Hydrogenation may also be employed as a product polishing step for oxidised waxes. Oxidised waxes with a high alcohol content find application in the production of nonionic wax emulsifiers and self-emulsifiable waxes. Alcohol-rich waxes may also be dehydrated to produce long-chain linear

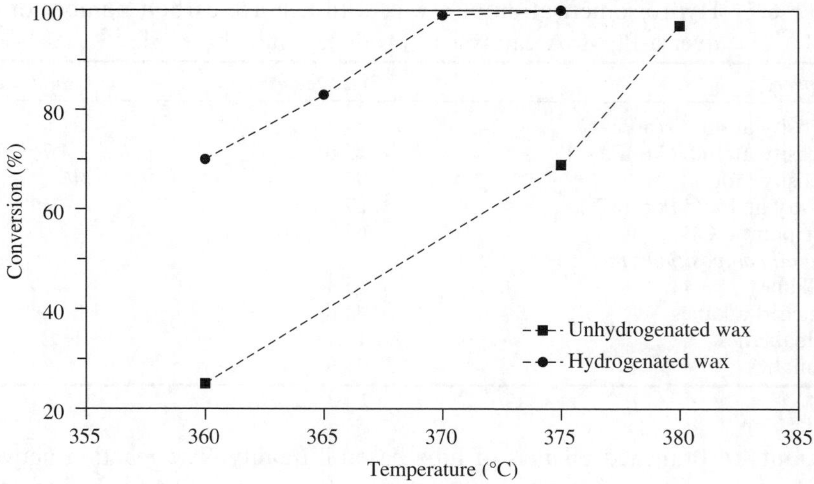

Figure 6.3 Hydrocracking of unhydrogenated (■) and hydrogenated (●) Fe-LTFT waxes over a commercially available sulfided $NiMo/SiO_2–Al_2O_3$ catalyst at 7 MPa, LHSV $0.55\,h^{-1}$ and H_2:wax ratio 1500:1.

alkenes. The HYD of the oxidised waxes over copper chromite and Ru/C catalysts has been reported.[34] An alcohol yield of 50–55% at 180–190 °C, 10 MPa hydrogen pressure and LHSV $0.4\,h^{-1}$ could be obtained during the HYD over copper chromite. Hydrotreating at temperatures above 200 °C resulted in significant over-hydrogenation of the oxidised wax to alkanes. The combination of copper chromite with Ru/C increased carboxylic acid hydrogenation, but it did not improve the overall alcohol yield.

6.3.2 Hydroisomerisation of Waxes

The upgrading of the Fischer–Tropsch waxes to lube base oils involves the partial conversion of long-chain *n*-alkanes to branched alkanes to improve cold flow properties (Figure 5.14). Hydroisomerisation affects the viscosity index of lube base oil and there is a trade-off between the decrease in viscosity index and an improvement in the cold flow properties of the products. This trade-off between viscosity index and cold flow properties is illustrated by the study of Calemma *et al.* (Table 6.7).[43] A decrease in the pour point, from + 63 to − 21 °C should be noted. This was achieved at the expense of a decrease in the viscosity index from 194 to 146. With increasing conversion, the yield of lube base oil ultimately decreased due to an increase in the yield of cracking products. Under optimal operating conditions, a base oil yield of up to 60% per pass could be obtained.

Long-chain linear alkanes such as *n*-octacosane (*n*-C_{28}), *n*-hexatriacontane (*n*-C_{36}) and *n*-tetratetracontane (*n*-C_{44}) were investigated using a 0.3% Pt/MSA catalyst in a stirred micro-autoclave.[44] The objective was to convert these

Table 6.7 Hydroisomerisation of a wax with average carbon number of C_{33} over a Pt/MSA catalyst to produce a lube base oil.

Property	*Wax feed*	*Lube base oil*
Viscosity at 40 °C (mPa · s)	–	21.6
Viscosity at 100 °C (mPa · s)	5.61	4.7
Viscosity index	194	146
Density at 15 °C (kg · m^{-3})	827	–
Pour point (°C)	63	–21
Base oil composition (mass%)		
n-Alkanes	35.4	–
Branched alkanes	44.6	86.2
Cycloalkanes	18.1	13.2
Aromatics	1.9	3.9

reactants to branched alkanes of lube base oil quality. The reaction network involved the conversion of *n*-alkanes *via* three competitive reactions that directly led to the formation of cracking products and two pseudo-components, namely '*i*-C_n lube', which is lump of branched alkanes with sufficiently low pour points to make them suitable for a base oil, and '*i*-C_n nolube' which is lump of branched alkanes with pour points unsuitable for a base oil. During reaction, the '*i*-C_n nolube' fraction was converted into '*i*-C_n lube' through subsequent HIS reactions.

Kobayashi and co-workers used a ^{13}C NMR method to investigate the molecular structures of the lube base oil and diesel fuel that can be prepared from Fischer–Tropsch waxes by HIS/HCR.[45–47] The aim was to determine the location and length of the branches. It was observed that the probability of methyl branching on the main carbon chain decreased in the order 2nd > 3rd > 4th and so on; the probability of methyl branching on the seventh, eighth and inner carbon atoms was almost equal. The catalysts used in this study were prepared by impregnating ammonium heptamolybdate solution and nickel nitrate solution separately on extrudates of an alumina, silica and mordenite mixture. Other catalysts were prepared by impregnating ammonium tungstate solution and nickel nitrate solution separately on extrudates of an alumina, silica–alumina and ultrastable Y zeolite mixture. The last type of catalyst gave higher conversion of the > 360 °C boiling fraction. As a consequence, the yield of diesel fuel obtained over this catalyst was greater. The experiments were conducted at a total pressure of 9 MPa and at temperatures between 340 and 370 °C.

Zhou *et al.* studied the effect of metal promoters on the activity and selectivity of tungstated zirconia (TZ) with 8 mass% W for the HIS of *n*-hexadecane in a trickle bed continuous flow reactor with the aim of designing an active catalyst for the conversion of Fischer–Tropsch waxes to fuels and lube base oil fractions.[48] It was found that Pt had a better promoting effect than either Ni or Pd. Pretreatment at temperatures between 300 and 400 °C for 3 h in H_2 slightly increased the yields of branched hexadecane isomers over Pt/TZ. Under the same conditions, the performance of sulfated zirconia (SZ) was compared with that of TZ catalysts; Pt/SZ was compared with the Pt/TZ. The former was a good cracking catalyst, whereas the Pt/TZ was suitable as a HIS catalyst. This observation was also confirmed during the HIS/HCR of two Fischer–Tropsch waxes. Thus, severe cracking was suppressed using the Pt/TZ catalyst to obtain branched isomers in the diesel fuel or lube base oil ranges.

A Pt/TZ catalyst was also investigated using n-C_{24} and n-C_{36} alkanes, and also a Fischer–Tropsch wax.[49] A Pt/TZ catalyst with 0.5 mass% Pt and 12.5 mass% W was used in conjunction with the addition of SZ, TZ and zeolites to increase its activity and selectivity at 200 °C to kerosene and distillate. The effect of improving the performance of Pt/TZ by adding the zeolite MOR revealed that an optimal mixing ratio exists for maximum conversion of n-C_{24} under certain reaction conditions. The hybrid catalysts consisted of physical mixtures of the solids. Hybrid catalysts based on Pt/TZ exhibited a higher catalytic activity and higher selectivity for transportation fuels when Fischer–Tropsch waxes were used as the feed.

6.3.3 Hydrocracking of Waxes

The studies conducted by Dry represent some of the early work on HCR to convert waxes from FTS into distillates for use as diesel fuel.[50,51] Under mild conditions and by recycling the fractions boiling above the distillate range to extinction, a final distribution of 80% distillate, 15% naphtha and 5% gas was obtained. The temperature required to achieve a specific conversion of Fischer–Tropsch waxes was about 30 °C lower compared with conventional vacuum gas oil.[52]

The composition of liquid products from the HCR of LTFT wax over a noble metal catalyst was investigated using HPLC and GC–MS techniques.[53] The focus was on the content and nature of aromatics formed. Low levels (<2 mass%) of aromatics were identified. Among them, short-chain alkylated benzenes were predominant. Small amounts of naphthalene and higher aromatics were also present. The bifunctional nature of the catalyst and the reaction conditions applied during the HCR of LTFT wax did not favour the formation of more aromatics. The liquid product had a low density, typically 760–780 kg m^{-3}. The low density of the distillate obtained from wax HCR makes it difficult to produce on-specification EN590-type diesel fuel in high yield from LTFT syncrude with current refining technology.[54]

The performances of some of the Pt/ASA catalysts in Table 5.27 were evaluated for the HCR of LTFT waxes and compared with that of a commercial sulfided base metal catalyst for the same (Figure 6.4).[55] The wax feed had a carbon number distribution ranging from C_{13} to C_{83}. The results show that at high conversions, the distillate selectivities of the Pt/Siral75 and PtW/Siral75 catalysts were higher than those of the PtW/Siral40 and sulfided base metal catalysts. Moreover, the cloud point and cetane number for the diesel produced with the PtW/Siral75 catalyst were − 11 °C and 77, respectively, compared with − 8 °C and 79 for the for the commercial sulfided base metal catalyst.

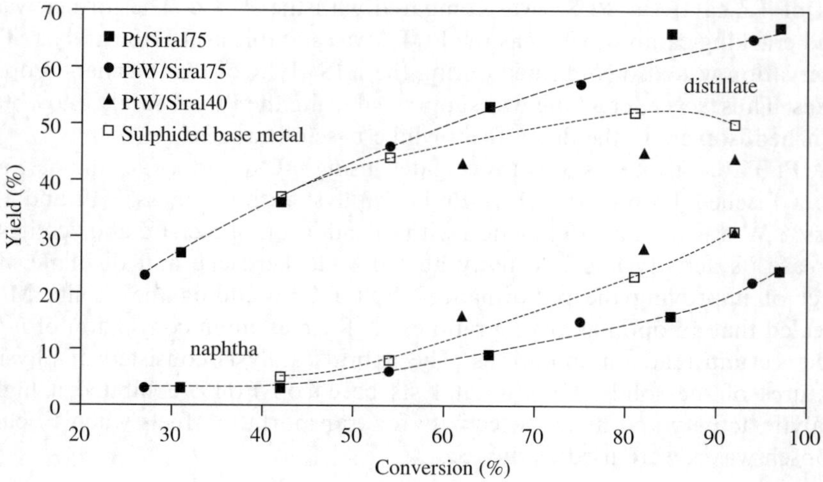

Figure 6.4 Hydrocracking of LTFT wax over Pt- and W-promoted silicated amorphous silica–alumina (Siral75) catalysts and a commercially available sulfided base metal hydrocracking catalyst at 7 MPa, LHSV 1 h^{-1} and H_2:wax ratio 1000:1.

Table 6.8 Influence of pressure on the hydrocracking of LTFT medium wax over a PtMo/Siral75 (silicated amorphous silica–alumina) catalyst at 370 °C, WHSV 1 h^{-1} and H_2:wax ratio 1200:1.

	Hydrocracked product		
Property	*3.5 MPa*	*5.0 MPa*	*7.0 MPa*
Conversion (%)	81	58	46
Distillate-to-naphtha ratio	3.1	4.3	5.3
Branched-to-linear alkane ratio	4.1	4	3.4
Cloud point (°C)	−17	−12	−10
Cetane number	72	72	74

The ability of unsulfided Pt-promoted amorphous silica–alumina catalysts with mild acidity to convert LTFT waxes to distillate with high selectivity at high conversion has also been pointed out by Calemma and co-workers.[56,57]

Fischer–Tropsch waxes can not only be hydrocracked at lower temperature, but also at lower pressure than in conventional HCR, as indicated before (Table 5.23).[58] The effect of pressure on the HCR of LTFT waxes at 370 °C is illustrated by the data in Table 6.8.[59] Similar data for HCR at 380 °C have also been published.[60] Conversion and HIS of the products increased with decrease in operating pressure. This is in line with the bifunctional HCR mechanism (Sections 5.2.1 and 5.3.1), where the first step involves the formation of alkenes (deHYD) at the metal site followed by protonation, rearrangement and cracking on a Brønsted acid site. Consequently, an increase in the H_2 pressure should lead to a lower steady-state concentration of alkenes and of carbocations on the catalyst surface. A lower hydrogen pressure should lead to increased HIS followed by HCR. The negative effect of higher H_2 pressure on HIS and HCR can be offset by increasing the temperature. For example, over a sulfided NiMo/SiO_2–Al_2O_3 catalyst an increase of around 15 °C was necessary to maintain the same HCR conversion when the pressure was increased from 3.5 to 7 MPa.[60] However, catalyst deactivation due to coke formation was observed when the H_2 pressure was too low. Stable operation over a 100 day test period has been reported for HCR of waxes at 3.5 MPa, but operation at pressures below 1 MPa definitely led to catalyst deactivation.[60] A threshold pressure has not been indicated, however.

The distillate selectivity during HCR of LTFT waxes is also influenced by the liquid hourly space velocity and H_2:wax ratio, mainly as result of a change in wax conversion.[4] Thus, the diesel selectivity increased with increasing LHSV and thereby decreasing contact time of the feed with the catalyst. An increase in the H_2:wax ratio resulted in an increase in conversion. For example, increasing the H_2:wax ratio from 500:1 to 1500:1 almost doubled the conversion.

Recycling of the 'unconverted' wax from once-through operation to the reactor increased the overall conversion, but resulted in a reduced distillate-to-naphtha ratio. The optimum that is observed for one catalyst is not necessarily the same for other catalysts, as can be seen from Figure 6.4. The 'unconverted' wax may not be hydrocracked, but this does not imply that it has not been hydroisomerised. The distillate selectivity will deteriorate when recycling 'unconverted' wax with the fresh wax feed, since the recycle is isomerised and more reactive than the fresh feed. This follows from the fundamentals of the wax hydrocracking mechanism and, unless care is taken in the commercial design to compensate for this, the distillate selectivity will deteriorate compared with the once-through values. Designs employing wax HCR with recycle to extinction should therefore feed the wax recycle at a point closer to the bottom of the catalyst bed and not at the top of the catalyst bed.

6.3.4 Catalytic Cracking of Waxes

The catalytic cracking of FT wax has been investigated by a number of groups.[61–69] An economic comparison of FCC and HCR for the upgrading of Fischer–Tropsch wax indicated that the former, with its more olefinic product slate, is more economical than one based on HCR.[66] This is contrary to the perception that has been created by the exclusive use of HCR technology for the upgrading of wax in new LTFT facilities.[19] However, it is understandable, since these facilities do not produce transportation fuels as in a normal fuels refinery, but naphtha and a high cetane number distillate blending stock.

Catalytic cracking of the Fischer–Tropsch wax under conditions approaching FCC over several acidic catalysts produced a high octane number gasoline, except for ASA, which is not shown, which only produced gaseous products (Table 6.9).[69] Over a mesoporous Al–MCM-41 catalyst, having a

Table 6.9 Catalytic cracking of Fischer–Tropsch wax over several acidic catalysts at 560 °C, contact time 12 s and catalyst:wax ratio 2:1.

		Product yield (%)		
Cracking catalyst	*Conversion (%)*	*LPG*	*Naphtha*	*RON*
Al-MCM-41	42	21	18	91
H-ZSM-5 (3% crystalline)	78	43	30	83
ASA + H-Y	86	30	52	85
ASA + H-ZSM-5	91	46	37	91

Table 6.10 Catalytic cracking of Fe-LTFT medium wax (commercial fixed bed synthesis) and crude oil-derived gas oil over equilibrium HY catalyst in a micro activity testing unit at 520 °C and catalyst:feed ratio 3:1.

	FCC products	
Description	*Crude oil gas oil*	*LTFT wax*
Conversion (mass%)	61.6	88.1
Product distribution (%)		
C_2 and lighter	2.6	1.8
C_3–C_4	11.6	31.4
C_5–220 °C	43.1	52.7
220 °C and heavier	38.4	11.9
Coke	4.3	2.2
Naphtha properties		
RON	90.4	85.8
MON	79.8	77.6

similar number of acid sites as ASA, a cracking conversion of about 40% was achieved with 20% selectivity to gasoline. The higher cracking activity of the Al–MCM-41 catalyst was attributed to stronger acid sites than those present in ASA. It has been noted that H-Y and H-ZMS-5 catalysts were very active and in the reported work these catalysts were diluted with ASA.

Wax is more easily cracked than crude oil-derived residues. The catalytic cracking work performed at Amoco indicated that at 520 °C and a catalyst:oil ratio of 3:1 the conversion of Fischer–Tropsch medium wax was 88% compared with 62% of conventional crude oil-derived gas oil (Table 6.10).[65] The higher reactivity of wax compared with gas oil can be explained by the differences in molecular composition. The wax consists almost entirely of long-chain alkanes that are easy to crack. In contrast, crude oil-derived gas oil also contains molecules that consist of heteroatom-containing aromatics linked by aliphatic side-chains. Although the bridging aliphatic side-chains can easily be cracked, the intermediate products from cracking are the aromatic fragments, which are more difficult to crack.[67]

In the case of typical FCC operation with H-Y, it was found that the condition of the H-Y catalyst had only a minor influence on the conversion and

Table 6.11 Catalytic cracking of Fe-LTFT wax (obtained from Mobil slurry bubble column FTS at 250 °C and 2.6 MPa) over a rare earth-exchanged H-Y (Engelhard HEZ-53) catalyst in a riser unit at hydrocarbon partial pressure 0.11 MPa, residence time 1 s and catalyst:feed ratio 4.2:1–4.4:1.

Property	*Equilibrium H-Y*	*Equilibrium H-Y*	*Coked H-Y*
Temperature, top/maximum (°C)	465/478	504/523	505/524
Conversion (mass%)	91.4	93	91.1
Product distribution (%)			
C_2 and lighter	2.0	3.5	4.3
C_3–C_4	17.2	21.7	19.5
C_5–194 °C	56.5	56.3	57.0
194–344 °C	23.2	17.6	19.8
Coke	1.1	0.9	−0.6
Naphtha properties			
RON	89.8	91.5	91.6
Distillate (unhydrogenated) properties			
Cetane index	53	51	49
Pour point (°C)	−23	−23	−34

product yields obtained from wax cracking (Table 6.11).[63] Likewise, catalytic cracking of wax was found to be fairly insensitive to the catalyst:feed ratio.[65]

6.3.5 Co-catalysts for Wax Conversion During FTS

The use of cracking catalysts (HCR and catalytic cracking) in combination with FTS has been considered by a number of researchers. The idea behind this concept is to break the Anderson–Schulz–Flory distribution of products that is inherent to FTS by introducing a different catalytic functionality. This requires matching the operating windows of the catalyst for FTS with that of the co-catalyst. The most challenging and direct approach is to design a catalyst that can perform FTS and cracking conversion of the Fischer–Tropsch products all on a single catalyst, as pioneered by Chang *et al.* with Fe/H-ZSM-5.[70] A similar approach was followed by Egiebor *et al.*, who investigated Fe/H-Y catalysts.[71] Although neither of these catalysts produced wax, the same principle has been investigated in conjunction with lower temperature FTS that produces wax.[72,73] These studies employed a mixture of catalysts in the same reactor, rather than a single catalyst.

A major challenge in performing iron-based FTS together with co-conversion of the Fischer–Tropsch products in the same reactor is to prevent migration of the alkali promoters from the Fischer–Tropsch catalyst to the acidic co-catalyst. When that happens, the alkali promoters neutralise the acid sites on the co-catalyst, leading to co-catalyst deactivation. One way of overcoming this obstacle is to physically separate FTS and the co-catalyst, but without the intermediate product cooling and separation steps usually associated with the Fischer–Tropsch gas loop (as discussed in Section 4.2.1). This approach has been studied for the conversion of wax-containing LTFT products by a number of groups.[74–77] In these investigations, FTS and co-conversion are not fully segregated. Depending on the level of integration between the two steps (similar operating window or not), it can arguably no longer be considered as co-catalysis during FTS.

References

1. M. E. Dry, *Catal. Rev. Sci. Eng.*, 1981, **23**, 265.
2. D. O. Leckel, *Energy Fuels*, 2009, **23**, 38.
3. M. E. Dry, in *Methane Conversion*, ed. D. M. Bibby, C. D. Chang, R. F. Howe and S. Yurchak, Elsevier, Amsterdam, 1988, p. 447.
4. D. O. Leckel, *Energy Fuels*, 2005, **19**, 1795.
5. A. de Klerk, *Ind. Eng. Chem. Res.*, 2007, **46**, 5516.
6. R. J. O'Brien, L. Xu, S. Bao, A. Raje and B. H. Davis, *Appl. Catal. A*, 2000, **196**, 173.
7. D. Wei, J. G. Goodwin Jr., R. Oukaci and A. H. Singleton, *Appl. Catal. A*, 2001, **210**, 137.
8. D. B. Bukur, *Catal. Today*, 2005, **106**, 275.
9. Anon., *Pet. Econ.*, 2008, **75** (6), 36.
10. G. W. Roberts, J. M. Biales, E. L. Dodge and P. K. Kilpatrick, *Prepr. Pap. Am. Chem. Soc. Div. Pet. Chem.*, 1999, **44** (1), 65.
11. R. R. Oder, *Stud. Surf. Sci. Catal.*, 2007, **163**, 337.
12. F. Asinger, *Paraffins Chemistry and Technology*, Pergamon Press, Oxford, 1968.
13. R. Freerks, presented at the AIChE Spring National Meeting, New Orleans, 2003, paper 86d.
14. J. H. Le Roux and S. Oranje, *Fischer–Tropsch Waxes*, Sasol, Sasolburg, 1984.
15. R. L. Espinoza, A. P. Steynberg, B. Jager and A. C. Vosloo, *Appl. Catal. A*, 1999, **186**, 13.
16. F. J. M. Schrauwen, in *Handbook of Petroleum Refining Processes*, ed. R. A. Meyers, McGraw-Hill, New York, 2004, p. 15.25.
17. J. Ansorge, *Prepr. Pap. Am. Chem. Soc. Div. Fuel Chem.*, 1997, **42** (2), 654.
18. L. P. Dancuart, R. de Haan and A. de Klerk, *Stud. Surf. Sci. Catal.*, 2004, **152**, 482.
19. A. de Klerk, *Prepr. Pap. Am. Chem. Soc. Div. Pet. Chem.*, 2008, **53** (2), 105.
20. K. Fraser, in *Fundamentals of Gas to Liquids*, 2nd edn, ed. E. Soutar Petroleum Economist, London, 2005, p. 15.
21. N. Fabricius, in *Fundamentals of Gas to Liquids*, 2nd edn, ed. E. Soutrar Petroleum Economist, London, 2005, p. 12.
22. M. Guisnet, F. Alvarez, G. Giannetto and G. Perot, *Catal. Today*, 1987, **1**, 415.
23. E. Furimsky, *Catalysts for Upgrading Heavy Petroleum Feeds*, Elsevier, Amsterdam, 2007.

24. T. Higgins, *World Refining*, 2004, **41**, 4.
25. C. S. Snodgrass and M. Perrin, *J. Inst. Pet. Technol.*, 1938, **24**, 289.
26. G. H. Dazeley and C. C. Hall, *Fuel*, 1948, **27**, 50.
27. H. G. Könnecke and G. Gawalek, *Chem. Tech. Berlin*, 1956, **8** (10), 603.
28. D. T. A. Huibers and H. I. Waterman, *Brennst.-Chem.*, 1960, **41**, 297.
29. L. B. Bragg, *Ind. Eng. Chem.*, 1941, **33**, 376.
30. H. H. Voge and G. M. Good, *J. Am. Chem. Soc.*, 1949, **71**, 593.
31. A. Kossiakoff and F. O. Rice, *J. Am. Chem. Soc.*, 1943, **65**, 590.
32. B. H. Rosen, *Ind. Eng. Chem.*, 1960, **52**, 14.
33. Z. Jisheng, C. Languang, S. Shuhe and C. Shaoxin, in *Proceedings of the 12th Annual International Pittsburgh Coal Conference*, 1995, p. 681.
34. F. H. A. Bolder, A. de Klerk and J. L. Visagie, *Ind. Eng. Chem. Res.*, 2009, **48**, 3755.
35. E. L. J. Breet, A. S. Luyt and S. Oranje, *S. Afr. J. Chem.*, 1990, **43**, 83.
36. E. L. J. Breet and A. S. Luyt, *S. Afr. J. Chem.*, 1991, **44**, 101.
37. A. de Klerk, *Ind. Eng. Chem. Res.*, 2003, **42**, 6545.
38. A. de Klerk, *Ind. Eng. Chem. Res.*, 2004, **43**, 6898.
39. R. A. Sheldon and J. K. Kochi, *Metal-catalyzed Oxidations of Organic Compounds*, Academic Press, New York, 1981.
40. F. H. A. Bolder, *Energy Fuels*, 2007, **21**, 1396.
41. D. O. Leckel, in *Proceedings of the 7th European Congress on Catalysis*, Sofia, 2005, paper O5-05.
42. D. O. Leckel, *Energy Fuels*, 2007, **21**, 662.
43. V. Calemma, S. Peratello, C. Perego, A. Moggi and R. Giardino, *Prepr. Pap. Am. Chem. Soc. Div. Pet. Chem.*, 1999, **44** (3), 241.
44. V. Calemma, S. Peratello, F. Stroppa, R. Giardino and C. Perego, *Ind. Eng. Chem. Res.*, 2004, **43**, 934.
45. M. Kobayashi, M. Saitoh, S. Togawa and K. Ishida, *Energy Fuels*, 2009, **23**, 513.
46. M. Kobayashi, M. Saitoh, K. Ishida and H. Yachi, *J. Jpn. Petr. Inst.*, 2005, **48**, 365.
47. M. Kobayashi, S. Togawa and K. Ishida, *J. Jpn. Petr. Inst.*, 2006, **49**, 194.
48. Z. Zhou, Y. Zhang, J. W. Tierney and I. Wender, *Fuel Process. Technol.*, 2003, **83**, 67.
49. S. Zhang, Y. Zhang, J. W. Tierney and I. Wender, *Fuel Process. Technol.*, 2001, **69**, 59.
50. M. E. Dry, *Catal. Rev. Sci. Eng.*, 1981, **23**, 265.
51. M. E. Dry, *Hydrocarbon Process.*, 1982, **61** (8), 121.
52. M. E. Dry, in *Catalysis Science and Technology*, Vol. **1**, ed. J. R. Anderson and M. Boudart, Springer, Berlin, 1981, p. 159.
53. D. O. Leckel, *Energy Fuels*, 2009, **23**, 32.
54. A. de Klerk, *Energy Fuels*, 2009, **23**, 4593.
55. D. O. Leckel, *Ind. Eng. Chem. Res.*, 2007, **46**, 3505.
56. V. Calemma, S. Peratello, S. Pavoni, G. Clerici and C. Perego, *Stud. Surf. Sci. Catal.*, 2001, **136**, 307.

57. V. Calemma, C. Gambaro, W. O. Parker Jr., R. Carbone, R. Giardino and P. Scorletti, *Catal. Today*, 2010, **149**, 40.
58. C. Bouchy, G. Hastoy, E. Guillon and J. A. Martens, *Oil Gas Sci. Technol. Rev. IFP*, 2009, **64**, 91.
59. D. O. Leckel and M. Liwanga-Ehumbu, *Energy Fuels*, 2006, **20**, 2330.
60. D. O. Leckel, *Energy Fuels*, 2007, **21**, 1425.
61. I. Kobolakis and B. W. Wojciechowski, *Can. J. Chem. Eng.*, 1985, **63**, 269.
62. J. Abbot and B. W. Wojciechowski, *Ind. Eng. Chem. Prod. Res. Dev.*, 1985, **24**, 501.
63. T. M. Leib, J. C. W. Kuo, W. E. Garwood, D. M. Nace, W. R. Derr and S. A. Tabak, in *Proceedings of the AIChE Annual Meeting*, Washington, DC, 1988, paper 61d.
64. W. J. Reagan, *Prepr. Pap. Am. Chem. Soc. Div. Fuel Chem.*, 1994, **39** (2), 337.
65. M. M. Schwartz, *DOE Contract No. DE-AC22-91PC90057 – Final Report*, 1995.
66. G. N. Choi, S. J. Kramer, S. S. Tam, J. M. Fox and W. J. Reagan, *Prepr. Pap. Am. Chem. Soc. Div. Fuel Chem.*, 1996, **41** (3), 1079.
67. X. Dupain, R. A. Krul, M. Makkee and J. A. Moulijn, *Catal. Today*, 2005, **106**, 288.
68. X. Dupain, R. A. Krul, C. J. Schaverien, M. Makkee and J. A. Moulijn, *Appl. Catal. B*, 2006, **63**, 277.
69. K. S. Triantafyllidis, V. G. Komvokis, M. C. Papapetrou, I. A. Vasalos and A. A. Lappas, *Stud. Surf. Sci. Catal.*, 2007, **170**, 1344.
70. C. D. Chang, W. H. Lang and A. J. Silvestri, *J. Catal.*, 1979, **56**, 268.
71. N. O. Egiebor, W. C. Cooper and B. W. Wojciechowski, *Appl. Catal.*, 1989, **55**, 47.
72. N. Guan, Y. Liu and M. Zhang, *Catal. Today*, 1996, **30**, 207.
73. X. Song and A. Sayari, *Energy Fuels*, 1996, **10**, 561.
74. J. C. W. Kuo, *DOE Contract No. AC22-80PC30022 – Final Report*, 1983.
75. R. L. Varma, N. N. Bakhshi, J. F. Mathews and S. N. Ng, *Ind. Eng. Chem. Res.*, 1987, **26**, 183.
76. Z. W. Liu, X. Li, K. Asami and K. Fujimoto, *Energy Fuels*, 2005, **19**, 1790.
77. Z. W. Liu, X. Li, K. Asami and K. Fujimoto, *Appl. Catal. A*, 2006, **300**, 162.

CHAPTER 7

Upgrading of Fischer–Tropsch Oxygenates

Oxygenates are ubiquitous in Fischer–Tropsch syncrude (Table 1.1). The distribution of oxygenates between the different product fractions from FTS depends on the polarity and boiling point of the oxygenates, and also the efficiency of the stepwise cooling after FTS (Section 4.2.1). A large portion of the lighter boiling more polar oxygenates ends up in Fischer–Tropsch aqueous product (reaction water), but a significant fraction remains in the oil product. The information presented thus far highlighted the influence of oxygenates on the catalysis of some upgrading steps.

There is a rich chemistry associated with the conversion of FTS oxygenates to chemicals. Many oxygenates in Fischer–Tropsch syncrude can be extracted and sold as chemicals (see also Chapter 9).[1] The further beneficiation of the purified oxygenates will not be discussed, since there is little difference between the conversion of purified oxygenates from FTS and that from other sources.

Three aspects of oxygenate conversion in the products from FTS will be considered in more detail. First, the acid-catalysed reactions of oxygenates in general: this is pertinent to all processes involving acid catalysis. Acid catalysis is employed in many refinery conversion processes, for example, alkylation, OLI, etherification, HIS, HCR and FCC. Second, the refining of the FTS aqueous product will be explored, which is a topic that has not received much attention in the literature. This product stream is composed almost entirely of oxygenates and water. The discussion consequently also has some bearing on the upgrading of biomass and aqueous effluent from oil sands processing. Lastly, some processes will be considered that deal specifically with oxygenate conversion in the FTS oil and gaseous products.

RSC Catalysis Series No. 4
Catalysis in the Refining of Fischer–Tropsch Syncrude
By Arno de Klerk and Edward Furimsky

Published by the Royal Society of Chemistry, www.rsc.org

7.1 Acid-catalysed Reactions of Oxygenates

The acid-catalysed conversion of oxygenates pertinent to OLI, HIS, HCR and catalytic cracking has been touched upon in Chapter 5. Of these reactions, the most influential in the context of Fischer–Tropsch refining is the formation of carboxylic acids over acid catalysts (Section 5.1.6). However, it is not the only acid-catalysed oxygenate conversion.

The acid-catalysed conversion of oxygenates pertinent to FTS has been investigated with SPA as acid catalyst.[2] The reaction of different oxygenates was studied in isolation and in the presence of alkenes. It was noted that under industrially relevant conditions, interactions between different oxygenate classes are also possible. In another study, the influence of various oxygenates was investigated over an acidic resin catalyst, also in the presence of alkenes.[3] There are also reports dealing with the influence of different oxygenate classes in Fischer–Tropsch-derived feed on conversion over bifunctional acidic catalysts.[4,5]

In addition to the studies dealing with specific compound classes, the influence of mixed oxygenates in Fischer–Tropsch feed materials on the conversion of hydrocarbons over acidic catalysts has been investigated.[6–9] Many of these findings have already been discussed (Chapter 5).

Most of the acid-catalysed chemistry of oxygenates involves transformation via an alcohol or a carbonyl functionality. These two chemistries will therefore be discussed. In addition to these, there are various other acid-catalysed oxygenate conversions and reference to any organic chemistry text book will show that oxygenates in general have a rich chemistry on their own and in reaction with other compounds.

7.1.1 Acid-catalysed Alcohol Conversion

The alcohols are one of the most abundant oxygenate classes and they are present at percentage levels in both HTFT and LTFT syncrudes. Alcohols are dehydrated over acid catalysts to produce the corresponding ethers and/or alkenes, depending on the severity of dehydration. This is an endothermic reaction and it is equilibrium limited.

Superficially, alcohol dehydration is a straightforward acid-catalysed reaction, but in reality the mechanism is influenced by the nature of the catalyst. Some of the complexity introduced by the catalyst can be seen from the side-products during dehydration.

Over SPA, the dehydration of 2-propanol at 140 °C in a batch reactor for 20 h resulted in the formation of mostly of propene, at better than 75% conversion.[2] Less than 5% of the 2-(1-methylethoxy)propane (diisopropyl ether) and OLI products of propene were produced. Some 2-propoxypropane was found, indicating that in some way the catalyst allowed reaction on the α-carbon, either through the reverse reaction to hydrate the propene or direct alcohol etherification with the propene. The conversion of 1-butanol was only about 30% and resulted in a more complex product spectrum (Figure 7.1). In

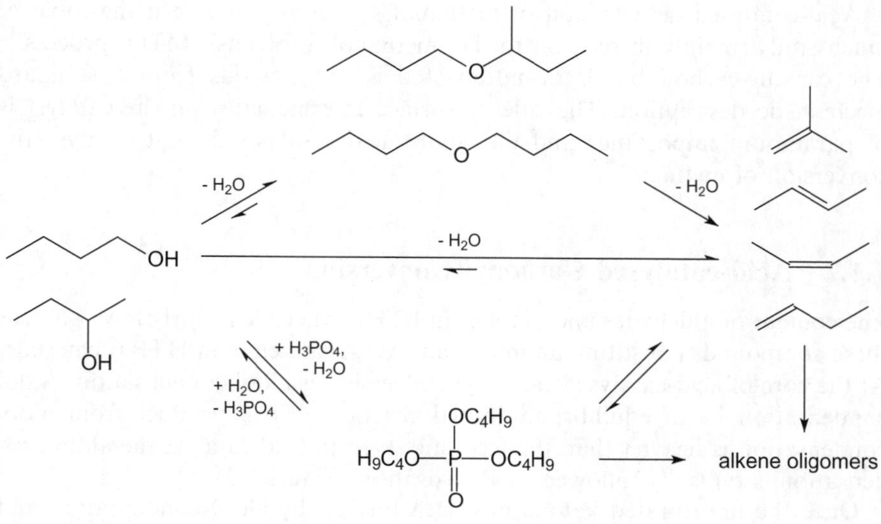

Figure 7.1 Reaction network of alcohol dehydration over SPA.

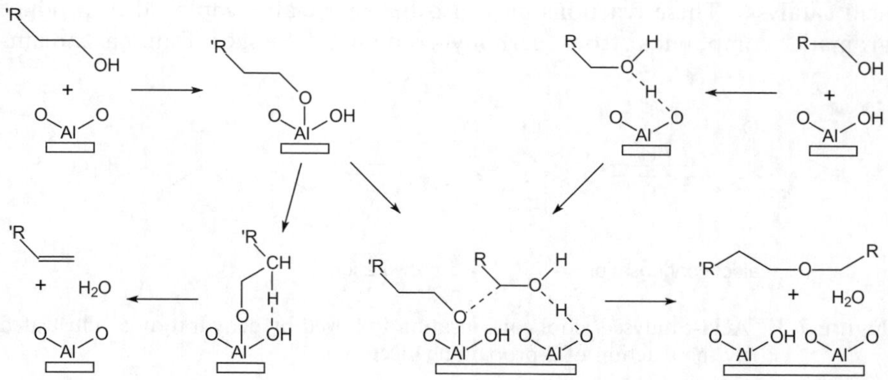

Figure 7.2 Mechanism of alcohol dehydration over alumina.

addition to butoxybutane and 2-butoxybutane (<5% selectivity), all four butene isomers and their oligomers were found in the product. Hydrogen transfer reactions took place, as evidenced by the presence of *n*-butane and 2-methylpropane, also trace amounts of aromatics, in the product. Some tributyl phosphoric acid ester was also found in the product.

Over alumina, the dehydration mechanism is somewhat different, as can be seen from the mechanism proposed by Shi and Davis (Figure 7.2).[10] In addition to the dehydration steps, alumina is known to catalyse dehydrogenation. The propensity of alumina to catalyse dehydrogenation is strongly influenced by the catalyst pretreatment.[11]

Acid-catalysed dehydration of methanol specifically has been the topic of much study, mainly in relation to the 'methanol-to-olefins' (MTO) process.[12] The carbon–carbon bond formation step is not obvious from a standard mechanistic description. The role of surface intermediates on the catalyst is of paramount importance and ultra-pure acid catalysts do not catalyse the conversion of methanol.[13]

7.1.2 Acid-catalysed Carbonyl Conversion

The content of aldehydes and ketones in LTFT syncrude is fairly low, whereas these compounds constitute an important oxygenate class in HTFT syncrude. At the core of acid-catalysed carbonyl conversion is aldol condensation. Aldol condensation is an equilibrium-limited reaction.[14] The product from aldol condensation is heavier than the feed and over an acid catalyst the aldol condensation is typically followed by dehydration (Figure 7.3).

Once the unsaturated ketone has been formed by aldol condensation and dehydration, two subsequent acid-catalysed reactions may follow. The first is a repetition of the aldol condensation and dehydration that may ultimately lead to aromatisation of the product (Figure 7.4). It is in this way that carbonyl compounds can easily form heavy carbonaceous deposits and aromatic coke on acid catalysts. These reactions may also be beneficially employed to produce aromatic compounds from carbonyl-containing Fischer–Tropsch streams

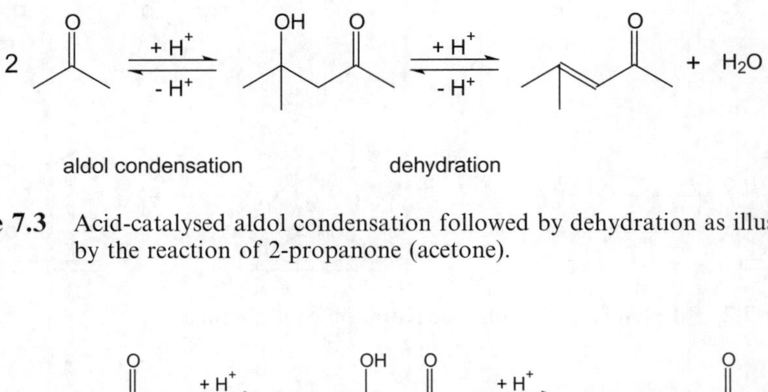

Figure 7.3 Acid-catalysed aldol condensation followed by dehydration as illustrated by the reaction of 2-propanone (acetone).

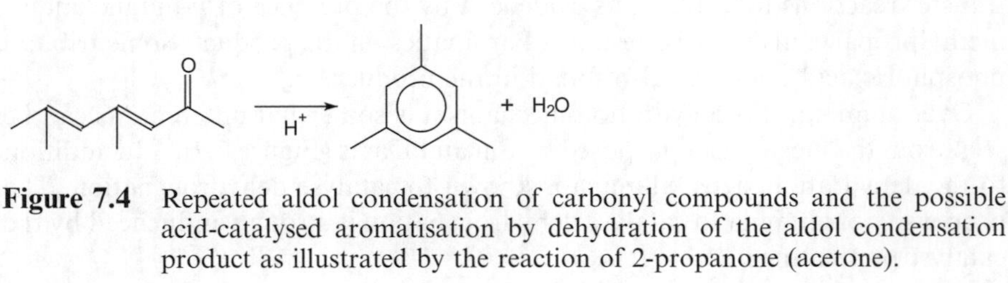

Figure 7.4 Repeated aldol condensation of carbonyl compounds and the possible acid-catalysed aromatisation by dehydration of the aldol condensation product as illustrated by the reaction of 2-propanone (acetone).

without resorting to catalytic reforming. The reaction network is more complex than suggested by Figure 7.4, and a more detailed discussion can be found in a paper by Salvapati *et al.*[15]

The second of the reactions following on aldol condensation is carboxylic acid formation by hydrolytic cleavage of the aldol condensation product (Figure 5.13). This reaction has already been discussed (Section 5.1.6).

Over SPA, more than 90% conversion of propanal was reported after 20 h in a batch reactor at 140 °C.[2] The products were mainly due to aldol condensation, with the main products being approximately equal amounts of 2-methyl-2-pentenal and 1,3,5-trimethylbenzene. The conversion of 2-pentanone under similar conditions was around 40%. In both instances carboxylic acids were detected in the product, in addition to alkenes, as one would expect from the hydrolytic cleavage of aldol condensation products.

Acid-catalysed ketone rearrangement reactions have been reported by Fry and co-workers.[16–20] Not all catalysts are equally active for such rearrangement reactions. In particular, SPA has been reported to have a propensity for ketone rearrangement.

7.2 Oxygenate Conversion in the Fischer–Tropsch Aqueous Product

The composition of the water-soluble oxygenates depends on the nature and operation of the FT process. Fused iron-based high-temperature Fischer–Tropsch synthesis yields a product containing mainly alcohols, carbonyl compounds and carboxylic acids (Table 4.7). The organic products in the aqueous stream are about 7–10% of the total HTFT product. The product from precipitated iron-based low-temperature Fischer–Tropsch synthesis contains less water-soluble oxygenates, about 3% of the total LTFT product, and is richer in alcohols, especially methanol. Cobalt-based LTFT synthesis generally produces less water-soluble oxygenates (Table 1.1).

The oxygenates in the FTS aqueous product can be either recovered and purified or they can be converted to products that can be refined with the Fischer–Tropsch gaseous and oil product fractions. Recovery of the oxygenates by separation is a difficult task because of the complex liquid–liquid-vapour equilibria and numerous azeotropes.[21] The decision to extract the oxygenates will typically be determined by market demands for such speciality chemicals and the cost/complexity of the production facility. When the water-soluble oxygenates are not recovered, the aqueous product from FTS has to be treated before disposal to reduce its environmental impact.[22] In the latter case, the conversion of the mixed oxygenate product to alkenes may be an appropriate way to simplify the refinery.[23,24]

Three mixtures of alcohols were used in a dehydration study carried out by Nel and de Klerk.[23] The alcohol mixtures had the following compositions: a light C_2–C_3 alcohol fraction, a heavy alcohol fraction consisting mostly of C_3–C_6 alcohols and an intermediate alcohol fraction consisting mostly of C_4 and C_5

alcohols. All of these alcohol mixtures were obtained from the HTFT aqueous product. Most of the dehydration experiments were carried out with a commercial η-alumina catalyst that was low in metal impurities, but contained about 0.9% silica. Two other catalysts were also used, namely a C84/3-type SPA and Amberlyst 15 sulfonic acid resin catalyst.

Conversions of alcohols to the corresponding alkenes of >95% were observed above 360 °C (Figure 7.5).[23] The ease of alcohol dehydration catalysed by alumina increased with increasing chain length for primary linear C_2–C_6 alcohols. Dehydration of the secondary and tertiary alcohols was more facile than that of primary alcohols. It was possible to selectively dehydrate secondary and tertiary alcohols in mixtures containing primary alcohols. A selectivity of non-primary to primary alcohol dehydration of >50:1 was achieved with Amberlyst 15, whereas SPA gave a selectivity ratio of 10:1 at best and η-alumina less than 2:1. The non-olefinic dehydration products consisted mainly of ethers. At low temperatures, primary alcohols are dehydrated to form mainly ethers. Ether formation also benefits from higher pressures.[25] With increasing temperature, ether formation passes through a maximum before dehydration becomes dominated by alkene formation.[26,27]

In a separate experiment, the stability of an η-alumina catalyst was evaluated for the dehydration of heavy alcohol mixtures.[23] A 1:1 mixture of alcohols and water was passed over the catalyst at 350 °C and near atmospheric pressure. After 30 days of continuous operation, the alcohol conversion decreased by less than 1%, indicating that η-alumina is a stable catalyst for the conversion of alcohols in the Fischer–Tropsch aqueous product. The catalysis involved in alcohol dehydration will be discussed further in Section 8.3.

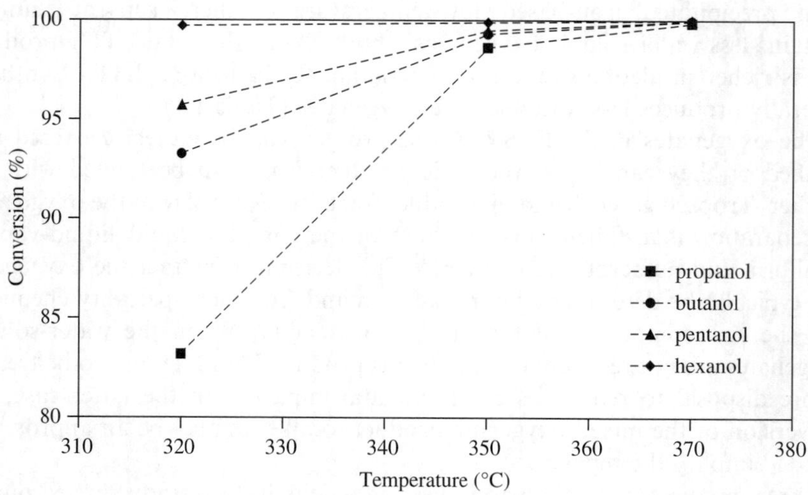

Figure 7.5 Dehydration of an aqueous C_3–C_8 alcohol mixture from HTFT synthesis over an η-alumina catalyst at near atmospheric pressure and LHSV $0.2\,h^{-1}$ on an alcohol basis. The feed mixture contained 50% water.

Another approach that has been suggested for Fischer–Tropsch aqueous product refining is to employ acid catalysis to convert the mixed oxygenates to alkenes and aromatics.[24] The repeated aldol condensation of aldehydes and ketones, followed by dehydration, produces, amongst others, mononuclear aromatics (Figure 7.4). The aromatics thus produced are all alkylated benzenes, with the exception of the product from ethanal, which is just benzene.

7.3 Oxygenate Conversion in the Fischer–Tropsch Oil Product

In general, it has been concluded that alumina is a good catalyst for the conversion of alcohols in a typical Fischer–Tropsch matrix of products that may include other oxygenate classes.

The dehydration of alcohols in the presence of carboxylic acids, ketones and aldehydes in oil fractions from FTS was investigated by Bolder and Mulder.[28] They observed that the alcohols could be readily dehydrated to alkenes at 380 °C over pure η-alumina. Almost quantitative dehydration of the alcohols was achieved, whereas reactions of carbonyl compounds and acids were incomplete and short-lived. At the same time, at least 70% of the 1-alkenes in the feed were initially isomerised to internal alkenes. Over a period of 8 days at 350 °C and an LHSV of 6 h^{-1}, the IS of alkenes gradually diminished until the fraction of 1-alkenes equalled the original concentration in the feed plus the fraction formed from the dehydration of 1-alcohols. The conversion of carbonyl compounds and acids also decreased over this period to a negligible level. The IS activity was linked to strong acid sites that were selectively deactivated by the carbonyl compounds and acids. The rate of deactivation increased with increasing concentration of the carbonyl compounds and acids in the feed. The loss of these catalytic sites did not diminish the catalyst activity towards alcohol dehydration. Oxidative regeneration at 480 °C restored catalytic activity for the conversion of carbonyl compounds and carboxylic acids, and also the IS of alkenes. The pattern of decreasing conversion of carbonyls, carboxylic acids and double bond IS with time on-stream was repeated after each regeneration.

The dehydration of alcohols in LTFT naphtha over η-alumina has been suggested as a way to improve distillate production and quality.[29]

The presence of alcohols in products from FTS could be beneficially employed for the removal of carboxylic acids by esterification. Esterification was catalysed by conventional acidic catalysts (liquid acids and acidic resins), and also metal oxide catalysts, such as MoO_3–Al_2O_3 and WO_3–Al_2O_3.[30] The catalysis involved in this type of esterification will be discussed further in Section 8.5.1.

The deoxygenation of FTS naphtha over alumina and alumina-rich materials, such as bauxite, has been employed in a number of commercial refineries associated with FTS.[31] The deoxygenation is carried out in such a way that it is accompanied by double bond IS.[32,33] Combined deoxygenation and IS has two advantages in a FT fuels refinery. The first is the improvement of the quality of the naphtha fraction by double bond IS of the linear 1-alkenes to internal alkenes that have higher octane numbers. After such catalytic treatment, the octane number is typically improved by about 10 octane units.[33] The second benefit is found in narrowing the carbon number distribution that is obtained from distillation. Oxygenates typically co-boil with hydrocarbons 2–4 carbon numbers apart; for example, 2-pentanone (102 °C) and 1-pentanol (138 °C) co-boil with C_8 and C_9 hydrocarbons, respectively. By deoxygenating the syncrude first and then fractionating the product, the molecules can be sent to the most appropriate refinery units.[34]

Bolder listed some advantages of retaining oxygenates in the Fischer–Tropsch oil product and specifically the distillate when it will be used as a transportation fuel.[30] In low concentrations, these oxygenates are beneficial to both gasoline and diesel combustion properties. Long-chain carboxylic acids in diesel reduce corrosion at low temperatures and improve fuel lubricity. Ethers and esters in diesel enhance the cetane number, improve the fuel lubricity properties and reduce noxious combustion products. Alcohols also reduce noxious combustion products and enhance the diesel cetane number. It is consequently more beneficial to convert carboxylic acids in diesel fractions to non-corrosive oxygen-containing compounds such as esters, rather than to alkanes. The beneficial effect of retaining oxygenates is illustrated by the lubricity improvement found by reducing the HDO severity of HTFT distillate (Figure 7.6).[35]

Not all of the beneficial effects of oxygenates observed in distillates are found in naphtha range products. With regard to oxygenates in motor gasoline, the

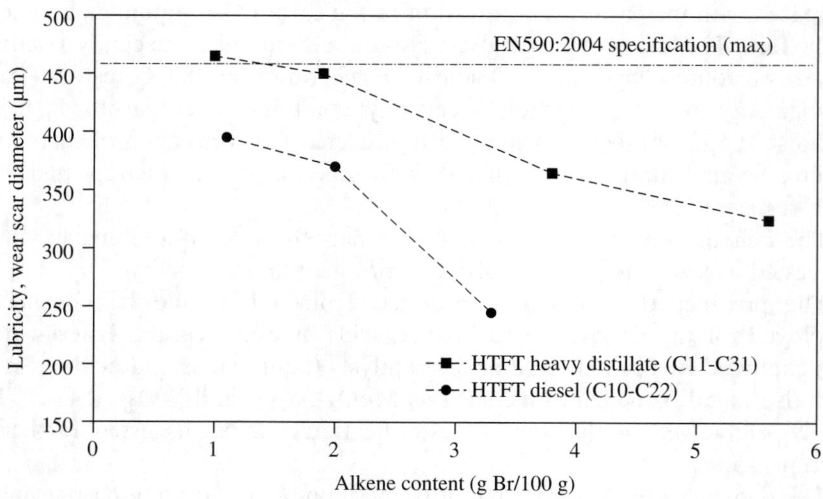

Figure 7.6 Beneficial effect of retaining oxygenates in distillate on the fuel lubricity as illustrated by the partial hydrogenation of two straight-run HTFT distillate fractions.

alcohols and ethers enhance the octane rating and improve combustion of the fuel. However, carboxylic acids boiling in the gasoline range are corrosive and need to be removed. Esters produced during carboxylic acids removal have low octane numbers and would likewise be undesirable in motor gasoline.

References

1. A. Redman, in *Proceedings of the 18th World Petroleum Congress*, Johannesburg, 2005, cd179.
2. A. de Klerk, R. J. J. Nel and R. Schwarzer, *Ind. Eng. Chem. Res.*, 2007, **46**, 2377.
3. D. Smook and A. de Klerk, *Ind. Eng. Chem. Res.*, 2006, **45**, 467.
4. D. O. Leckel, in *Proceedings of the 7th European Congress on Catalysis*, Sofia, 2005, paper O5-05.
5. D. O. Leckel, *Energy Fuels*, 2007, **21**, 662.
6. C. T. O'Connor, S. T. Langford and J. C. Q. Fletcher, in *Proceedings of the 9th International Zeolite Conference*, Montreal, 1992, p. 467.
7. M. Cowley, *Energy Fuels*, 2006, **20**, 1771.
8. A. de Klerk, *Energy Fuels*, 2007, **21**, 625.
9. T. N. Mashapa and A. de Klerk, *Appl. Catal. A*, 2007, **332**, 200.
10. B. Shi and B. H. Davis, *J. Catal.*, 1995, **157**, 359.
11. B. H. Davis, *J. Catal.*, 1972, **26**, 348.
12. C. D. Chang, *Catal. Rev. Sci. Eng.*, 1983, **25**, 1.
13. J. F. Haw, W. Song, D. M. Marcus and J. B. Nicholas, *Acc. Chem. Res.*, 2003, **36**, 317.
14. J. P. Guthrie, *Can. J. Chem.*, 1978, **56**, 962.
15. G. S. Salvapati, K. V. Ramanamurthy and M. Janardanarao, *J. Mol. Catal.*, 1989, **54**, 9.
16. W. H. Corkern and A. Fry, *J. Am. Chem. Soc.*, 1967, **89**, 5888.
17. A. Fry and W. H. Corkern, *J. Am. Chem. Soc.*, 1967, **89**, 5894.
18. F. E. Juge and A. Fry, *J. Org. Chem.*, 1970, **35**, 1876.
19. M. Oka and A. Fry, *J. Org. Chem.*, 1970, **35**, 2801.
20. A. Fry and M. Oka, *J. Am. Chem. Soc.*, 1979, **101**, 6353.
21. T. Q. Elliot, C. S. Goddin Jr and B. S. Pace, *Chem. Eng. Prog.*, 1949, **45**, 532.
22. A. C. Vosloo, L. P. Dancuart and B. Jager, presented at the 11th World Clean Air and Environment Congress, Durban, 1998, paper 6F-2.
23. R. J. J. Nel and A. de Klerk, *Ind. Eng. Chem. Res.*, 2007, **46**, 3558.
24. R. J. J. Nel and A. de Klerk, *Prepr. Pap. Am. Chem. Soc. Div. Fuel Chem.*, 2009, **54** (1), 118.
25. H. Feilchenfeld, *Ind. Eng. Chem.*, 1953, **45**, 855.
26. H. J. Solomon, H. Bliss and J. B. Butt, *Ind. Eng. Chem. Fundam.*, 1967, **6**, 325.
27. H. Knözinger, *Angew. Chem. Int. Ed. Engl.*, 1968, **7**, 791.
28. F. H. A. Bolder and H. Mulder, *Appl. Catal. A*, 2006, **300**, 36.

29. R. J. J. Nel and A. de Klerk, *Ind. Eng. Chem. Res.*, 2009, **48**, 5230.
30. F. H. A. Bolder, *Prepr. Pap. Am. Chem. Soc. Div. Pet. Chem.*, 2009 **54** (1), 1.
31. A. de Klerk, *Prepr. Pap. Am. Chem. Soc. Div. Pet. Chem.*, 2008, **53** (2), 105.
32. C. J. Helmers, A. Clark and R. C. Alden, *Oil Gas J.*, 1948, **47** (26), 86.
33. F. H. Bruner, *Ind. Eng. Chem.*, 1949, **41**, 2511.
34. A. de Klerk, *Prepr. Pap. Am. Chem. Soc. Div. Fuel Chem.*, 2009, **54** (1), 116.
35. A. de Klerk and M. J. Strauss, *Prepr. Pap. Am. Chem. Soc. Div. Fuel. Chem.*, 2008, **53** (1), 313.

CHAPTER 8

Catalysis in the Refining of Fischer–Tropsch Syncrude

In Chapter 4 (Section 4.3), the properties of Fischer–Tropsch (FT) syncrude and conventional crude oil were compared and some differences were listed out (Table 4.9).[1] Based on these compositional differences, one may suspect that the refining processes of FT syncrude and crude oil are dissimilar, although the extent of the differences may not be clear. It should emphatically be stated that the differences in composition between FT syncrude and crude oil are meaningful differences and that they significantly influence the catalysis, catalyst selection and refining technologies that can be used.[2] It is possible to refine FT syncrude using a crude oil refining approach, but it results in suboptimal refining and refinery design.[3,4]

The design of a Fischer–Tropsch refinery differs markedly from that of a conventional crude oil refinery. This is illustrated by comparing a generic modern crude oil refinery (Figure 8.1) and a generic modern HTFT refinery (Figure 8.2).[2]

The conversion units that are employed in a Fischer–Tropsch refinery depend on the product slate that is being targeted. From an analysis of commercial FT refineries it has been pointed out that:[3]

1. FT syncrude is best refined to transportation fuels with co-production of chemicals, although it is possible to refine it to only fuels or only chemicals.
2. Refining of HTFT and LTFT syncrude requires different refinery designs, although the same type of conversion units may be applicable.
3. Oxygenates present in FT syncrude have to be dealt with specifically to avoid processing problems.
4. Alkenes give FT syncrude synthetic capability and alkene OLI is a key refining technology.

RSC Catalysis Series No. 4
Catalysis in the Refining of Fischer–Tropsch Syncrude
By Arno de Klerk and Edward Furimsky

Published by the Royal Society of Chemistry, www.rsc.org

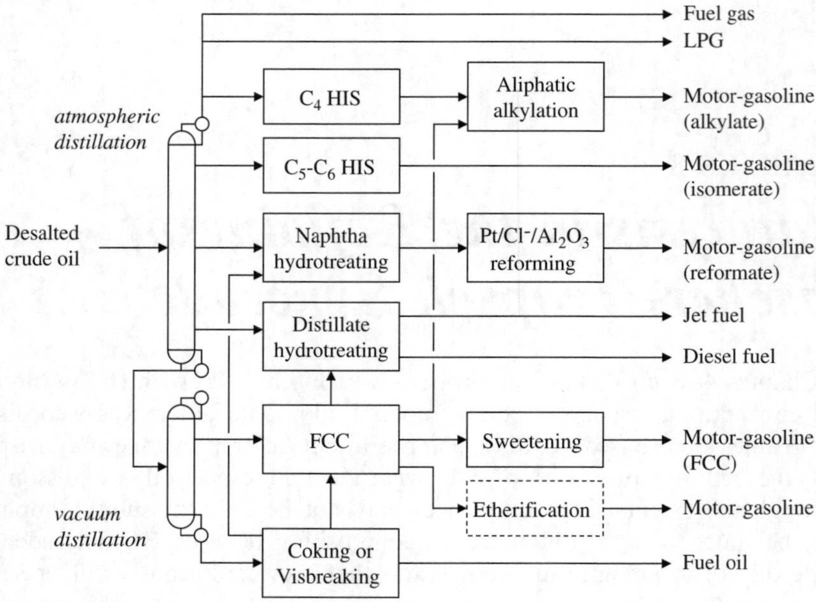

Figure 8.1 Conventional modern crude oil refinery. This is a generic design that does not represent any specific refinery. It illustrates the refining pathways typically needed for the refining of crude oil to on-specification transportation fuels.

Many of the feed peculiarities of FT syncrude and how these influence catalysis have already been discussed. In this chapter, the discussion is broadened to include some of the other types of conversion processes that have not yet been discussed and that may find application in Fischer–Tropsch fuel refineries. More specifically, four catalyst types were identified that play a key role in the production of transportation fuels from FTS:[5]

1. alumina and alumina-rich catalysts, such as bauxite, which are used for double bond IS and deoxygenation and also alcohol dehydration to alkenes and ethers;
2. solid and liquid phosphoric acid catalysts that are used for the OLI of alkenes, alkylation of aromatics with alkenes and ethene hydration;
3. nonacidic Pt/L-zeolite-based catalysts for catalytic reforming of C_6–C_8 naphtha;
4. mildly acidic Pt/SiO_2–Al_2O_3 catalysts for HIS and HCR of distillate, residue and waxes to produce fuels and lubricating oils.

It is important to realise that some of the conversion processes that are ubiquitous in crude oil refineries[6] have poor compatibility with materials from FTS (Table 8.1).[2] Efficient FT refining requires an alternative catalyst selection or even a different type of technology. This does not imply that conventional

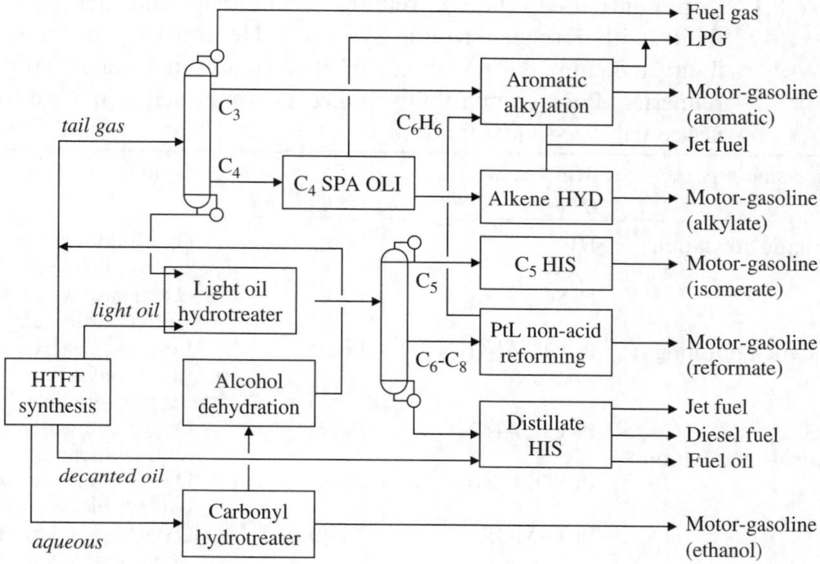

Figure 8.2 High temperature Fischer–Tropsch refinery. This is a generic design that does not represent any specific refinery. It illustrates the refining pathways typically needed for the refining of HTFT syncrude to on-specification transportation fuels. In this design, the tail gas is not cryogenically separated.

crude oil refining technologies cannot be used in conjunction with FT refining, but in that case the designs would be suboptimal.

Good compatibility indicates that the technology and catalyst combination has a benefit of being used with feed materials from FTS. Neutral compatibility indicates that it can be used with FT feed, but that FT feed has no advantage or drawbacks compared with crude oil-derived feed for the application. Poor compatibility reflects either FT feed incompatibility or a mismatch between the aim of the technology and nature of the feed, even though the technology may be compatible with the feed. For example, FCC can be applied successfully for the conversion of LTFT waxes to produce light alkenes (Section 6.3.4), but it is wasteful to select LTFT technology in combination with FCC, because it performs carbon rejection on an already hydrogen-rich feed.

A different set of refining technologies and catalysts can be found that have good compatibility with FT syncrude (Table 8.2).[2,5] With some exceptions, these technologies would not normally be considered for crude oil refining. The four catalyst types highlighted earlier figure prominently in this list.

Generally, the good compatibility with FT feed as opposed to a crude oil-derived feed can be ascribed to one or more of the following factors:

1. compatibility with oxygenates;
2. special alkene transformations;

Table 8.1 Commonly used crude oil refining technologies and their compatibility with Fischer–Tropsch syncrude. Neutral to good compatibility indicates the possibility of efficient use in Fischer–Tropsch refineries. Poor compatibility indicates some inefficiency and not necessarily feed incompatibility.

Conversion process	*Main catalysts for crude oil refining*	*FT compatibility*	*Comments*
Aliphatic alkylation	HF	Poor	Oxygenate/water sensitive; little butanes
	H_2SO_4	Poor	Oxygenate/water sensitive; little butanes
Catalytic reforming	$Pt/Cl^-/Al_2O_3$-based	Poor	Low $N+2A$ feed content; oxygenate/water sensitive
C_5/C_6 hydroisomerisation	$Pt/Cl^-/Al_2O_3$	Poor	Oxygenate/water sensitive; olefinic feed
	$Pt/SO_4^{2-}/ZrO_2$	Poor	Oxygenate/water sensitive; olefinic feed
	Pt/H-MOR	Neutral	Oxygenate tolerant; compatible with FT feed
Alkene etherification	Acidic resin	Neutral	Some oxygenate inhibition
Alkene oligomerisation	SPA	Good	Mechanism favours FT *n*-alkenes
	ASA	Good	Oxygenate tolerant
Sweetening	Co phthalocyanine	Irrelevant	No sulfur in FT syncrude
Hydrotreating	Sulfided NiMoW/Al_2O_3	Neutral	Adding S to S-free feed; oxygenate tolerant
Hydrocracking	Sulfided NiMoW/SiO_2–Al_2O_3	Neutral	Adding S to S-free feed; oxygenate tolerant
Fluid catalytic cracking	USY	Poor	C-rejection of H-rich feed; compatible with FT feed
Visbreaking of residue	No catalyst	Irrelevant	Residue viscosity comparatively low
Coking	No catalyst	Poor	C-rejection of H-rich feed; low Conradson carbon

3. benefit of feed linearity;
4. sulfur-free nature of the feed.

Many of the key technologies listed in Tables 8.1 and 8.2 have already been discussed in detail in Chapter 5, namely HYD, HIS, IS, OLI and HCR. The remainder of the chapter will focus on those conversion technologies and

Table 8.2 Conversion processes and catalysts that have been identified with good Fischer–Tropsch compatibility.

Conversion process	*Catalyst*	*Application*
Double bond isomerisation	Alumina	Usefulness limited in modern FT refineries
Pentene skeletal isomerisation	Alumina	Feed for etherification, but 10–15% side-products
Alkene di-/oligomerisation	H-ZSM-5	High cetane number, low-density distillate
	ASA	Low cetane number, high-density distillate
	SPA	Good 'alkylate' from *n*-butenes; good jet fuel component
	Thermal	Lubricating oil from *n*-alkenes
Aromatic alkylation	SPA	Benzene reduction; synthetic jet fuel with olefinic feed
Hydrocracking	Unsulfided $Pt/SiO_2–Al_2O_3$	High cetane number distillate; jet fuel component
Thermal cracking	Thermal	Limited use in modern FT refineries, except for chemicals and lubricating oil
Catalytic reforming	Nonacidic Pt/L-zeolite	Aromatic motor gasoline; platform for aromatics chemicals
Alcohol dehydration	Alumina	Aqueous product refining; fuel ethers

catalysts that are relevant to FT refineries, but that have not yet been covered in detail, namely catalytic reforming, aromatic alkylation, alcohol dehydration and etherification. These technologies are not less important, but they are only relevant when a Fischer–Tropsch refinery is designed to produce final on-specification transportation fuels. These technologies are not found in refineries of the SMDS type or upgraders such as Oryx GTL, which produce blending components rather than final fuels.[3]

8.1 Catalytic Reforming

Catalytic reforming technology was developed to upgrade low-octane naphtha to a high-octane motor gasoline blending component that is rich in aromatic compounds.[7] Hydrogen is co-produced during this process and hydrogen production has become equally important,[8] due to the increasing pressure on refineries to increase their hydroconversion severity.[9]

There are two classes of catalytic reforming catalysts that have markedly different responses to the nature of the feed material, namely Pt-promoted chlorinated alumina ($Pt/Cl^-/Al_2O_3$) and nonacidic Pt-promoted L-zeolite (Pt/L-zeolite). The former is a bifunctional catalyst, containing acid and metal sites, whereas the latter is a monofunctional catalyst, containing only metal sites.

The reaction network is fairly complex and multiple reaction pathways are possible.[10] The main reaction classes found during catalytic reforming are as follows:

1. DeHYD–HYD, which is the addition or removal of hydrogen by metal sites.
2. HIS, which results in a rearrangement of the skeletal structure and requires both metal and acid sites.
3. Aromatisation that involves deHYD of a cycloalkane or dehydrocyclisation of an alkane, whereby the ring structure is created during deHYD. Both processes occur on metal sites.
4. Cracking, HCR and hydrogenolysis, which are various ways of reducing the carbon chain length of the product by either acid- or metal-catalysed C–C bond scission. These are undesirable side-reactions during catalytic reforming.

The rate-limiting step in catalytic reforming is alkane activation, which is an endothermic process.[11] Catalytic reforming is conducted at high temperature in order to provide energy for alkane activation and because aromatics formation is thermodynamically favoured by high temperature. By increasing the temperature (severity of operation), the octane number of the final product can be increased and operation of catalytic reformers is typically controlled in such a way that the product (reformate) is of sufficiently high octane number to meet motor gasoline octane demand in the refinery.

8.1.1 Reforming Over $Pt/Cl^-/Al_2O_3$ Catalysts

Catalytic reformers found in conventional crude oil refineries employ $Pt/Cl^-/Al_2O_3$-based bifunctional catalysts. The first Pt-based reforming process to be used for refining was the UOP Platforming-process that came on-stream in 1949.[12] In industry, the term 'platforming' is often colloquially used to refer to catalytic naphtha reforming. Typical operating ranges are 490–525 °C and 1.4–3.5 MPa for semi-regenerative reforming and 525–540 °C and 0.3–1.0 MPa for reformers with continuous catalyst regeneration (CCR).[7,8,13]

In $Pt/Cl^-/Al_2O_3$-based catalysts, the Pt metal can be stabilised by the addition of a second metal. In most cases, the second metal is rhenium, tin or iridium. The support material is acidified by co-feeding chloroalkanes, such as CCl_4 or C_2Cl_4, which also retards Pt agglomeration and aids Pt redispersion during regeneration.[8] Acidity is required to catalyse IS reactions, such as the conversion of alkylcyclopentanes to cyclohexane species, which can then readily be converted into aromatics by deHYD. The chlorination of the catalyst necessitates removal of all water and oxygenates from the feed. Water and oxygenates can react with the chlorided alumina support to produce hydrochloric acid. Hydrochloric acid is corrosive. Dechlorination is accompanied by loss of strong acidity, and it is therefore a source of catalyst deactivation.

Fischer–Tropsch-derived feed invariably contains oxygenates, which is a drawback when employing chlorinated catalysts.

As with all noble metal catalysts, the $Pt/Cl^-/Al_2O_3$-based reforming catalysts are also sensitive to sulfur poisoning and the feed should preferably contain no sulfur. In this respect, feed derived from FTS has an advantage, although some of this advantage may be eroded when a sulfided base metal HYD catalyst is employed for feed pretreatment.

The composition of the feed plays an important role in determining the severity of operation that will be required to achieve a desired reformate octane number. Cycloalkanes (naphthenes) react much faster than acyclic alkanes, since cycloalkanes require fewer reaction steps to form a six-membered ring structure that can be directly dehydrogenated to produce an aromatic. Feed materials that contain a high concentration of cycloalkanes are called rich naphthas. The richness of a naphtha feed is expressed by the number $N+2A$, where N refers to the percentage of naphthenes in the feed and A refers to the percentage of aromatics in the feed. Rich naphthas require less severe conditions than lean naphthas to obtain the same reformate octane number. FTS-derived naphtha contains little cycloalkanes and aromatics, with HTFT naphtha having more cyclic material than LTFT naphtha that essentially contains mainly acyclic material. HTFT naphtha, which contains some aromatics and cycloalkanes, typically has an $N+2A$ number of less than 30,[14] whereas that of LTFT naphtha approaches zero. Such lean naphthas make very poor feed material for $Pt/Cl^-/Al_2O_3$-based reforming. The reforming of FTS-derived naphtha therefore has a higher gas make and lower aromatics yield compared with the reforming of crude oil-derived feed at similar conversion.[7] It has been reported that reforming of HTFT naphtha with an end point of 180 °C resulted in a liquid yield of only 70–75% at a research octane number (RON) of around 80,[14] clearly not attractive values.

The carbon chain length of the hydrocarbons in the feed also affects reforming. Alkane reactivity for catalytic reforming over $Pt/Cl^-/Al_2O_3$ catalysts increases in the order $C_6 < C_7 \ll C_8 \approx C_9$ and heavier. In general, C_6 and C_7 compounds are not considered a desirable feed for standard catalytic reforming due to their low reactivity. Inclusion of C_6 material in the feed is also undesirable due to its high benzene selectivity. It should therefore be clear that catalytic reforming over $Pt/Cl^-/Al_2O_3$ catalysts has poor compatibility with Fischer–Tropsch syncrude. Although standard catalytic reforming is employed industrially with feed derived from FTS, it is not the reforming technology of choice for FT refining.

8.1.2 Reforming Over Nonacidic Pt/L-Zeolite Catalysts

Nonacidic Pt/L-zeolite reforming catalysts are a more recent development than $Pt/Cl^-/Al_2O_3$ catalysts.[15] The two main technologies based on nonacidic Pt/L-zeolite catalysts are the Aromax process of Chevron Phillips Chemical Company[16,17] and the RZ-Platforming-process of UOP.[18]

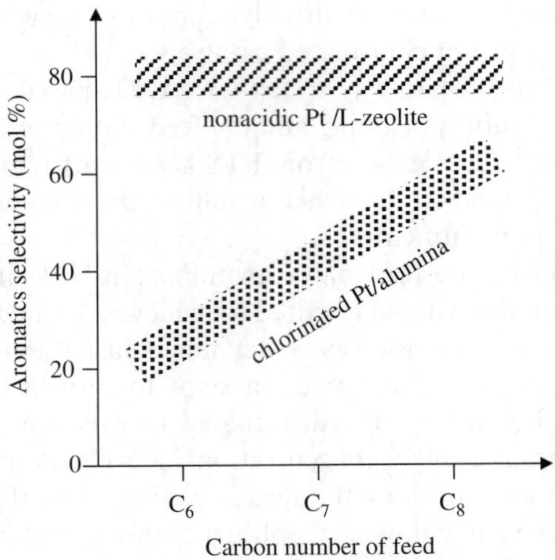

Figure 8.3 Reforming selectivity to aromatics under comparable conditions for different feed carbon numbers over nonacidic Pt/L-zeolite- and chlorinated Pt/alumina-based reforming catalysts.

On account of the very high aromatics selectivity of nonacidic Pt/L-zeolite-based catalysts with C_6–C_8 naphtha and especially with C_6–C_7 *n*-alkane feed, this type of reforming technology is mainly employed for chemicals production. It is also immediately apparent that linear hydrocarbon-rich naphtha from FTS should have good compatibility with this type of reforming. With increasing carbon number, the performance advantage of nonacidic Pt/L-zeolite over standard $Pt/Cl^-/Al_2O_3$ reforming becomes less (Figure 8.3).[18] Nonacidic Pt/L-zeolite-based reforming is consequently not normally used with C_9 and heavier feed materials.

Pt/L-zeolite reforming catalysts have no acidity and any residual acidity in the L-zeolite structure is typically removed by ion exchange with potassium and/or barium. By doing so, acid-catalysed side-reactions are eliminated and the mechanism is dependent on metal site-catalysed conversion only. The very high (around 90%) aromatics selectivity of linear hydrocarbon conversion over L-zeolite is ascribed to the shape selectivity of the zeolite structure, which ensures end-on attachment of the molecule (Figure 8.4).[19] End-on attachment is a prerequisite for 1,6-ring closure to selectively produce benzene from *n*-hexane and toluene from *n*-heptane.

The operating conditions of Pt/L-zeolite reforming are similar to those of reforming over chlorinated Pt/alumina-based catalysts, but it requires no chloroalkane addition. Nonacidic Pt/L-zeolite catalysts are extremely sensitive to sulfur poisoning and sulfur in the feed must be removed to levels below $0.05\ \mu g\ g^{-1}$.[20] Sulfur presents no difficulty when this technology is employed with Fischer–Tropsch-derived feed, since it is already sulfur free.

The effect of oxygenates on a PtK/L-zeolite has been studied and it was reported that oxygenates and CO suppressed conversion, whereas water had no

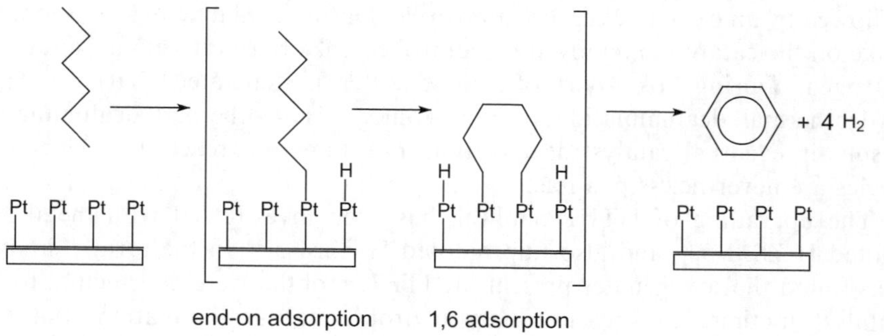

Figure 8.4 End-on adsorption of *n*-alkanes on Pt/L-zeolite that leads to 1,6-ring closure and aromatisation.

effect.[21] This indicated that FT feeds, even feed materials containing some oxygenates, can be used in conjunction with Pt/L-zeolite reforming. Overall, Fischer–Tropsch syncrude has good compatibility with nonacidic Pt/L-zeolite reforming on account of its linearity and the absence of sulfur.[22,23]

8.1.3 Aromatisation Over Metal-promoted ZSM-5 Catalysts

The aromatisation of C_3–C_5 hydrocarbons is related to reforming, but such units are generally not associated with refineries. The aim of aromatisation is mainly to convert normally gaseous alkanes to aromatic-rich liquid hydrocarbons for chemical production. Like reforming, an added advantage is the co-production of hydrogen.

It has been shown that light alkanes can be activated and aromatised on H-ZSM-5, without the rapid catalyst deactivation that is seen on many other acidic zeolites. The geometric constraints imposed by the ZSM-5 zeolite structure cause it to have a lower coking tendency than Beta- and Y-zeolites. In addition, ZSM-5 has a much larger coke capacity than less coking zeolites. More coke lay-down is therefore required before complete deactivation occurs.[24]

Over H-ZSM-5, hydrogen rejection occurs by hydrogen transfer to alkenes (forming alkanes), which limits the aromatics yield that can be obtained. When a metal is added to produce a bifunctional catalyst, the hydrogen can be desorbed as molecular hydrogen (H_2) and the aromatics yield is substantially increased.[25] Commercial aromatisation processes therefore employ bifunctional catalysts that are either based on Zn/ZSM-5 (for example, the Alpha process of Asahi) or Ga/ZSM-5 (for example, the Cyclar process of BP).

The operating conditions of metal promoted ZSM-5-based aromatisation are similar to those of catalytic naphtha reforming and are in the range 450–520 °C and < 1 MPa pressure. Aromatisation processes are characterised by periodic operation, with each production cycle (in the order of 2 days) being followed by an oxidative regeneration cycle. During oxidative regeneration, the coke on the catalyst is removed by controlled coke burn-off with air diluted in nitrogen. During coke burn-off, some water is generated that can cause hydrothermal dealumination of the zeolite.[26] Hydrothermal dealumination results in eventual catalyst deactivation, but numerous reaction–regeneration cycles are nevertheless possible.

The upgrading of HTFT naphtha has been investigated with metal-promoted H-ZSM-5,[27] and also unpromoted H-ZSM-5.[28] In the former study it was found that oxygenates present in HTFT naphtha were detrimental to the catalyst lifetime, causing not only hydrothermal dealumination, but also selective loss of the metal. The use of metal-promoted ZSM-5 with straight-run naphtha range material from FTS is therefore not recommended. Conversely, the light hydrocarbons from FTS are substantially oxygenate free and aromatisation of alkanes and alkane–alkene mixtures may be considered, as was suggested previously.[29]

8.2 Aromatic Alkylation

Aromatic alkylation is not normally associated with refining, but rather with petrochemical production. However, in a Fischer–Tropsch refinery it becomes an indispensable technology when nonacidic Pt/L-zeolite-based reforming technology is employed. It is likewise a valuable technology if FTS is used in a coal-to-liquids facility with low-temperature coal gasification technology. In both instances the refinery has to process benzene-rich materials. In the future, benzene alkylation may also become a more prominent crude oil refining technology. The increasingly stringent regulation of the benzene content in motor gasoline is likely to necessitate some refinery intervention to reduce benzene in final motor gasoline. Of the technologies for refinery benzene reduction, benzene alkylation has the advantage that it retains the octane value of benzene in the final motor gasoline.[30,31]

There are various commercial processes for the acid-catalysed alkylation of benzene with either ethylene or propylene. The catalysts most often used are solid phosphoric acid (SPA) and zeolite-type materials such as H-ZSM-5 (Mobil-Badger), H-MCM-22 (Mobil-Ratheon/Mobil-Badger), HY-zeolite (CDTech) and modified H-Beta-zeolite (Enichem).[32,33] These processes all operate at high aromatic-to-alkene ratios to minimise alkene OLI as a side-reaction.

In a Fischer–Tropsch refinery, where alkenes are more abundant, a different operating philosophy is possible. Alkene OLI and benzene alkylation can be combined into a single refining step to reduce refinery benzene levels and produce motor gasoline and jet fuel blending components. It has been reported that >80% conversion of benzene to alkylated benzenes was obtained over SPA during industrial testing (Table 8.3).[34] It has also been reported that combined propene OLI and aromatic alkylation is able to produces a synthetic Jet A-1 jet fuel.[35] Propene is the preferred alkene feed for alkylation over SPA,

Table 8.3 Aromatic alkylation combined with alkene oligomerisation. Test results obtained during industrial operation in an SPA-catalysed OLI process at 180–210 °C, 3.8 MPa and LHSV 1.3 h^{-1}, operated in 'diesel mode' with olefinic naphtha recycle.

	Industrial HTFT operation	
Description	*OLI only*[a]	*OLI and alkylation*[b]
Conversion of alkenes in feed (%)		
Propene	>95	98
Butenes	>95	97
Benzene	–	85
Toluene	–	81
Unhydrogenated motor gasoline		
RON	96	95.2
MON	82.3	81.7
Hydrogenated motor gasoline		
RON	71	73.0
MON	74.4	73.0

[a]Typical values obtained in an industrial unit operated with mixed C_3–C_4 HTFT alkene feed.
[b]Hydrotreated coal tar naphtha (52% aromatics) co-fed in a 1:8 mass ratio with fresh feed.

although butene-rich feeds can also be employed. When butene is used as the alkylating alkene, the per pass benzene conversion is lower.

Aromatic alkylation over SPA is recommended over zeolite-catalysed aromatic alkylation in Fischer–Tropsch refineries, and also in conventional crude oil refineries, for the following reasons:

1. Benzene alkylation over SPA requires a lower operating temperature than that over zeolites.
2. Low aromatic-to-alkene ratio operation is possible without affecting OLI performance, which enables benzene alkylation to be performed in existing SPA-based OLI units.
3. Multiple alkylation is low over SPA even when operating at low aromatic-to-alkene ratios.
4. No subsequent transalkylation reactor is required.
5. SPA is more resistant to feed impurities than zeolite catalysts.

8.3 Alcohol Dehydration to Alkenes

The dehydration of alcohols to alkenes is an important synthetic fuels refining technology. In many synthesis gas conversion technologies, alcohols are primary products, *e.g.* syngas-to-methanol[36] and FTS.

'Methanol-to-olefins' (MTO) conversion is a well-known application of H-ZSM-5 catalysts.[37] It is a key refining step in synthetic fuel facilities based on methanol and may also have application in FTS for upgrading the light alcohols in the aqueous product (Section 7.2). In this respect, alcohol dehydration can simplify the aqueous product refinery,[38,39] as can be seen from the generic HTFT refinery design shown in Figure 8.2. By converting the aqueous alcohol mixture into alkenes, the alkenes can be easily separated from the water and co-processed with the rest of the FT alkenes.

Industrially, the catalyst that is most often employed for alcohol dehydration is alumina.[40] Alumina is stable in the presence of large amounts of water under the operating conditions required for dehydration. It has also been pointed out that alumina is well suited for the conversion of a wide range of materials from FTS (see also Chapter 7).[5]

Alcohol dehydration is endothermic and reversible, with the equilibrium favouring dehydration over hydration.[41,42] The water that is produced during the dehydration reaction has a small impact on the conversion and water is often co-fed with the alcohols to reduce the adiabatic temperature decrease during dehydration. Co-feeding water with the alcohols has the additional benefit of diluting the surface concentration of the alcohols and the alkenes formed during the conversion, thereby reducing side-reactions.[43] This type of operation has been practised on industrial scale with mixed HTFT alcohols employing η-alumina as catalyst, and also with HTFT-derived 1-octanol using γ-alumina as catalyst. Some studies in support of these applications and other possible Fischer–Tropsch applications of alcohol dehydration to alkenes have been published.[38,44–47]

It should be noted that the reverse reaction of alcohol dehydration, namely alkene hydration, is also relevant to Fischer–Tropsch refining. Ethene hydration to ethanol is a useful way to convert ethene into a transportable product when the FT refinery is not close to a petrochemical market. The ethanol thus produced can also be used as a motor gasoline blending component.

The hydration of ethene to ethanol is a commercial process with high ethene recycle due to the unfavourable hydration equilibrium. It is a phosphoric acid-catalysed process.[48] The application of ethene hydration in a FT refinery has one additional advantage, namely that the side-products can be co-processed with the FT aqueous product, making it considerably cheaper than a stand-alone process.

Another application of alcohol dehydration is the partial dehydration of alcohols to ethers, but this will be discussed separately (Section 8.4.2).

8.4 Etherification

8.4.1 Etherification of Alkenes with Alcohols

Ethers produced by the reaction of short-chain alcohols with short-chain branched alkenes exhibit good motor gasoline properties (Table 8.4).[49] In countries that allow the use of fuel ethers, these compounds are commonly employed as high-octane blending components that provide a convenient way to improve motor gasoline quality. Etherification of C_5 and C_6 alkenes with methanol is practised industrially using HTFT feed.[50]

Table 8.4 Blending octane numbers and vapour pressure (P_{vap}) at 37.8 °C of commonly considered fuel ethers.

Compound	*RON*	*MON*	P_{vap} *(kPa)*
2-Methoxy-2-methylpropane (MTBE)	118	101	55
2-Ethoxy-2-methylpropane (ETBE)	118	101	40
2-Methoxy-2-methylbutane (TAME)	115	100	25
2-(1-Methylethoxy)propane (DIPE)	110	97	34

Etherification is acid catalysed and conversion is equilibrium limited, with ether formation being favoured by low temperature.[51] The catalyst that is most often used for etherification is Amberlyst 15, a sulfonic acid-exchanged divinylbenzene–styrene copolymer resin. Other acidic resin catalysts and zeolites are also used.[49,52]

The etherification process has to be operated with an excess of alcohol in order to reduce alkene OLI, which is an acid-catalysed side-reaction (see also Section 5.1.1.7). Various oxygenates can inhibit the etherification reaction and can also participate in side-reactions, often forming water.[53] Water is known to inhibit conversion over acidic resin catalysts.[54] Etherification of alkenes with alcohols over silica–alumina-based materials is less common, although some work in this field has been reported.[55,56]

8.4.2 Etherification of Alcohols

Acidic resin catalysts have been successfully employed for ether synthesis from 1-pentanol and 1-hexanol.[57,58] These longer chain ethers can be employed as cetane enhancers in diesel fuel. The use of silica–alumina-based catalysts for alcohol etherification has also been reported. Ether yields ranging from 30 to 75% were obtained at 200 °C from C_2–C_8 alcohols over ion-exchanged montmorillonites.[59] Various zeolites have likewise been tested for alcohol etherification reactions.[60,61] The complete dehydration of the alcohols to the corresponding alkenes is usually the dominant side-reaction.

It has been suggested that the alcohols in LTFT naphtha can be converted into linear fuel ethers to improve the overall yield and the quality of the distillate from LTFT refining. The reaction network for the conversion of C_5–C_{12} alcohols over η-alumina was studied in the operating rang 250–350 °C, 0–4 MPa and WHSV 1–4 h^{-1}.[62] The main products were the corresponding linear ethers and linear 1-alkenes. Under unoptimised conditions, the highest ether yield was 54% and it was obtained by conversion at 300 °C, 1 MPa and WHSV 1 h^{-1}. The main side-products were aldehydes and alkene dimers.

Dehydration over alumina occurred predominantly on Lewis acid sites, with acid-catalysed side-reactions, such as dimerisation, taking place over strong acid sites. Dehydrogenation took place over basic and/or redox sites. It was reported that dehydration to produce 2-alkenes was *cis*-selective and did not occur by Brønsted acid-catalysed double bond IS, but rather by dehydration–hydration–dehydration over Lewis acid sites.[62]

8.5 Other Fischer–Tropsch-related Oxygenate Conversions

Some oxygenate conversions were investigated with the purpose of resolving specific FT refining challenges, but have not yet found their way into conceptual refinery designs or chemical production processes. These cannot be classified as commercial conversion technologies yet, but it is worthwhile discussing the catalysis that may find application in the future.

8.5.1 Esterification of Carboxylic Acids

Esterification is a well-known and commercially practised conversion technology in which both homogeneous and heterogeneous catalysts are employed.[63] The subsequent discussion is limited to the application of esterification for the conversion carboxylic acids at low concentration (less than 2%) in Fischer–Tropsch syncrude.

Aliphatic carboxylic acids are primary products from FTS. In the naphtha range, the short-chain carboxylic acids may cause corrosion in processing units and in chemical extraction processes. For chemicals products, carboxylic acids must be removed before solvent extraction with basic compounds can be performed.[64] When employing FTS-derived material as feed for hydroformylation, carboxylic acids must also be removed, because they facilitate unwanted reactions and affect the activity of Rh-based hydroformylation catalysts.[65] Short-chain carboxylic acids also increase the corrosiveness of fuels and must be removed from motor gasoline and jet fuel and in a crude oil refining context esterification has been suggested for the removal of acids.[66]

Two classes of catalysts were evaluated for esterification of carboxylic acids in a Fischer-Tropsch mixture, namely metal oxide catalysts and strong acids:[67]

1. The metal oxide catalysts tested were WO_3 precipitated on Al_2O_3, containing 20–30 mass% WO_3, and MoO_3 on alumina, containing about 15 mass% MoO_3 and 3% sulfate. It was reported that maximum carboxylic acid conversion was reached at about 210 °C and decreased slightly at higher temperatures. No catalyst deactivation was observed over a period of 7 days of continuous operation and the conversion was close to the equilibrium conversion.
2. The strong acid catalysts tested were Nafion NR50, a perfluorinated sulfonic acid resin, and the homogeneous catalyst *p*-toluenesulfonic acid. The strong acids were able to esterify the carboxylic acids at 80 °C (compared with the 210 °C of the metal oxide catalysts).

By employing this type of conversion, the carboxylic acid content in HTFT distillate could be reduced from 12 to less than 1 mg KOH g^{-1}.

8.5.2 Aromatisation of Carbonyls

Repeated aldol condensation of aldehydes and ketones followed by dehydration leads to the formation of aromatics.[68] About one-third of the oxygenates in the aqueous product from fluidised bed Fe-HTFT synthesis are carbonyl compounds. Acid catalysis can be employed to convert the mixed oxygenates (alcohols and carbonyls) to alkenes and aromatics in a single step.[39] This has been discussed in Section 7.2.

References

1. A. de Klerk, *Green Chem.*, 2007, **9**, 560.
2. A. de Klerk, *Green Chem.*, 2008, **10**, 1249.
3. A. de Klerk, *Prepr. Pap. Am. Chem. Soc. Div. Pet. Chem.*, 2008, **53** (2), 105.
4. A. de Klerk, in *Advances in Fisher Tropsch Synthesis, Catalysts and Catalysis*, ed. B. H. Davis and M. L. Occelli, Taylor & Francis (CRC Press), Boca Raton, FL, 2009, p. 331.
5. A. de Klerk, *Prepr. Pap. Am. Chem. Soc. Div. Fuel Chem.*, 2009, **54** (1), 116.
6. P. Leprince (ed.), *Petroleum Refining. Volume 3. Conversion Processes*, Technip, Paris, 2001.
7. M. Lapinski, L. Baird and R. James, in *Handbook of Petroleum Refining Processes*, ed. R. A. Meyers, McGraw-Hill, New York, 2004, p. 4.3.
8. G. Martino, in *Petroleum Refining. Volume 3. Conversion Processes*, ed. P. Leprince, Technip, Paris, 2001, p. 101.
9. J. J. Alves and G. P. Towler, *Ind. Eng. Chem. Res.*, 2002, **41**, 5759.
10. B. H. Davis, *Catal. Today*, 1999, **53**, 443.
11. F. Zaera, *Appl. Catal. A*, 2002, **229**, 75.
12. M. J. Sterba and V. Haensel, *Ind. Eng. Chem. Prod. Res. Dev.*, 1976, **15**, 2.
13. R. L. Peer, R. W. Bennett, D. E. Felch and E. von Schmidt, *Catal. Today*, 1993, **18**, 473.
14. J. S. Swart, G. J. Czajkowski and R. E. Conser, *Oil Gas J.*, 1981, **79** (35), 62.
15. J. R. Bernard and J. Nury, *US Pat.*, 4104320, 1978.
16. T. R. Hughes, R. L. Jacobson, P. W. Tamm and W. C. Buss, *Stud. Surf. Sci. Catal.*, 1986, **28**, 725.
17. T. R. Hughes, R. L. Jacobson and P. W. Tamm, *Stud. Surf. Sci. Catal.*, 1988, **38**, 317.
18. J. D. Swift, M. D. Moser, M. B. Russ and R. S. Haizmann, *Hydrocarbon Technol. Int. Q.*, 1995, Autumn, 86.
19. S. J. Tauster and J. J. Steger, *J. Catal.*, 1990, **125**, 387.
20. T. Fukunaga and V. Ponec, *J. Catal.*, 1995, **157**, 550.
21. M. E. Dry, R. J. Nash and C. T. O'Connor, in *Proceedings of the 12th International Zeolite Conference*, Baltimore, 1998, p. 2557.
22. M. E. Dry, *Appl. Catal. A*, 1996, **138**, 319.
23. M. E. Dry, *Appl. Catal. A*, 2004, **276**, 1.
24. P. C. Mihindou-Koumba, H. S. Cerqueira, P. Magnoux and M. Guisnet, *Ind. Eng. Chem. Res.*, 2001, **40**, 1042.

25. T. Mole, J. R. Anderson and G. Creer, *Appl. Catal.*, 1985, **17**, 141.
26. A. de Lucas, P. Canizares, A. Durán and A. Carrero, *Appl. Catal. A*, 1997, **154**, 221.
27. A. de Klerk, in *Catalysis in Application*, ed. S. D. Jackson, J. S. J. Hargreaves and D. Lennon, Royal Society of Chemistry, Cambridge, 2003, p. 24.
28. N. N. Madikizela-Mnquanqeni, *Catal. Today*, 2009, **143**, 126.
29. L. P. Dancuart, R. de Haan and A. de Klerk, *Stud. Surf. Sci. Catal.*, 2004, **152**, 482.
30. A. R. Goezler, A. Hernandez-Robinson, S. Ram, A. A. Chin, M. N. Harandi and C. M. Smith, *Oil Gas J.*, 1993, **91** (13), 63.
31. A. de Klerk and R. J. J. Nel, *Energy Fuels*, 2008, **22**, 1449.
32. T. F. Degnan Jr, C. M. Smith and C. R. Venkat, *Appl. Catal. A*, 2001, **221**, 283.
33. C. Perego and P. Ingallina, *Catal. Today*, 2002, **73**, 3.
34. T. M. Sakuneka, R. J. J. Nel and A. de Klerk, *Ind. Eng. Chem. Res.*, 2008, **47**, 7178.
35. T. M. Sakuneka, A. de Klerk, R. J. J. Nel and A. D. Pienaar, *Ind. Eng. Chem. Res.*, 2008, **47**, 1828.
36. P. J. A. Tijm, F. J. Waller and D. M. Brown, *Appl. Catal. A*, 2001, **221**, 275.
37. C. D. Chang, *Catal. Rev. Sci. Eng.*, 1983, **25**, 1.
38. R. J. J. Nel and A. de Klerk, *Ind. Eng. Chem. Res.*, 2007, **46**, 3558.
39. R. J. J. Nel and A. de Klerk, *Prepr. Pap. Am. Chem. Soc. Div. Fuel Chem.*, 2009, **54** (1), 118.
40. H. Knözinger, *Angew. Chem. Int. Ed. Engl.*, 1968, **7**, 791.
41. C. S. Cope and B. F. Dodge, *AIChE J.*, 1959, **5**, 10.
42. C. S. Cope, *AIChE J.*, 1964, **10**, 277.
43. M. E. Winfield, in *Catalysis. Volume VII. Oxidation, Hydration, Dehydration and Cracking Catalysts*, ed. P. H. Emmett, Reinhold, New York, 1960, p. 93.
44. K. McGurk, in *Proceedings of the South African Chemical Engineering Congress*, Sun City, 2003, paper P082.
45. F. H. A. Bolder and H. Mulder, *Appl. Catal. A*, 2006, **300**, 36.
46. N. P. Makgoba, T. M. Sakuneka, J. G. Koortzen, C. van Schalkwyk, J. M. Botha and C. P. Nicolaides, *Appl. Catal. A*, 2006, **297**, 145.
47. F. H. A. Bolder, A. de Klerk and J. L. Visagie, *Ind. Eng. Chem. Res.*, 2009, **48**, 3755.
48. L. M. Elkin, W. S. Fong and P. L. Morse, *Synthetic Ethanol and Isopropanol: Process Economics Program Report 53A*, SRI, Menlo Park, CA, 1979.
49. P. Travers, in *Petroleum Refining. Volume 3. Conversion Processes*, ed. P. Leprince, Technip, Paris, 2001, p. 291.
50. A. de Klerk, *Ind. Eng. Chem. Res.*, 2003, **43**, 6349.
51. J. F. Izquierdo, F. Cunill, M. Vila, J. Tejero and M. Iborra, *J. Chem. Eng. Data*, 1992, **37**, 339.

52. J. G. Goodwin Jr, S. Natesakhawat, A. A. Nikolopoulos and S. Y. Kim, *Catal. Rev. Sci. Eng.*, 2002, **44**, 287.
53. D. Smook and A. de Klerk, *Ind. Eng. Chem. Res.*, 2006, **45**, 467.
54. R. Bringué, J. Tejero, M. Iborra, J. F. Izquierdo, C. Fité and F. Cunill, *Top. Catal.*, 2007, **45**, 181.
55. S. Wang and J. A. Guin, *Energy Fuels*, 2001, **15**, 666.
56. S. Wang and J. A. Guin, *React. Kinet. Catal. Lett.*, 2002, **75**, 169.
57. R. Bringué, J. Tejero, M. Iborra, J. F. Izquierdo, C. Fité and F. Cunill, *Ind. Eng. Chem. Res.*, 2007, **46**, 6865.
58. R. Bringué, J. Tejero, M. Iborra, C. Fité, J. F. Izquierdo and F. Cunill, *J. Chem. Eng. Data*, 2008, **53**, 2854.
59. J. A. Ballantine, M. Davies, I. Paterl, J. H. Purnell, M. Rayanakorn, K. J. Williams and J. M. Thomas, *J. Mol. Catal.*, 1984, **26**, 37.
60. I. Hoek, T. A. Nijhuis, A. I. Stankiewicz and J. A. Moulijn, *Appl. Catal. A*, 2004, **266**, 109.
61. J. Tejero, C. Fité, M. Iborra, J. F. Izquierdo, F. Cunill and R. Bringué, *Microporous Mesoporous Mater.*, 2009, **117**, 650.
62. R. J. J. Nel and A. de Klerk, *Ind. Eng. Chem. Res.*, 2009, **48**, 5230.
63. Y. Liu, E. Lotero and J. G. Goodwin Jr, *J. Catal.*, 2006, **242**, 278.
64. T. Hahn, in *Proceedings of the South African Chemical Engineering Congress*, Sun City, 2003, paper P013.
65. E. Mieczynska, A. M. Trzeciak and J. J. Ziolkowski, *J. Mol. Catal.*, 1993, **80**, 189.
66. Y. Z. Wang, X. Y. Sun, Y. P. Liu and C. G. Liu, *Energy Fuels*, 2007, **21**, 941.
67. F. H. A. Bolder, *Prepr. Pap. Am. Chem. Soc. Div. Pet. Chem.*, 2009, **54** (1), 1.
68. G. S. Salvapati, K. V. Ramanamurthy and M. Janardanarao, *J. Mol. Catal.*, 1989, **54**, 9.

CHAPTER 9

Commercial Products from Fischer–Tropsch Syncrude

Various products made from Fischer–Tropsch syncrude were discussed in relation to catalysis in previous chapters. The discussions were organised based on catalysis and feed transformations, not products. This chapter provides an overview of the products that are produced in conjunction with FTS.

Transportation fuels are the most important products from FTS. Usually several upgrading or refining steps are necessary before specifications of commercial products are attained. In addition to transportation fuels, related products and chemicals may be produced. High-quality lubricating base oil that is used for the preparation of lubricants may be prepared from waxy oil and wax. Various chemicals may be directly recovered and purified from the syncrude or co-produced from processes involved in syngas preparation.[1,2] There is a wide range of applications for the FT oxygenates, alkenes and *n*-alkanes, which cover the spectrum from commodities to niche market applications. The aim of the chapter is to provide an overview in the context of catalysis and opportunities for catalysis to upgrade and refine Fischer–Tropsch syncrude and associated co-products.

9.1 Transportation Fuels

High-quality transportation fuels (motor gasoline, aviation fuels and diesel fuel) can be prepared from the gaseous and liquid streams obtained from FTS. The molecular properties of FT syncrude that influence the conversion into fuels the most are the alkene content, the oxygenate content and the linearity of the molecules.

Differences in composition between Fischer–Tropsch syncrudes and conventional crude oils result in different challenges during refining to produce transportation fuels. Despite the high linear hydrocarbon content of syncrude,

RSC Catalysis Series No. 4
Catalysis in the Refining of Fischer–Tropsch Syncrude
By Arno de Klerk and Edward Furimsky

Published by the Royal Society of Chemistry, www.rsc.org

Table 9.1 Comparison of straight run (unrefined) properties of HTFT and LTFT naphtha fractions with those of different Arabian Light crude oil naphtha fractions.

	HTFT	*LTFT*	*Arabian Light crude oil*	
Property	*20–105 °C*	*20–100 °C*	*20–80 °C*	*80–180 °C*
Yield of total (mass%)	30	10	5	15
Density (kg m^{-3})	680	680	660	750
RON clear	68	43	61	24
Sulfur (mass%)	0	0	0.02	0.04
Alkenes (mass%)	85	55	0	0
Aromatics (mass%)	2	0	2	14

from a refining perspective the properties of straight run HTFT syncrude compares favourably with those of a good-quality crude oil (Table 9.1).[3]

The recent literature provides an overview of commercial fuels production from FTS and some of the challenges faced specifically when producing diesel fuel.[4–6]

Specifications for transportation fuels and the local market situation where the FTS facility is located determine the refining requirements. The market requirements will also determine the preferred product distribution, for example, North American markets are more in favour of motor gasoline than diesel fuel, whereas the converse is true in Europe. The subsequent discussion is somewhat biased towards fuels production in South Africa, because it is the only country at present where a substantial volume (20–30%) of the transportation fuels is produced by FTS.[7] The two main South African producers of fuels from FTS are Sasol and PetroSA.

9.1.1 Motor Gasoline

There are four compound classes allowed in motor gasoline, namely alkanes (including cycloalkanes), alkenes, aromatics and oxygenates. Fuel specifications place a limit on all of these compound classes, except for the alkanes. Typical limits are alkenes (maximum 18 vol.%), aromatics (maximum 35 vol.%) and oxygenates (maximum 2.7 mass% as oxygen; 10–15 vol.% as oxygenates). The molecular composition within each class is not regulated, except for the aromatics and oxygenates. The maximum benzene content in the aromatics is regulated and the oxygenate classes that are allowed in motor gasoline are limited to alcohols and ethers. In addition to these limitations, specific limits have been set for sulfur and lead.

Comparing these specifications with the composition of straight run naphtha from both HTFT and LTFT synthesis (Table 9.2),[8] it is clear that the alkene content is much higher and that of aromatics is much lower than allowed. The oxygenates also include compound classes that are undesirable for motor gasoline, such as aldehydes, ketones and short-chain carboxylic acids.

Table 9.2 Composition of the straight run naphtha fractions (C_5–C_{10}) of different Fischer–Tropsch technologies.

	Composition of naphtha (mass%)			
	Fe-HTFT	*Fe-LTFT*		*Co-LTFT*
Compound class	*Fluidised bed*	*Slurry bed*	*Fixed bed*	*Slurry bed*
Alkanes	13	29	60	54
Alkenes	70	64	32	35
Aromatics	5	0	0	0
Oxygenates	12	7	8	11

Table 9.3 Properties of straight run naphtha from German Co-LTFT synthesis.

	Straight run naphtha			
Property	*Normal pressure Co-LTFT*		*Medium pressure Co-LTFT*	
Distillation end point (°C)	150	200	150	200
RON	57	43	38	25
Alkene content (%)	33	–[a]	19	–[a]

[a]Not reported.

Furthermore, the aliphatic hydrocarbons are mainly *n*-alkanes and linear 1-alkenes, which affect the properties of the motor gasoline.

Due to the differences in the compositions of the naphtha fractions from the main types of FTS, motor gasoline production from each type will be considered individually.

9.1.1.1 *Motor Gasoline from Co-LTFT Synthesis*

The bulk of the naphtha fraction obtained from German normal-pressure and medium-pressure Co-LTFT operation was used as blending components in motor gasoline. Due to the hydrogenating nature of the Co-based catalyst, the major compound class in the straight run naphtha was alkanes and the octane number was correspondingly low (Table 9.3).[9] In spite of this fact, the straight run product was not further refined. The only type of catalysis employed for upgrading the lighter fractions to fuels was liquid phosphoric acid OLI of the C_3–C_4 gaseous products. The benefit of further catalytic upgrading was investigated and demonstrated,[10] but not applied industrially.

With the more demanding motor gasoline specifications at present, producers of Co-LTFT-derived naphtha have opted not to refine the low-octane naphtha to motor gasoline. The naphtha resembles the straight run naphtha from a paraffinic crude oil, being rich in linear hydrocarbons. However, with the absence of aromatics and cycloalkanes, the heavy naphtha from Co-LTFT makes a poor feed material for motor gasoline refining. The naphtha obtained from Co-LTFT synthesis in the Shell Middle Distillate Synthesis process in Bintulu, Malaysia, is hydrotreated and sold as *n*-alkanes. The Co-LTFT-derived naphtha from the Sasol Slurry Bed process as applied in the Oryx GTL facility in Ras Laffan, Qatar, is sold as cracker feed material. Good performance has been demonstrated with Co-LTFT naphtha as cracker feed.[11] The current commercial Co-LTFT facilities therefore consider the naphtha as a chemical intermediate and not suitable for fuels refining.

9.1.1.2 Motor Gasoline from Fe-LTFT Synthesis

When comparing the naphtha fractions from LTFT synthesis produced with the same reactor technology (Table 9.2), the higher alkene content of Fe-LTFT naphtha makes it better suited than Co-LTFT naphtha for motor gasoline production. The only commercial facility where Fe-LTFT syncrude was refined to motor gasoline was at Sasol 1 in Sasolburg, South Africa. The original refinery design for Sasol 1 combined the light gases (C_3–C_4) from Fe-LTFT synthesis with those from Fe-HTFT synthesis and co-refined them to motor gasoline.[12,13] The light Fe-LTFT naphtha (C_5–C_7) was the only Fe-LTFT cut that was refined separately to a motor gasoline component. After bauxite treatment, the light Fe-LTFT naphtha had a sufficiently high octane number due to its high alkene content to be employed as a motor gasoline blending component.[14] Bauxite treatment, and how the catalysis of bauxite treatment improves octane number, will be discussed in the next section (see also Section 7.3).

9.1.1.3 Motor Gasoline from Fe-HTFT Synthesis

Generally, the FT primary products such as stabilised light oil (SLO) have a higher octane number than conventional straight run crude oil-derived naphtha having a similar boiling range (Table 9.1). Refining Fe-HTFT naphtha to motor gasoline should therefore in principle be easier than refining crude oil-derived naphtha to motor gasoline.[3]

The Hydrocol process in Brownsville, TX, USA was the first commercial Fe-HTFT-based technology. It employed a fixed fluidised bed as reactor for FTS and the facility had motor gasoline production as its main objective.[15] The naphtha range product was rich in linear 1-alkenes and the straight run octane number could easily be improved by mild catalytic treatment over alumina-rich material that caused double bond IS of the alkenes and some deoxygenation (Table 9.4).[16]

In the final motor gasoline, the bauxite-treated motor gasoline was blended with 'polymer'-gasoline and *n*-butane (Table 9.5).[16] The 'polymer' gasoline was produced by the SPA OLI of the gaseous (C_3–C_4) fraction from Hydrocol FTS. A similar refining strategy was employed for the Fe-HTFT naphtha in the

Table 9.4 Research octane number (RON) and motor octane number (MON) of straight run and bauxite-treated Hydrocol naphtha without and with addition of tetraethyllead (TEL).

Hydrocol naphtha	*TEL (ml gal^{-1})*[a]	*RON*	*MON*
Straight run	0	68.4	62.0
	3	84.5	74.4
Bauxite-treated straight run	0	86.7	75.9
	3	94.2	82.1

[a]1 ml gal^{-1} = 0.2642 ml l^{-1}.

Table 9.5 Properties of the final motor gasoline produced by the Hydrocol Fe-HTFT process after refining. Properties before and after addition of tetraethyllead (TEL) are given.[a]

	Hydrocol motor gasoline	
Property	*Clear*	*+ 3 ml gal^{-1} TEL*[b]
Density at 15 °C (kg m^{-3})	717	717
Reid vapour pressure (kPa)	63	63
RON	91.4	97.2
MON	80.2	84.1

[a]The Hydrocol refinery contained only two conversion units, bauxite-catalysed double bond IS and SPA-catalysed OLI.
[b]1 ml gal^{-1} = 0.2642 ml l^{-1}.

original Sasol 1 design. The olefinic product from OLI was not hydrogenated, since there was no limit on the alkene content of motor gasoline. The SPA-derived olefinic motor gasoline typically had a RON in the range 95–97 and MON in the range 81–83. It is worth pointing out that by employing only two of the key catalyst types for FT refining, namely alumina- and phosphoric acid-based catalysts, the refinery could produce a leaded product with a 91 road octane number ($\frac{1}{2}$RON + $\frac{1}{2}$MON). In this respect, the Hydrocol refinery (and original Sasol 1 refinery design) performed as well as or even better than the more complex Sasol 2 and 3 refineries that were constructed 30 years later, but employed a crude oil refinery design.[13,17]

The crude oil refinery design approach followed for Fe-HTFT naphtha upgrading at the Sasol 2 and 3 refineries (currently known as Sasol Synfuels) mainly employed a combination of standard catalytic reforming using a chlorinated Pt/alumina-based catalyst (Section 8.1.1) and SPA OLI (Section 5.1.3.1). As was pointed out earlier, standard catalytic reforming with chlorinated Pt/alumina-based catalyst systems are poorly suited for the refining of very linear naphtha. Since the Fe-HTFT naphtha contained little cyclic material, the $N+2A$ of the reformer feed was very low. In order to obtain a RON 90 product, the volume yield was only around 82%.[17]

Table 9.6 Properties of the final motor gasoline produced by refining of Fe-HTFT syncrude at Sasol Synfuels in 2009. The Fe-HTFT motor gasoline also contains hydrotreated coal tar naphtha from low-temperature coal gasification.

Property	*Fe-HTFT-derived lead-replacement gasoline*	*Fe-HTFT-derived unleaded gasoline*	*Typical crude oil-derived gasoline in South Africa*
Density at 20 °C ($kg\,m^{-3}$)	723	729	728
Reid vapour pressure (kPa)	66	67	72
RON	93	93	93
MON	83	83	83
Aromatics content (vol.%)	25	29	27
Alkene content (mass%)	30	30	~12
Oxygenates (mass% O)	0.05	0.14	0.09
Sulfur content ($\mu g\,g^{-1}$)	<10	<10	150

With the phase-out of leaded gasoline, it was necessary to improve the octane number of the motor gasoline. Pentene IS and etherification technologies were added to the Sasol Synfuels refinery to generate high-octane components, and chemical production units (linear 1-alkene extraction and catalytic cracking) were employed to remove low-octane components from the naphtha. The latter reduced the overall volume of the motor gasoline pool by converting naphtha range material into chemicals. This ultimately yielded a product that is very similar to the motor gasoline produced by crude oil refining (Table 9.6).[4]

It will be noted that the RON 93 Fe-HTFT-derived motor gasoline (Table 9.6) contains little oxygenates. The fuel ethers (TAME) are mainly employed in the RON 95 motor gasoline that is co-produced in a smaller volume.[18]

Motor gasoline is also produced commercially from Fe-HTFT syncrude in the PetroSA GTL facility in Mossel Bay, South Africa. The Fe-HTFT naphtha is co-refined with associated natural gas liquids (mainly alkanes) to produce a RON 95 unleaded motor gasoline (Table 9.7).[4]

9.1.2 Jet Fuel

The main compound class in jet fuels (aviation turbine fuels) that is regulated is aromatics, since aromatic compounds are soot precursors that can affect combustion. Particulate matter generated during soot formation is not only detrimental to the environment, but can also damage the turbine. The maximum aromatic content of Jet A-1 is limited to 25 vol.% and less than 3 vol.%

Table 9.7 Properties of the final motor gasoline produced by refining of Fe-HTFT syncrude and natural gas liquids at the PetroSA gas-to-liquids facility in South Africa.

Property	*Fe-HTFT-derived unleaded gasoline*
Density at 20 °C (kg m^{-3})	748
Reid vapour pressure (kPa)	72
RON	95
MON	85
Aromatics content (vol.%)	37
Alkene content (mass%)	8
Oxygenates (mass% O)	–
Sulfur content (μg g^{-1})	<10

naphthalenes (bicyclic aromatics). Synthetic Jet A-1 from FTS must conform to more stringent specifications, which require the aromatic content to be in the range 8–22 vol.%.[19]

In addition to limitations imposed on the acidity and sulfur content of jet fuels, alkenes and heteroatom-containing compounds are not directly regulated, but are indirectly regulated by thermal stability requirements. Thermal stability is a key property, since the jet fuel is used as a heat exchange fluid in the engine and airframe. Any precursor to gum formation (for example, aldehydes and ketones) must be excluded from the fuel. For Jet A-1 fuel, a freezing point of less than –47 °C is required to ensure that the fuel remains pumpable in the low-temperature conditions during high-altitude flight. This places a limitation on the concentration of *n*-alkanes. Properties such as the distillation curve, energy content, density and viscosity are also important.

9.1.2.1 Jet Fuel from HTFT Synthesis

Since 1999, the international airport in Johannesburg, South Africa, routinely made use of semi-synthetic jet fuel. In 2008, fully synthetic jet fuel produced from HTFT products were also approved under Defence Standard 91–91, Issue 6.[19]

It can be generally stated that it is easy to refine FT syncrude to jet fuel. It is therefore ironic that marketing attempts by FT fuel producers to differentiate FT-derived fuels from petroleum fuels had some unintended consequences for the use of FT-derived kerosene as jet fuel. An extensive testing programme had to be undertaken in order to qualify material from FTS for use in jet fuel, despite the fact that it is possible to produce a jet fuel from FTS that falls well within the composition range of jet fuels produced from crude oil. Even at the time of writing, the semi-synthetic jet fuel (mixture of Fe-HTFT-derived kerosene and crude oil-derived kerosene) and fully synthetic jet fuel (specific Fe-HTFT kerosene components) that are allowed by international jet fuel specifications[19] are very restrictive in terms of their origin and refining pathways.

Single unit production of fully synthetic jet fuel involving combined OLI and aromatic alkylation over an SPA catalyst has been suggested (see also Section 8.2).[20] Although such a jet fuel complies to all the Jet A-1 specifications, the specific refining pathway has not yet been qualified.

9.1.2.2 Jet Fuel from LTFT Synthesis

According to the Defence Standard 91–91, Issue 6 international jet fuel specifications,[19] LTFT-derived material is not allowed in Jet A-1. This is unfortunate, as one can in principle produce on-specification Jet A-1 of comparable quality from crude oil, HTFT syncrude and LTFT syncrude. A discussion of the literature dealing with LTFT syncrude refining to produce jet fuel for commercial aviation use is consequently somewhat academic. However, the international Jet A-1 fuel specifications do not govern military applications. There is a strong correlation between civil and military jet fuel specifications, but the military is more pragmatic about the origin of the feed.

The US Army is considering the use of a single kerosene-type fuel for all of its gas turbine and (tactical) diesel engine applications. The fuel must comply with 'Jet Propulsion 8' (JP-8) specifications and is termed 'Battlefield Use Fuel of the Future' (BUFF). The exception is the fuel for use on aircraft carriers, which requires conformity with JP-5 specifications. JP-5 is essentially the same as JP-8, except that it has a higher flash point. The higher flash point provides an additional degree of safety in handling fuels on aircraft carriers. Due to supply security issues, Fischer-Tropsch-derived fuels are of specific interest.[21]

The introduction of FT fuels into the military fleet faces several challenges, for example, the interchangeability of FT fuels with conventional crude oil-derived kerosene-type fuels. Specifically, there is a concern about the elastomer compatibility of fuel systems already conditioned using conventional-type fuels with subsequent exposure to FT fuels containing no aromatics.[22] This is not a concern, of course, if the LTFT-derived jet fuel complies with synthetic jet fuel specifications requiring a minimum of 8% aromatics.

The ability to produce a jet fuel blend component from LTFT wax is very dependent on the nature of the hydrocracking catalyst. It was demonstrated that a kerosene fuel conforming to anticipated BUFF specifications can be produced from LTFT products.[22] This was achieved by fractionation of the LTFT hydrocarbons to remove the light fractions to comply with the volatility requirements (flash point). In addition, a heavy portion of the feed had to be removed to achieve low-temperature fluidity requirements (viscosity, pour point and freezing point). The yield of the JP-8 fuel complying with specifications, namely a freezing point of –47 °C and a minimum flash point of 38 °C, was about 31 vol.% of the fractionator feed. For JP-5 fuel, the freezing point and flash point requirements, namely –47 °C and 60 °C, respectively, could be achieved at a yield of about 22 vol.%. These parameters could be attained by setting a target final boiling point of the product. A heavy fraction from the fractionator accounted for more than 60% of the feed. The yield was limited by the degree of HIS over the HCR catalyst. Employing a sulfided base metal HCR catalyst hampered jet fuel production.

The suitability of the Syntroleum FT S-5 product for jet fuel applications was evaluated by Muzzell *et al.* by comparing its properties with specifications of the commercial JP-5 fuel.[23] Most properties of the S-5 fuel conformed to the specifications for JP-5 fuels, except for density (Table 9.8).[23,24] The sulfur, nitrogen, oxygen and aromatics contents were below the detection limit. The FT fuel was produced over a highly isomerising HCR catalyst and had a low *n*-alkane content, with a branched-to-linear alkane ratio of 14:1.5. The branched alkanes were mostly methyl-branched. Consequently, the freezing point of the S-5 fuel was well beyond specifications requirements.

Comparing the Sasol and Syntroleum processes for the production of 'isoparaffinic kerosene' (IPK) for use in jet fuel (Table 9.8) highlights the critical nature of the HCR catalyst selection. The S-5 kerosene was obtained by HCR over a noble metal catalyst, whereas the SSPD kerosene was obtained by HCR over a sulfided base metal catalyst. There is clearly value in using Pt/SiO_2–Al_2O_3-based catalysts for this type of application and specifically catalysts where the metal-to-acid site balance is tuned to give a more isomerised product.

9.1.3 Diesel Fuel

The key performance measure of diesel fuel is its cetane number and on a molecular level it is a measure of the ease with which the molecule can be thermally decomposed in the presence of air at high temperature and pressure. The cetane number is therefore a measure of the inherent thermal stability of the molecule and its autoxidation propensity. Cetane number improves with increasing carbon number and in the order aromatics < cycloalkanes < alkanes.[25] Since *n*-alkanes have high cetane numbers, distillates (170–360 °C) from FTS generally have a cetane number exceeding that required by diesel fuel specifications.

Sulfur and polynuclear aromatics are regulated in diesel fuel, but neither affects the refining of products from FTS. Fischer–Tropsch syncrude is practically free of sulfur and only HTFT syncrude contains some (<0.5% of distillate) polynuclear aromatic material.

Other diesel fuel properties that are important include density, viscosity, cold flow, lubricity, flash point and distillation range. Density and viscosity influence the volume and energy value of material that is injected with each engine stroke. LTFT syncrude has a lower density, which translates into a higher volumetric fuel consumption for the same power delivery. FT syncrude also has poor cold flow properties on account of its high *n*-alkane content and, as in the case of jet fuel, some branching must be introduced by HIS. Boundary layer lubricity is related to the polarity of the compounds present in the diesel fuel.[26] Material from FTS inherently has good lubricity, which is provided by 1-alcohols and long-chain carboxylic acids, but it may be destroyed during too severe hydroprocessing (see also Section 7.3).

Table 9.8 Synthetic and semi-synthetic jet fuels produced from the kerosene obtained by hydrocracking of LTFT syncrude from the Sasol Slurry Phase Distillate (SSPD) and Syntroleum (S-5) processes.

	Synthetic		*Semi-synthetic*		*Specifications*	
Property	*SSPD*	*S-5*	*SSPD/Merox*[a]	*SSPD/DHC*[b]	*Jet A-1*	*JP-5*
Density at 15 °C ($kg\,m^{-3}$)	747	764	776	784	775–840	788–845
Flash point (°C)	45	64	48	53	<38	<60
Freezing point (°C)	–48	–51	–51	–50	<–47	<–46
Viscosity at –20 °C (mPa s)	4.2	6.1	4.4	4.6	<8.0	<8.5
Smoke point (mm)	>50	>43	36	37	>25	>19
Net combustion heat ($MJ\,kg^{-1}$)	44.1	44.1	–	–	>42.8	>42.6
Composition						
Aromatics (vol.%)	0	0.4	9.9	6.5	8–25	<25
Sulfur (mass%)	<0.01	<0.0001	0.07	<0.01	<0.3	<0.4
Thiol content (mass%)	0.0002	<0.0001	0.0004	0.0003	<0.003	<0.002
Acidity (mg KOH g^{-1})	0.009	0.0014	0.009	0.01	<0.015	<0.015
Distillation (°C)						
IBP	154	183	152	156		Report[c]
T10	168	194	169	179	<205	<206
FBP	267	267	267	278	<300	Report[c]

[a]Blend of SSPD material and Merox-sweetened crude oil-derived kerosene in a 1:1 ratio.
[b]Blend of SSPD material and crude oil-derived kerosene from distillate hydrocracking (DHC) in a 1:1 ratio.
[c]Value must be stated on the jet fuel analysis, but the value is not subject to regulation.

9.1.3.1 Diesel Fuel from LTFT Synthesis

The perception has been created that hydrocracked LTFT waxes yield good-quality diesel fuel. This perception is based mainly on the high cetane number of the hydrocracked products, which consist of mainly linear and branched aliphatic hydrocarbons, with little cyclic or aromatic compounds being present. Conventional unmodified diesel engines have been used for evaluation of a variety of FTS-derived fuels and good emission performance has been reported for such fuels during engine testing.[27–29] The work performed by Syntroleum illustrates the point (Table 9.9).[27] Except for hydrocarbon emissions that remained invariant, a reduction in all other emissions was observed using LTFT diesel fuel compared with conventional crude oil-derived diesel. Similar reductions in emissions were also observed for a light-duty diesel engine.

The interchangeable colloquial use of the terms 'distillate' (referring to boiling range) and 'diesel fuel' (distillate that meets legislated fuel specifications) can result in misleading perceptions about the suitability of FTS for on-specification diesel fuel production. A high cetane number, low-density distillate from FTS may not necessarily conform to diesel fuel specifications. Furthermore, meeting diesel fuel specifications without resorting to blending with crude oil-derived distillate may not be easy.[6]

It is significant to point out that historically, Co-LTFT-derived distillate was not considered a good diesel fuel: '... straight FT fractions, in spite of their high cetane numbers, do not make the most satisfactory Diesel fuels ...'.[30] The properties of the distillate fraction from German Co-LTFT synthesis depended on its distillation range and origin, but generally it had a high cetane number and low density (Table 9.10).[30,31] The distillate was employed as a blending component with low cetane number, high-density material, such as coal tar, brown coal tar and crude oil residue cuts. In this way, a product was obtained that had a cetane number of 40–45 and good energy density. The distillate cut point that was employed was determined by the climate, with summer diesel fuel being a 150–320 °C cut and winter fuel essentially a kerosene 150–250 °C cut.[9]

Shell made it very clear during the development and commercialisation of the SMDS process that the intention was to use the Co-LTFT-derived distillate as a blending component with a distillate of crude oil origin.[32] This does not

Table 9.9 Comparison of emissions from a 5.9 l Cummings B engine on a test stand operated with LTFT distillate (Syntroleum S-2) and crude oil-derived diesel fuel.

	Emissions ($g\ bhp^{-1}\ h^{-1}$)[a]			
Test fuel	*Hydrocarbons*	*CO*	*NO_x*	*Particulates*
EPA No. 2 diesel fuel	0.10	1.3	4.0	0.10
Syntroleum S-2	0.10	0.8	3.2	0.06

[a] $1\ g\ bhp^{-1}\ h^{-1} = 0.3725\ \mu g\ J^{-1}$.

Table 9.10 Properties of straight run distillates from German Co-LTFT synthesis.

	Straight run Co-LTFT distillate	
Property	*Weil and Lane*[30]	*Ward et al.*[31]
Density (kg m^{-3})	760	772
Cetane number	96	80
Flash point (°C)	49	78
Pour point (°C)	–20	–1

imply that LTFT-derived distillate cannot be used as a neat diesel fuel, but it would not meet the requirements of some diesel fuel specifications. It has been reported that on a molecular level LTFT syncrude is unsuitable for the production of on-specification EN 590:2004 diesel fuel in high yield. In Fischer–Tropsch refining there is a trade-off between distillate density, cetane number and distillate yield, called the 'FT density–cetane–yield triangle'.[6] With respect to the EN 590:2004 diesel fuel specifications, it is possible to meet any two of these three requirements without too much refining effort, but meeting all three with FT syncrude as feed material is difficult. Leckel reached the same conclusion in his review paper on FT diesel fuel.[5]

Elastomers found in fuel injection system of engines will swell when in contact with diesel fuel. The extent of swelling depends on the aromatic content of the fuel. This may lead to some problems when fuels with reduced aromatic content are being gradually introduced into the market. Elastomers that have been exposed to high-aromatic fuel and then to low-aromatic fuel may cause leaching of absorbed aromatics, causing them to shrink. These effects have been studied in detail by Lamprecht.[33]

Moreover, hydroprocessed LTFT diesel generally has poor lubrication properties, because the surface-active compounds (oxygenates) are destroyed during severe hydroprocessing. Boundary layer lubrication has to be improved by additives, which is generally necessary for all severely hydroprocessed fuels.

In South Africa, Fe-LTFT synthesis has been employed since the 1950s for the production of diesel fuel in combination with Fe-HTFT.[1] Although it was not the original design intention to improve the diesel fuel properties by combining Fe-LTFT, Fe-HTFT and coal pyrolysis distillates, the outcome was definitely synergistic. More recently, it has been reported that a combined LTFT and HTFT synthesis configuration is considered for the new 80 000 bpd Mafutha coal-to-liquids facility in South Africa.[34]

Some of the benefits of combining LTFT and HTFT distillates have been highlighted (Table 9.11).[35] These results show that the HTFT DHT diesel improves the density, volumetric heating value and viscosity of the FT diesel blends. On the other hand, LTFT diesel improves the cetane number and cold flow properties of the FT diesel blends. Moreover, engine tests of the blends indicated a beneficial effect of the LTFT diesel fraction on the NO_x and CO_x emissions.

Table 9.11 Selected fuel properties of hydroprocessed HTFT and LTFT distillate blends.

	Ratio of HTFT to LTFT distillate in blend						
Property	*100:1*	*85:15*	*70:30*	*50:50*	*30:70*	*15:85*	*0:100*
Density at 15 °C ($kg\,m^{-3}$)	809	803	797	789	781	775	769
Cetane number	57	59	61	66	67	69	73
Alkene content (g Br per 100 g)	9.4	8.2	6.7	5.4	3.2	1.9	0.6
Aromatics content (mass%)	23.9	20.3	16.8	12	7.3	3.7	0.1
Sulfur content ($\mu g\,g^{-1}$)	3	2	2	<1	<1	<1	<1
Flash point (°C)	78	74	72	66	63	60	58
Viscosity at 40 °C (mPa s)	2.14	2.11	2.10	2.07	2.03	2.01	2.00
CFPP (°C)	0	–1	3	–6	–11	–19	–19
Lubricity, HFRR wear (μm)	547	549	552	556	560	612	617
Distillation (°C)							
IBP	184	180	166	159	153	152	151
T10	208	205	200	195	189	184	182
T50	239	242	242	243	245	246	249
T95	363	359	351	343	336	330	325
FBP	385	385	379	367	358	345	334

The severity of hydroprocessing, whether it is hydrotreating of straight run distillate, HIS or HCR, influences the final product properties. In order to have acceptable cold flow properties, some branching must be introduced, although this results in some cetane number loss. Catalyst selection is important in determining the product quality. For example, all of the sulfur found in the LTFT distillate from the Sasol Slurry Phase Distillate process is due to sulfur addition in the refinery. This is a direct consequence of employing a sulfided base metal HCR catalyst. The LTFT distillate from the Shell Middle Distillate Synthesis process is sulfur-free, since it employs a noble metal HCR catalyst.

9.1.3.2 *Diesel Fuel from HTFT Synthesis*

It is possible to produce a hydroprocessed straight run Fe-HTFT distillate that meets European EN590:2004 diesel fuel specifications.[6,36] However, the ability to do so is very dependent on the catalyst selection and operating conditions employed during hydroprocessing.

The combined distillate and residue fractions from Fe-HTFT synthesis contain around 27% aromatics and the fraction of aromatics in each distillation cut increases with increasing boiling point. The aromatics are mainly alkyl mono-aromatics and the cetane number of the unrefined material is 55.[36] The heavier fraction of the Fe-HTFT syncrude therefore contributes positively to the cetane number and density, which is important to overcome the FT density–cetane number–yield triangle for diesel fuel production.[6] When only the lighter straight run distillate is refined, the density is correspondingly lower (Table 5.31).

Table 9.12 Properties of straight run bauxite-treated Hydrocol (Fe-HTFT) distillate before and after hydrogenation.

	Hydrocol bauxite-treated distillate	
Property	*Unhydrogenated*	*Hydrogenated*
Density ($kg\,m^{-3}$)	806	806
Cetane number	56[a]	71
Pour point (°C)	–9	–1
T90 distillation (°C)	304	327

[a]Reportedly an estimated value.

The Hydrocol process produced an olefinic distillate that was refined only by bauxite-treatment, which resulted in partial deoxygenation. It has been shown that the cetane number of the distillate can be improved by HYD (Table 9.12),[37] but this was not applied in the commercial Hydrocol plant. The distillate and residue fractions from Fe-HTFT synthesis constitute less than 10% of the total syncrude. In order to improve the distillate yield from HTFT refineries, additional distillate can be produced by the OLI of gaseous and naphtha-range alkenes.

In the PetroSA HTFT refinery, H-ZSM-5 is employed as an OLI catalyst in the COD process for the production of distillate from light alkenes. This catalyst is well suited to distillate production and its pore-constrained geometry limits branching in the distillate range material (Table 5.2) to yield a distillate with good cetane number. The hydrogenated distillate from H-ZSM-5 OLI is therefore not much different from the product obtained during the HCR of LTFT wax. This material is then blended with hydrotreated straight run HTFT distillate, C_3 and heavier alcohols recovered from the HTFT aqueous product and straight run distillate from natural gas liquids (Table 9.13).[38]

In the Sasol Synfuels HTFT refinery, distillate fractions are produced in a number of units that are blended to yield a final diesel fuel (Table 9.14).[4,36] The main contributor to the final diesel fuel is the light distillate obtained by hydrotreating the straight run SLO. There is also a heavy SLO-derived distillate that is produced by catalytic dewaxing of the residue from the SLO distillate hydrotreater. Although distillate is produced commercially from the SPA-based OLI of gaseous alkenes, this distillate fraction is a kerosene cut. This material has a low cetane number (typically less than 35) and low density;[39] it is generally not included in the diesel fuel. The coal pyrolysis liquids that are co-produced during the coal-to-syngas conversion are also hydrotreated and included in the diesel fuel to increase the density of the blend. It will be noted from Table 9.14 that the coal tar distillate is severely hydrotreated. The coal tar hydrotreating conditions employed commercially are 280–380 °C, 18.5 MPa and LHSV 0.25 h^{-1}.[40,41]

9.1.4 Other Fuel Types

Liquid petroleum gas (LPG) is employed in some countries as a transportation fuel. The amount of straight run propane and butanes from FTS is around

Table 9.13 Different distillates and distillate blends produced at the Fe-HTFT GTL facility of PetroSA.

Property	*Hydrotreated PetroSA HTFT distillates* Mossgas 1	Mossgas 2	Mossgas COD
Composition of distillate			
Distillate from ZSM-5 OLI	63	60	100
Straight run SLO	30	28	0
Straight run gas liquids[a]	7	7	0
Heavy FT alcohols	0	5	0
Hydrotreated distillate			
Density at 20 °C (kg m^{-3})	808.8	806.5	800.7
Cetane number	53.0	49.3	51.4
Aromatics content (vol.%)	16.4	15.9	10.1
Sulfur content (mass%)	<0.001	<0.001	<0.001
Distillation (°C)			
IBP	222	81	229
T90	322	318	323
FBP	360	363	361

[a]Associated natural gas liquids that are co-recovered with the natural gas feed.

Table 9.14 Distillate blending components and final diesel fuel produced at the Fe-HTFT CTL facility of Sasol.

Property	*Hydroprocessed distillates* Light SLO distillate	Heavy SLO distillate	Coal tar distillate	*Final diesel fuel*
Blending volume (%)	75 ± 5	4 ± 2	22 ± 6	–
Density at 20 °C (kg m^{-3})	812	860	870	829
Cetane number	55	38	53	55
Aromatics content (vol.%)	20	25	13	*ca.* 25
Polynuclear aromatics (mass%)	0.05	0.32	0.05	<1
Sulfur content (μg g^{-1})	1	5	1	<5
Acidity (mg KOH g^{-1})	0.02	0.04	0.0001	–[a]
Flash point (°C)	94	111	60	77
Viscosity at 40 °C (mPa s)	2.4	8.8	2.3	2.2
CFPP (°C)	–30	16	–7	–6
Lubricity, HFRR wear (μm)	345	340	508	<460
Distillation (°C)				
IBP	190	190	150	192
T95	249	469	355	348[b]
FBP	370	490	370	394

[a]Not reported.
[b]T90 distillation.

2–3% of the syncrude. Many refining processes, such as HCR and catalytic reforming, produce some additional LPG, making LPG a meaningful commercial product in all FT facilities.

Heavier transportation fuels, such as fuel oils, are generally not produced commercially in FT facilities. However, a small volume of waxy oil is produced in the HTFT refinery.[36] Some heavy cuts can also be employed as cracking materials, such as the slack wax cuts in the German Co-LTFT refineries.[30]

9.2 Lubricating Oils

The residual wax from FTS is a suitable feed for the preparation of lubricating base oils. Fischer–Tropsch wax can readily be hydroisomerised (catalytically dewaxed) to produce lube base oils possessing properties similar to those of products derived from petroleum feeds (Section 6.3.2).

The same or slightly modified catalytic processes that are suitable for dewaxing of the conventional vacuum gas oil and deasphalted oil can also be used for dewaxing FT waxes. If the final boiling point of the feed wax exceeds that of the lube base oil, the dewaxing catalyst must also exhibit good HCR activity. When co-production of lube base oil with transportation fuels is considered, a catalyst possessing high HCR and HIS activity is desirable.

Lubricants can be distinguished from transportation fuels by their high viscosity and high boiling range, typically >400 °C. The final lubricants are prepared from lube base oil by mixing in various additives. An important characteristic of lube base oil and the final lubricant product is their viscosity index. This index is an indication of the change in viscosity with increasing temperature. A higher viscosity index indicates a smaller change in the viscosity of the lube base oil with an increase in temperature. Among the different hydrocarbon groups, linear hydrocarbons exhibit the highest viscosity index, and aromatics exhibit the lowest viscosity index. Based on the criteria of viscosity and boiling range, wax from LTFT synthesis appears to be an ideal feed for lube base oil preparation.

The cold flow properties of lube base oils having a high content of linear hydrocarbons have to be adjusted to meet performance specifications. Lube base oils prepared from LTFT wax must therefore be subjected to dewaxing before it can be used for the preparation of lubricants. The conversion of linear to branched alkanes affects the viscosity index. This penalty is offset by a significant improvement in the cold flow properties of the lube base oil.

Lubricants must also be resistant to oxidation. The oxidation stability of lubricants can be enhanced by the addition of various antioxidants. It was reported that oxidation inhibitors suitable for lubricants prepared from FT-derived lube base oils are not necessarily equally suitable for lubricants of petroleum origin.[42] For example, for a lubricant prepared from LTFT wax, the best antioxidant was a combination of 0.5 mass% triphenyl phosphite and

0.5 mass% chromium oleate. When tested under the same conditions, tricresyl phosphite was the best antioxidant for a lubricant of petroleum origin.

Waxes from LTFT synthesis can readily be converted into lubricating oils using HIS catalysts, such as mildly acidic $Pt/SiO_2–Al_2O_3$. This type of conversion will be commercially applied in Shell's Pearl GTL facility in Las Raffan, Qatar, for the production of lubricating base oil from Co-LTFT waxes.[43] This will be the first large-scale commercial application of lubricating oil production from LTFT waxes since German lubricating oil production from FTS during the Second World War.

A review of German synthetic lubricant production routes by Horne stated that most lubrication oils were produced by polymerisation in the presence of $AlCl_3$.[44] The LTFT waxes were thermally cracked to produce linear 1-alkenes that were then oligomerised over $AlCl_3$ to produce 'polyalphaolefin' (PAO)-type lubricating oils. Lubricating oils were also produced by chlorinating LTFT distillate and then performing Friedel–Crafts alkylation of naphthalene in the presence of $AlCl_3$. Other production methods were mentioned by Weil and Lane,[30] but seem not to have been applied commercially.

9.3 Chemicals

Fischer–Tropsch syncrude is an attractive feedstock for the production of chemicals, which are generally higher value products than transportation fuels. The types of chemicals that can readily be produced depend on the type of FTS. Some chemicals are present in high concentration in syncrude, for example the light alkenes and oxygenates in HTFT syncrude (Tables 4.1 and 4.8). Various process configurations have been outlined in the literature and include both extractive and synthetic approaches.[1,2,45–48] Further discussion will be limited to those chemicals that are produced commercially in Fischer–Tropsch facilities.

9.3.1 Oxygenates

The oxygenates that can typically be recovered from FT syncrude are those contained in the aqueous product (Table 4.7). Among them, alcohols are of main commercial interest due to their abundance in HTFT and LTFT syncrude, but carbonyl compounds and carboxylic acids are also of interest in HTFT syncrude (see also Chapter 7). Apart from the oxygenates that can be directly recovered, alcohols and aldehydes may also be produced from alkenes and synthesis gas by hydroformylation.[49]

9.3.1.1 Alcohols from Separation

Alcohols are primary products from FTS. The light alcohols (C_1–C_4) on condensation dissolve in the aqueous product phase. The alcohols are mainly linear 1-alcohols and can be recovered from the FT aqueous product by distillation.

Such recovery is more profitable from Fe-HTFT syncrude due to the higher concentration of light alcohols,[50] but light alcohols can in principle also be recovered from LTFT syncrude.

Historically, the extraction of LTFT alcohols was applied commercially at Sasol 1, where the aqueous products from Fe-HTFT and Fe-LTFT synthesis were combined.[14] However, none of the industrial LTFT-based facilities constructed since the 1990s include aqueous product refining. Light alcohols are at present recovered from FT syncrude only at the Fe-HTFT-based facilities of PetroSA and Sasol.

The yield of light alcohols from HTFT synthesis can be increased by selective HYD of the carbonyls (aldehydes and ketones) dissolved in the FT aqueous product. The use of an Ni/SiO_2–Al_2O_3 catalyst, such as the Süd-Chemie G-134, performs well in this application.[50]

Carbonyl to alcohol HYD at the PetroSA facility converts all carbonyl compounds to alcohols and the alcohols are sold as mixtures under the trade name Mosstanol.[51] At Sasol Synfuels, only the ethanal is hydrogenated to ethanol and pure alcohols are recovered and sold, in addition to alcohol mixtures.[2] Some of the light alcohols are processed further to other chemicals, for example the conversion of ethanol into ethyl acetate.[52] Heavier alcohols are present in the oil product from FTS, but they are not commercially recovered.

9.3.1.2 Alcohols from Hydroformylation

Aldehydes can be synthesised from alkenes and synthesis gas (CO and H_2) by hydroformylation, with subsequent HYD to produce the corresponding alcohols. There is consequently a natural synergy between hydroformylation technology and FTS, since both alkenes and synthesis gas are readily available.

Sasol at present has three Rh-based hydroformylation processes in commercial operation making use of material from FTS: C_{12}–C_{13} detergent alcohols produced from HTFT distillate, 1-butanol synthesis from propene and the production of 1-octanol as an intermediate product in the synthesis of 1-octene from 1-hexene.[47,53]

9.3.1.3 Carbonyls from Separation

Short-chain carbonyl compounds (C_2–C_5), similarly to light alcohols, dissolve in the FT aqueous product on condensation. Depending on the aqueous product refining strategy, these compounds may be recovered as mixtures or pure compounds.[14] Propanone (acetone) and 2–butanone (methyl ethyl ketone) are recovered commercially from the HTFT aqueous product; the heavier ketones are also recovered, but not separated into individual compounds. Some of these products are used for further processing, for example, for the production of 4–methyl–2–butanone (methyl isobutyl ketone) from propanone over a Pd/acidic resin catalyst.

9.3.1.4 Carboxylic Acids from Separation

Carboxylic acids are primary FTS products. The aqueous product after light oxygenate recovery contains about 1–2% of carboxylic acids and is fairly corrosive. It was found that the carboxylic acids could be selectively extracted with MTBE. A carboxylic acid recovery pilot plant was built in the chemical work-up section of the Sasol Synfuels facility to recover ethanoic acid (acetic acid) and propanoic acid from the HTFT aqueous product. However, corrosion problems and equipment failures, resulting in poor on-stream times, plagued the pilot plant. Furthermore, to scale this process up to a commercial scale would have required a large MTBE inventory, and also the largest diameter extractor in the world, which made it a very energy-intensive process. Acid recovery was therefore never taken beyond the pilot plant stage, although product was delivered commercially to the market.[2]

9.3.1.5 Other Oxygenates

At the Sasol 1 facility, oxidised waxes are produced commercially by batch autoxidation of LTFT wax (Section 6.2.2).[54] Various grades of oxidised and saponified oxidised waxes are marketed, with differing degrees of oxidation and different ratios of oxygenate functionalities (Table 6.5).[55,56]

Autoxidation was also used for carboxylic acid and soap production from Co-LTFT products in Germany during the Second World War.[57]

9.3.2 Alkenes

9.3.2.1 Ethene

Ethene (ethylene) is one of the 10 most abundant products from Fe-HTFT (Table 4.1). Ethane is almost equally abundant (Table 4.1) and can be cracked with high selectivity to ethene. Both of these compounds can be recovered from the HTFT tail gas. However, in order to benefit from the significant C_2 fraction in the Fe-HTFT syncrude, the FT gas loop design must include a cryogenic separation section. Ethene is commercially produced at Sasol Synfuels, but cryogenic C_2 separation was not included in the designs of the Hydrocol, Sasol 1 and PetroSA HTFT gas loops.

The amount of ethene produced by LTFT technologies has not yet warranted commercial recovery. However, it should be noted that Fe-LTFT catalysts deactivate in such a way that the selectivity of light alkenes, including ethene, increases with time on-stream.[58] The chemical potential of Fe-LTFT processes therefore benefits from FT catalyst deactivation, and with appropriate reactor technology the activity and selectivity level of Fe-LTFT synthesis can be controlled to maximise this benefit. The same does not hold true for Co-LTFT synthesis.

Downstream processing of ethene by polymerisation is the main commercial application of FTS-derived ethene at present. Since ethene is not easily transportable (unless dedicated pipeline infrastructure is available), it implies that commercial HTFT-derived ethene production for use as a chemical should be in the proximity of ethene consumers. If not, downstream processing of the ethene should be included as part of the HTFT refinery design.

9.3.2.2 Propene

Propene (propylene) is the most abundant Fe-HTFT product (Table 4.1), and recovery of propene does not require cryogenic separation. Recovery of propene is not a unique benefit from FT refining. The propene production from crude oil refineries, mainly derived from FCC units, supplies around 25% of the European propene market, 50% of the North American market and 20% of the Asian market.[59] The advantage of Fe-HTFT synthesis for propene production is that it is a major primary product.

In the Hydrocol, Sasol 1 and PetroSA facilities, propene was employed for fuels production and the same was true for the Sasol Synfuels HTFT facility until the 1990s. Since then, propene was extracted commercially and used for polypropylene production, and also 1-butanol and acrylic acid production. Although the latter two facilities are located in Sasolburg close to the Sasol 1 site, the propene is supplied from Sasol Synfuels in Secunda.

The amounts of propene produced by LTFT technologies are less and have not been exploited commercially for chemicals production. As with ethene, Fe-LTFT has a better potential than Co-LTFT for propene production, due to the selectivity changes when the Fe-LTFT catalyst deactivates.[58]

9.3.2.3 Linear 1-Alkenes

Fischer–Tropsch syncrude is naturally rich in linear 1-alkenes (α-olefins), which are primary products from FTS. The carbon number distribution is such that only HTFT syncrude yields a significant fraction of the linear 1-alkenes in the range employed for chemicals production. In fact, half of the 10 most abundant chemicals in HTFT syncrude (Table 4.1) are C_4–C_8 linear 1-alkenes.

Although PetroSA considered 1-hexene extraction,[60] only Sasol Synfuels recovers 1-pentene, 1-hexene and 1-octene commercially as final products.[2,47,61,62] In addition to these, linear 1-alkenes are also recovered and used within the Sasol Synfuels facility as feed for hydroformylation.[53]

The concentration of linear 1-alkenes in LTFT syncrude is much lower and these compounds are not recovered commercially from LTFT syncrude.

The process flow diagram for 1-hexene recovery from HTFT condensate[61] is less complex than that required for 1-hexene recovery from HTFT stabilised light oil, where the oxygenate concentration is higher. The principles governing separation are nevertheless the same. In the case of 1-hexene, most of the close-boiling polar compounds can be removed from the 1-hexene-containing fraction by extractive distillation. However, it is not possible to remove the very close-boiling alkenes 2–methyl-1-pentene and 2–ethyl-1-butene by distillation.

In order to facilitate this difficult separation, the process includes an acidic resin-catalysed etherification step with methanol.[63] Commercial production of 1-pentene is conducted in the same unit on a campaign basis.

The recovery of 1-octene is more complex and different technologies have been developed for this purpose. The first process that was developed for the recovery of 1-octene from HTFT syncrude made use of a basic solvent for oxygenate extraction. In order to do so, the carboxylic acids were neutralised with potassium carbonate before oxygenate extraction with *N*-methylpyrrolidone (NMP), which was followed by super-fractionation to purify the final product. In the second process for 1-octene recovery, the neutralisation step was eliminated by applying azeotropic distillation for acid removal.[47,62,64]

Although there is clearly value in recovering linear 1-alkenes from HTFT syncrude, the global market is comparatively small. For example, in 2000–2001 Sasol Synfuels supplied around 25% of the global demand for 1-hexene.[2] Should FTS become more widely used in the future, it is unlikely that 1-alkene recovery will be as profitable, since the global market will quickly be saturated.

9.3.3 Alkanes

9.3.3.1 Aromatics-free n-Alkanes

The inherent low aromatic, high linear hydrocarbon content of LTFT naphtha and distillate makes it well suited for the production of aromatics-free or very low aromatics alkane solvents. For commercial solvent production, the LTFT material is hydrogenated and distilled into various cuts. Paraffinic solvents are produced commercially from Fe-LTFT syncrude at Sasol 1 and Co-LTFT syncrude at the SMDS facility in Bintulu, Malaysia.

Aromatics-free alkane solvents that are marketed under the trade names Mosspar and SloPar are also produced commercially from HTFT syncrude by PetroSA.[51] Some of the material is obtained by deep HYD of the product from HTFT alkene OLI over H-ZSM-5. Hydrogenation catalysis is very important for the production of these compounds, with unsulfided Ni-based or noble metal-based HYD catalysts being preferred.

9.3.3.2 Waxes

There is a wide range of applications for medium and hard waxes from LTFT (Chapter 6). The former is especially suitable for the production of candles. After pretreatment, hard wax can be used in products such as cosmetics, coatings, adhesives and plasticisers. Different grades of paraffin waxes are commercially produced from Fe-LTFT (Table 6.2) and Co-LTFT (Table 6.3).[55,65] Medium and hard waxes were also commercially produced from Co-LTFT synthesis in Germany.[66,67]

Despite the high degree of saturation of the straight run LTFT waxes, the waxes are typically hydrotreated to improve colour and stability.[68] This has been discussed earlier (Section 6.3.1).

The heavy fraction from HTFT syncrude is very aromatic and bears no resemblance to LTFT waxes. HTFT syncrude is therefore not suitable as feed for wax production.

9.3.4 Associated Chemical Products

In any FT facility, some by-products may be obtained from feed processing for synthesis gas production. The nature of these products depends on both the feed and the type of processing. Examples of such by-products found in industrial FT facilities are natural gas liquids that are condensed before the methane-rich gas is reformed to synthesis gas and the coal pyrolysis liquids that are co-produced during low-temperature coal gasification to produce synthesis gas. In addition to these products, there are also products that are co-produced during air separation and gas cleaning. Although none of these products are derived from FTS, they are associated with FTS and are chemical products that may be produced in the context of an FT facility.

9.3.4.1 Inert Gases

Unless an air-driven technology has been selected for gasification and/or reforming to produce synthesis gas, an air separation unit is required to produce oxygen for synthesis gas production. All current commercial FT facilities make use of oxygen-driven synthesis gas production processes and therefore contain air separation units. Depending on the scale, one or more of the following products may be co-produced in addition to oxygen: nitrogen, argon and less abundant rare gases, such as neon, krypton and xenon.

9.3.4.2 Coal Liquids

Coal liquids are co-produced during low-temperature gasification and metallurgical coke production from coal. These coal liquids are formed during thermal decomposition of coal in the temperature range 300–650 °C. Coal-to-liquids facilities that make use of low-temperature gasification will therefore have associated coal liquids.

The separation technologies associated with the recovery of the aromatic and phenolic products from coal liquids are well established.[69–71] Typical products that can be recovered are benzene, naphthalene, alkylated aromatics, phenol, cresols and xylenols. This type of recovery is applied commercially at the Sasol 1 and Sasol Synfuels sites. Although the basic refining and recovery technology has changed little since the Second World War, some new developments in phenol recovery have been devised and implemented at the Sasol 1 site.[72]

9.3.4.3 Nitrogen Compounds

Air separation provides nitrogen as a by-product and, in combination with hydrogen from reforming, it provides the raw materials for ammonia production. In addition to ammonia that can be produced synthetically, ammonia can also be recovered from low-temperature coal gasification.[73] There is consequently a good technology fit between ammonia production and coal-to-liquids facilities. Ammonia provides a platform for the production of other nitrogen-based chemicals, such as urea, nitric acid, ammonium nitrate and ammonium sulfate. Sasol produces fertilisers and explosives from their ammonia-based co-production in South Africa.[2]

9.3.4.4 Sulfur Compounds

Sulfur-containing compounds have to be removed during synthesis gas production, since sulfur is an FT catalyst poison. Depending on the nature of the synthesis gas cleaning technology employed, the sulfur compounds may be recovered as either hydrogen sulfide or sulfur oxides. These compounds can then be transformed into elemental sulfur or other sulfur-containing commodities, such as sulfuric acid.

References

1. J. Meintjes, *Sasol 1950–1975*, Tafelberg, Cape Town, 1975.
2. J. Collings, *Mind Over Matter. The Sasol Story: a Half-Century of Technological Innovation*, Sasol, Johannesburg, 2002.
3. A. de Klerk, *Green Chem.*, 2007, **9**, 560.
4. B. I. Kamara and J. Coetzee, *Energy Fuels*, 2009, **23**, 2242.
5. D. O. Leckel, *Energy Fuels*, 2009, **23**, 2342.
6. A. de Klerk, *Energy Fuels*, 2009, **23**, 4593.
7. N. Naidoo, presented at the 2nd Sub-Saharan Africa Catal. Symposium, Swakopmund, 2001, paper 1.
8. A. de Klerk, *Green Chem.*, 2008, **10**, 1249.
9. US Naval Technical Mission in Europe, *The Synthesis of Hydrocarbons and Chemicals from CO and H_2*, Technical Report 248-45, September 1945.
10. G. Egloff, E. S. Nelson and J. C. Morrell, *Ind. Eng. Chem.*, 1937, **29**, 555.
11. L. P. Dancuart, J. F. Mayer, M. J. Tallman and J. Adams, *Prepr. Pap. Am. Chem. Soc. Div. Pet. Chem.*, 2003, **48** (2), 132.
12. A. de Klerk, *Prepr. Pap. Am. Chem. Soc. Div. Fuel Chem.*, 2008, **53** (2), 105.
13. A. de Klerk, in *Advances in Fisher Tropsch Synthesis, Catalysts and Catalysis*, ed. B. H. Davis and M. L. Occelli, Taylor & Francis (CRC Press), Boca Raton, FL, 2009, p. 331.
14. J. C. Hoogendoorn and J. H. Salomon, *Br. Chem. Eng.*, 1957, **2**, 368.
15. P. C. Keith, *Oil Gas J.*, 1946, **45** (6), 102.
16. F. H. Bruner, *Ind. Eng. Chem.*, 1949, **41**, 2511.

17. J. S. Swart, G. J. Czajkowski and R. E. Conser, *Oil Gas J.*, 1981, **79** (35), 62.
18. A. M. van Wyk, M. A. K. Moola, C. J. de Bruyn, E. Venter and L. Collier, *Chem. Technol. (S. Afr.)*, 2007, April, 4.
19. United Kingdom Ministry of Defence, Defence Standard 91-91, Issue 6. *Turbine Fuel, Aviation Kerosine Type, Jet A-1. NATO Code: F-35. Joint Service Designation: AVTUR*, 8 April 2008.
20. T. M. Sakuneka, A. de Klerk, R. J. J. Nel and A. D. Pienaar, *Ind. Eng. Chem. Res.*, 2008, **47**, 1828.
21. C. A. Forest and P. A., Muzzell, *SAE Tech. Pap. Ser.*, 2005, 2005-01-1807.
22. D. Lamprecht, *Energy Fuels*, 2007, **21**, 1448.
23. P. A. Muzzell, R. L. Freerks, J. P. Baltrus and D. D. Link, *Prepr. Pap. Am. Chem. Soc. Div. Pet. Chem.*, 2004, **49** (4), 411.
24. D. Lamprecht and P. N. J. Roets, *Prepr. Pap. Am. Chem. Soc. Div. Pet. Chem.*, 2004, **49** (4), 426.
25. R. C. Santana, P. T. Do, M. Santikunaporn, W. E. Alvarez, J. D. Taylor, E. L. Sughrue and D. E. Resasco, *Fuel*, 2006, **85**, 643.
26. G. Knothe and K. R. Steidley, *Energy Fuels*, 2005, **19**, 1192.
27. P. F. Schubert, R. Freerks, H. L. Tomlison and B. Russell, *Prepr. Pap. Am. Chem. Soc. Div. Fuel Chem.*, 2000, **45** (3), 592.
28. T. Wu, Z. Huang, W. G. Zhang, J. H. Fang and Q. Yin, *Energy Fuels*, 2007, **21**, 1908.
29. K. Yehliu, O. Armas and A. L Boehman, *Prepr. Pap. Am. Chem. Soc. Div. Pet. Chem.*, 2009, **54** (1), 72.
30. B. H. Weil and J. C. Lane, *The Technology of the Fischer–Tropsch Process*, Constable, London, 1949.
31. C. C. Ward, F. G. Schwartz and N. G. Adams, *Ind. Eng. Chem.*, 1951, **43**, 1117.
32. S. T. Sie, M. M. G. Senden and H. M. W. van Wechem, *Catal. Today*, 1991, **8**, 371.
33. D. Lamprecht, *SAE Tech. Pap. Ser.*, 2007, 2007-01-0029.
34. Anon., *Oil Gas J.*, 2008, **106** (4), 10.
35. D. Lamprecht, L. P. Dancuart and K. Harrilall, *Energy Fuels*, 2007, **21**, 2846.
36. D. O. Leckel, *Energy Fuels*, 2009, **23**, 38.
37. J. A. Tilton, W. M. Smith and W. G. Hockberger, *Ind. Eng. Chem.*, 1948, **40**, 1269.
38. C. Knottenbelt, *Catal. Today*, 2002, **71**, 437.
39. A. de Klerk, *Energy Fuels*, 2006, **20**, 439.
40. D. O. Leckel, *Energy Fuels*, 2006, **20**, 1761.
41. D. O. Leckel, *Prepr. Pap. Am. Chem. Soc. Div. Fuel Chem.*, 2009, **54** (1), 125.
42. E. N. Givens, S. C. LeViness and B. H. Davis, *Prepr. Pap. Am. Chem. Soc. Div. Pet. Chem.*, 2005, **50** (1), 182.
43. N. Fabricius, in *Fundamentals of Gas to Liquids*, Petroleum Economist, London, 2005, p. 12.

44. W. A. Horne, *Ind. Eng. Chem.*, 1950, **42**, 2428.
45. M. E. Dry, *ACS Symp. Ser.*, 1987, **328**, 18.
46. A. P. Steynberg, W. U. Nel and M. A. Desmet, *Stud. Surf. Sci. Catal.*, 2004, **147**, 37.
47. A. Redman, in *Proceedings of the 18th World Petroleum Congress*, Johannesburg, 2005, cd179.
48. A. de Klerk, L. P. Dancuart and D. O. Leckel, in *Proceedings of the 18th World Petroleum Congress*, Johannesburg, 2005, cd185.
49. M. Beller, B. Cornils, C. D. Frohning and C. W. Kohlpaintner, *J. Mol. Catal. A.*, 1995, **104**, 17.
50. R. J. J. Nel and A. de Klerk, *Ind. Eng. Chem. Res.*, 2007, **46**, 3558.
51. R. Minnie, C. Knottenbelt, J. Clur, W. Grond, M. Karodia and H. de Wet, in *Fundamentals of Gas to Liquids*, Petroleum Economist, London, 2005, p. 22.
52. S. W. Colley and M. W. M. Tuck, in *Catalysis in Application*, ed. S. D. Jackson, J. S. J. Hargreaves and D. Lennon, Royal Society of Chemistry, Cambridge, 2003, p. 101.
53. K. McGurk, in *Proceedings of the South African Chemical Engineering Congress*, Sun City, 2003, P082.
54. A. de Klerk, *Ind. Eng. Chem. Res.*, 2003, **42**, 6545.
55. J. H. le Roux and S. Oranje, *Fischer–Tropsch Waxes*, Sasol, Sasolburg, 1984.
56. F. H. A. Bolder, A. de Klerk and J. L. Visagie, *Ind. Eng. Chem. Res.*, 2009, **48**, 3755.
57. F. Asinger, *Paraffins Chemistry and Technology*, Pergamon Press, Oxford, 1968.
58. M. J. Janse van Vuuren, J. Huyser, G. Kupi and T. Grobler, *Prepr. Pap. Am. Chem. Soc. Div. Pet. Chem.*, 2008, **53** (2), 129.
59. A. Zinger, presented at the World Petrochemical Conference, Houston, TX, 2006.
60. O. R. Minnie, F. W. Petersen and F. R. Samadi, in *Proceedings of the South African Chemical Engineering Congress*, Sun City, 2003, P083.
61. SRI, Process *Economics Program Report 12D, Linear Alpha Olefins*, SRI, Menlo Park, CA, 2001.
62. T. Hahn, in *Proceedings of the South African Chemical Engineering Congress*, Sun City, 2003, P013.
63. A. de Klerk, *Ind. Eng. Chem. Res.*, 2004, **43**, 6349.
64. D. Diamond, T. Hahn, H. Becker and G. Patterson, *Chem. Eng. Process.*, 2004, **43**, 483.
65. J. Ansorge, *Prepr. Pap. Am. Chem. Soc. Div. Fuel Chem.*, 1997, **42**(2), 654.
66. H. Koch and G. Ibing, *Pet. Refiner*, 1943, **22**, 89.
67. D. Gall, *Inst. Petrol. Rev.*, 1947, **1**, 336.
68. F. H. A. Bolder, *Energy Fuels*, 2007, **21**, 1396.
69. W. L. Glowacki, in *Chemistry of Coal Utilisation*, Vol. **2**, ed. H. H. Lowry, Wiley, New York, 1945, p. 1136.

70. A. R. Powell, in *Chemistry of Coal Utilisation*, Vol. **2**, ed. H. H. Lowry, Wiley, New York, 1945, p. 1232.
71. E. O. Rhodes, in *Chemistry of Coal Utilisation*, Vol. **2**, ed. H. H. Lowry, Wiley, New York, 1945, 1287.
72. G. Lund, W. Davis, E. Coogan, G. de Wit and R. Naidoo, *Chem. Technol. (S. Afr.)*, 2006, December, 10.
73. W. H. Hill, in *Chemistry of Coal Utilisation*, Vol. **2**, ed. H. H. Lowry, Wiley, New York, 1945, p. 1008.

CHAPTER 10
Patent Literature

The primary aim of patent literature is to protect the intellectual property of companies. Patents either prevent competitors from practising the protected technology or provide protection in order to license the technology. In the first instance, the aim is to generate income mainly by production and keeping other players out of the market. In the second instance, the aim is to generate income by technology transfer. The latter does not exclude the former.

The dearth of Fischer–Tropsch-specific refining technologies that can be licensed, despite significant patenting activity, indicates that at present companies prefer to prevent the use of their refining know-how by others. This may well be a strategy that is akin to the strategy employed by John D. Rockefeller to dominate the crude oil industry in the USA in the late 19th century.[1] He realised that crude oil has a price, but that crude oil fundamentally has no value. You have to refine the crude oil to produce useful products. By controlling refining capacity, he controlled the market.

The purpose of this overview of patent literature is to highlight the areas of catalysis and conversion chemistry that are of relevance to the field of upgrading and refining of Fischer–Tropsch syncrude. It should therefore be seen as an extension of the literature that was covered in the preceding chapters. By nature, the patent literature is often less rigorous in scientific method and proof, but by definition should be novel and inventive at the time of patenting.

It is common practice for the same patent to be filed in various countries. For the purpose of this review, and in order to avoid duplication, the information published by the United States Patent and Trademark Office (USPTO) was the primary source. The smaller number of patents cited from other jurisdictions is not indicative of a disparity in activity. In some cases patent applications are also cited to cover more recent developments. These patent applications may or may not ultimately become patents.

The increasing number of patents issued since 2000 reflects growing interest in the intellectual property associated with FTS. A portion of these patents deal with the upgrading of primary hydrocarbon products from FTS, such as gases,

RSC Catalysis Series No. 4
Catalysis in the Refining of Fischer–Tropsch Syncrude
By Arno de Klerk and Edward Furimsky

Published by the Royal Society of Chemistry, www.rsc.org

middle distillates, waxy liquids and hard wax. Pretreatment of primary products prior to their upgrading has also been attracting attention and various developments leading to catalysts with good activity and selectivity for conversion of Fischer–Tropsch syncrude can be found. However, it will also be clear from the range of topics presented in previous chapters that the patent landscape associated with Fischer–Tropsch refining catalysis is still sparsely populated.

10.1 Pretreatment of Primary Products Before Refining

10.1.1 Transportation of Syncrude

One of the main disadvantages of FTS compared with syngas-to-methanol is that it produces a wide boiling range of products spanning three or more phases. Due to the heterogeneity of FT syncrude, it requires some upgrading at the production site, whereas methanol is a single transportable liquid product. In this and some other respects,[2] the advantage of methanol over FT syncrude is clear.

In order to overcome this limitation, special procedures have to be devised for the transportation of FT syncrude if the syncrude is to be refined at an off-site facility. By doing so, the economy of scale of the FT refinery can be decoupled from that of FTS. It also allows the FT refinery to be designed in such a way that it can exploit co-refining with other carbon sources, including products from more than one FTS facility.

A process for converting the products from FTS into a pumpable FT syncrude has been proposed.[3] It involves the separation of the FT syncrude into a light fraction (boiling below 288 °C) and a heavy fraction (boiling above 288 °C). The heavy fraction is subjected to HCR/HIS over a fluorided Pt/Al_2O_3 catalyst and then recombined with the light fraction to yield a pumpable syncrude.

The transportation of unrefined wax from LTFT synthesis has also been addressed by patents awarded to Chevron. In one method,[4] granular particles of wax are coated with an inorganic powder that adsorbs the wax without being encapsulated by the wax during a hot drop wax test. In another method,[5,6] the wax particles are transported in a liquid containing 50% water having pH >5. The amount of wax in the liquid medium may vary between 20 and 90%. The size of wax particles is important to ensure stability of the mixture.

10.1.2 Contaminant Removal from Syncrude

A method for monitoring the content of solid carry over in FT primary products has been patented.[7] The product from FTS is irradiated with light and the transmitted light is measured to determine the solids content. On the basis of this analysis, the method and conditions for solids removal can be selected.

The problem of solid carryover is typically encountered with slurry bubble column (liquid–solid separation) and fluidised bed (gas–solid separation) reactor technologies. It is more of a problem in slurry bubble column reactors, where small catalyst particles are suspended in a liquid medium through which the synthesis gas is bubbled. Separating the catalyst particles from the liquid synthesis product is core to slurry bubble column-based Fischer–Tropsch technology.

The use of a magnetised filter element for the removal of catalyst from Fischer–Tropsch products was proposed by Mobil,[8] in a patent that predated the commercial use of slurry bed technology for FTS. Methods that employ a solvent in combination with density separation and optionally electromagnetic separation have also been suggested.[9–11]

Fine catalyst particles and dissolved metals in the FTS product caused serious problems with the startup of the Co-LTFT-based Sasol Slurry Phase Distillate process in the Oryx GTL facility.[12] A method based on inductively couple plasma (ICP) analysis to determine the dissolved metal content in LTFT wax has been proposed.[13] The detrimental effects of dissolved metals in FT syncrude is not limited to LTFT syncrude and the impact of metal carboxylates on commercial FT refining processes has been described.[14]

For the purpose of removing catalyst particles, methods of treating an FTS primary hydrocarbon stream with an active filtering catalyst was disclosed by Mayer *et al.*[15] and Johnson[16] of Chevron. These methods are capable of removing soluble and ultra-fine particulate contamination, fouling agents and/or plugging precursors to minimise plugging of the catalyst beds in downstream upgrading units. The use of a guard bed in order to thermally decompose and deposit dissolved material in products from FTS has also been proposed by Syntroleum.[17]

A process for removing aluminium contaminants from the FTS product was described by Kuperman *et al.* of Chevron.[18] In the proposed process, the contaminated product is treated with at least an equimolar amount of a dicarboxylic acid solution in water, allowing for the precipitation of aluminium-containing components.

Two-step processes for removing metal contaminants and insoluble matter from FTS-derived streams have also been devised by Sasol. The contaminants can be induced to form and grow particles by treating the hydrocarbons with an aqueous stream that may include an acid, which can then be removed by conventional means.[19] Another approach is to expose the syncrude to hydrothermal conditions, thereby converting the metal oxygenate species in the syncrude to products that can be removed by filtration.[20,21]

10.1.3 CO and CO_2 Removal from Syncrude

Light fractions recovered from FTS, especially from LTFT slurry bubble column reactors, contain significant quantities of carbon oxides. If not removed from the light fractions, these carbon oxides can consume a large amount of hydrogen during subsequent hydroprocessing. This may also result in increased catalyst deactivation due to localised hydrogen starvation and high local catalyst temperatures (CO hydrogenation is very exothermic).

Moore and Cambern of Chevron described a method that can be used to remove carbon oxides at appropriate points during processing.[22] Water that is usually present, although in small quantities, is another unwanted compound in the oil product from FTS that may be removed.

10.1.4 Deoxygenation of Syncrude

Oxygenate conversion in FT syncrude was discussed in Chapter 7. It has been pointed out that alumina-rich materials were beneficially used for upgrading Fischer–Tropsch naphtha by deoxygenation and double bond IS to increase its octane number for use as motor gasoline.[23] The importance of having a low cracking activity for catalytic deoxygenation is also described in the patent literature, for example, the use of deactivated cracking catalysts for this purpose.[24]

10.2 Refinery Configurations for Upgrading Syncrude

A refinery design (Figure 10.1) has been proposed by Mobil for the production of mainly motor gasoline from FTS.[25] In the proposed refinery configuration, the total light oil from HTFT synthesis is converted over an H-ZSM-5 catalyst

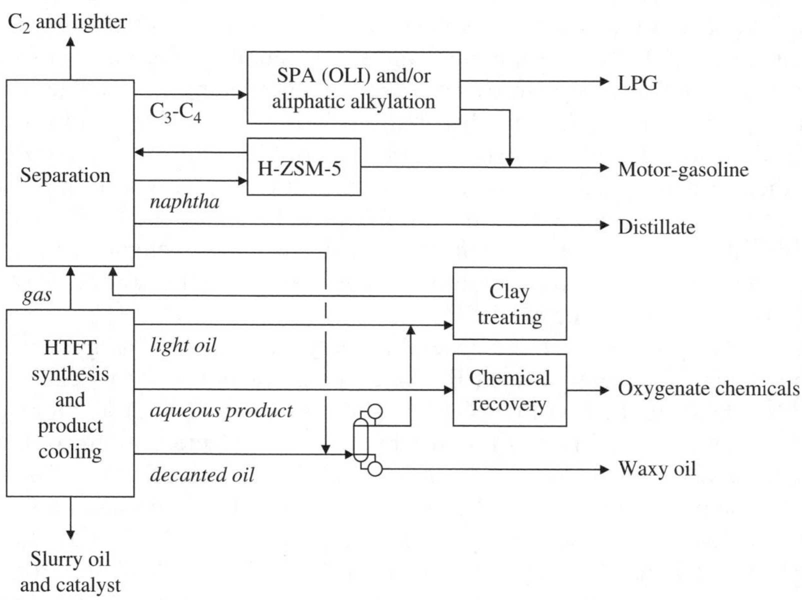

Figure 10.1 Refinery configuration for the upgrading of HTFT syncrude to mainly motor gasoline and some distillate.

to high-octane motor gasoline and some distillate. The proposed operating range for the H-ZSM-5 catalyst is 260–480 °C and 0–1.7 MPa and conversion may be conducted in a fixed bed or a fluidised bed reactor. It is further suggested that the oxygenates that can be recovered from the FT aqueous product may similarly be converted over H-ZSM-5. Since the proposed conversion co-produced a substantial amount of isobutane, it has further been suggested that the refinery configuration should include aliphatic alkylation in combination with solid phosphoric acid OLI to convert the C_3–C_4 fraction into motor gasoline. The examples and specifically the designation of decanted oil as a heavy product made it clear that the design in the patent was restricted to HTFT refining. It was also pointed out that there is benefit in separating the H-ZSM-5 conversion of C_5–C_6 naphtha from C_7 and heavier naphtha.[26] This is due to the difference in the cracking propensity of these two fractions.[27]

A refinement of the design that is shown in Figure 10.1 included HYD of the dienes in the light oil over a Pd- or Pt-containing HYD catalyst before H-ZSM-5 conversion.[28] It has also been claimed that it is advantageous to conduct the H-ZSM-5 conversion in the presence of hydrogen and that Ni/H-ZSM-5 (unsulfided or sulfided) is a beneficial modification of the catalyst. The use of Ni/H-ZSM-5 specifically has been described in a separate patent.[29] In practice, promoting the H-ZSM-5 with Ni would require sulfided operation, since metal leaching will take place under unsulfided conditions with HTFT light oil feed.[30,31]

The benefit of hydrogenating the heavier FT syncrude before fractionation and/or acid-catalysed conversion was realised and Mobil claimed a benefit of hydrogenating the FT material boiling above 150 °C before refining it.[32] This avoids the separation problems associated with atmospheric distillation of FT syncrude due to thermal decomposition in the reboiler (thermal cracking of oxygenates) and downstream carbon number broadening (oxygenate hydrogenation shifts the boiling point distribution). It is worth noting that the claims specified the use of a sulfided hydrotreating catalyst, thereby avoiding problems with metal leaching from the catalyst. Such sulfided hydrotreating operation has been described in another patent.[33] Taken together, the Mobil patents describe the technology and refinery design that were employed for the upgrading of the heavier than naphtha boiling material in the commercial Sasol Synfuels refineries in Secunda.[34,35]

Mobil also proposed a more general refinery configuration (Figure 10.2).[36] In this design, the water-washed FT syncrude is fractionated to yield C_3–C_4, naphtha, distillate, fuel oil (315–455 °C) and residue (>455 °C) fractions. The C_3–C_4 material is used for OLI, either by catalytic polymerisation (over solid phosphoric acid) or by conversion over H-ZSM-5. The possibility of making use of aliphatic alkylation has also been mentioned. The naphtha and fuel oil fractions are converted separately in units employing H-ZSM-5 catalysts. The straight run FT distillate fraction is hydrotreated with the distillate produced by dewaxing of the fuel oil fraction. However, despite the more general description, the refinery design has clearly been devised for HTFT syncrude and not for LTFT syncrude. Mobil also patented variations on the design

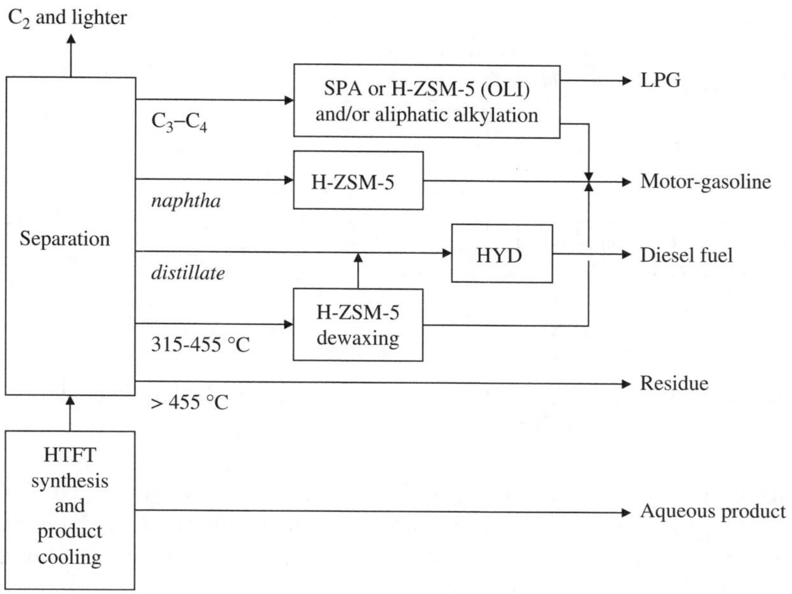

Figure 10.2 General refinery configuration for the upgrading of HTFT syncrude to transportation fuels.

shown in Figure 10.2.[37,38] The basic concept did not change, only the routing and positioning of units.

In what is a variation of the basic refinery design shown in Figure 10.2, the use of another catalyst type was suggested by Boersma and Sie of Shell in order to achieve separate upgrading of the naphtha and heavier than naphtha FTS products.[39] It was claimed that crystalline silicates can be employed for deoxygenation and aromatisation. The catalysts may include other metals (Fe, Ga, Ge) in low concentration, and also some alumina. It was stated that conversion can be conducted over the temperature range 200–500 °C and pressures below 10 MPa. The examples set more realistic operating criteria. FTS-derived naphtha (16.3% alcohols, 63.3% alkenes and 20.4% alkanes) was converted at 375 °C and 0.3 MPa over the silicate catalyst, obtaining a 71% yield of a C_5 and heavier naphtha. The product contained 62.2% alkanes, 18.6% cycloalkanes and 19.2% aromatics. Although not stated, the remaining 29% of the product mass was probably C_4 and lighter gaseous products and water from dehydration of the alcohols. FTS-derived distillate could be converted at 300 °C, but resulted in only moderate product improvement. Product improvement was mainly related to a lowering of the distillate pour point.

The design by Kuo of Mobil,[40,41] which involved pre-refining of the total product over H-ZSM-5, is an interesting departure from more conventional refinery designs. In this design, FTS was followed by hot separation to remove the FT catalyst fines and the remainder of the syncrude was then converted over H-ZSM-5 at elevated pressure and a temperature above 260 °C. Similar

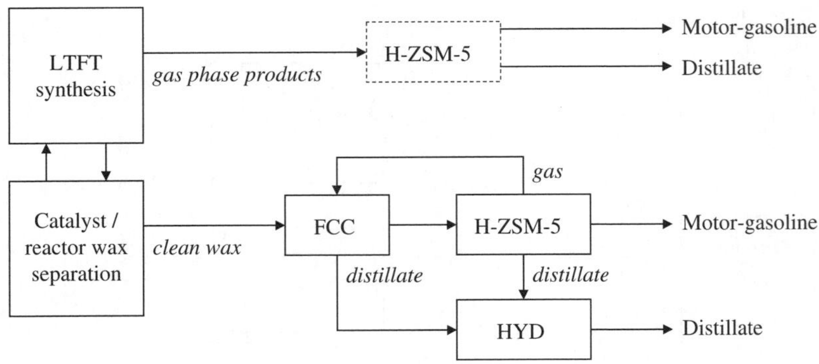

Figure 10.3 Refinery configuration for the upgrading of LTFT syncrude to transportation fuels.

inventions were proposed by Haag and others at Mobil that employed H-ZSM-5,[42,43] and also Beta-zeolite.[44] The concept of converting the total FT syncrude over H-ZSM-5 has since been revisited in the journal literature, for example in the work by Botes[45,46] (see also Section 6.3.5).

The preceding refinery configurations were mainly aimed at the upgrading of HTFT products. Mobil also suggested ways to refine LTFT syncrude. In addition to the hydrocracker-based LTFT design that was pioneered industrially by Shell, a fluid catalytic cracker-based design was proposed (Figure 10.3).[47] Generically stated, the patent specifically anticipated the use of slurry bubble column FTS, which was employed industrially for the first time a few years later.[48] The patent described the use of different H-ZSM-5-based units for converting the oxygenate- and alkene-rich primary products from FTS (in the gas phase) and that from fluid catalytic cracking (FCC) of wax (see also Sections 5.3.3.2 and 6.3.4). It was specifically mentioned that the wax from FTS is so reactive that a low residence time and a low-activity catalyst may be employed during fluid catalytic cracking. It has been suggested that discarded FCC catalyst (preferably faujasite) previously used for crude oil refining may be eminently suitable. It was further pointed out that the coke production from wax cracking is low and that the FCC regenerator would require additional fuel of about 2.1–2.4 MJ kg^{-1} wax to satisfy the heat balance.

It was found that the same conversion units could be used for the upgrading of HTFT and LTFT syncrude to produce on-specification EN228:2004 motor gasoline.[49] Although the units would differ in size depending on the syncrude composition, it has been claimed that a refinery design aimed at maximum motor gasoline production required only a combination of cracking, hydrotreating, aromatisation (reforming), HIS, OLI and aromatic alkylation. It was also noted that some of these units could be combined, for example OLI and aromatic alkylation (also see Section 8.2).[50] Analogous claims have been made for the production of synthetic jet fuel.[51] Irrespective of the syncrude, HTFT or LTFT, the only conversion units required for maximum synthetic jet fuel production complying with Defence Standard 91-91, Issue 6 Jet A-1 property requirements are HCR, hydrotreating, aromatisation (reforming), OLI and aromatic alkylation.

Refinery configurations that produce blending stocks have also been proposed, for example the configuration patented by Chevron.[52] This patent describes a process analogous to that employed in the commercial Oryx GTL facility. The description also includes extensions, such as oxygenate removal by conversion of the naphtha over alumina to produce a more olefinic, less oxygenate-rich naphtha product.

Syntroleum suggested a refinery configuration to produce linear alkylbenzenes and linear alkylbenzenesulfonates from FT syncrude.[53] The light distillate is dehydrated over alumina to increase its alkene content and the resulting product, which is an alkene and alkane mixture, is employed as alkylating feed. The aromatics needed for the alkylation are prepared by conventional catalytic reforming of the C_6–C_{10} naphtha. The heavier material (C_{20} and heavier) is hydrocracked to produce additional naphtha for catalytic reforming, in addition to kerosene and distillate cuts that are considered as product streams.

In an analogous proposal by Chevron,[54] catalytic dehydrogenation was proposed for the production of the alkenes. Aromatics production by catalytic reforming of the C_6–C_8 fraction over a nonacidic Pt/L-zeolite-based catalyst (Aromax technology) has been recommended.

10.3 Upgrading of Fischer–Tropsch Primary Products

The subsequent discussion of patents that describe refining processes for the upgrading of materials from FTS has been organised based on feed fraction, rather than by product or purpose. The patents cover upgrading of all hydrocarbon streams, namely light alkenes, naphtha, middle distillates, residues (wax) and aqueous products. It will be noticed that some patent literature paid specific attention to the FT oxygenates in these streams to differentiate them from analogous crude oil upgrading processes.

10.3.1 Light Alkene Conversion

It was indicated earlier that the content of light, normally gaseous alkenes depends on the type of FTS and its operating parameters. Generally, iron-based FTS yields a more olefinic product, but the patents that describe light alkene upgrading are also applicable to light alkenes derived from other refining processes, for example FCC of FT wax. Maximising the yield of middle distillates emerged as one of the main topics of invention.

Although solid phosphoric acid is not a good catalyst for distillate production, it is employed industrially in this role with light HTFT alkenes.[55,56] This awkward use of SPA with Fischer–Tropsch alkenes was anticipated by UOP.[57] The patent also claims applicability to feed with oxygenates, where the oxygenates have an oxygen content in the range 0.1–10 mass%.

Du Toit described a process for the production of diesel boiling range hydrocarbons by OLI of an olefinic stream containing branched short-chain

(C_3–C_8) alkenes using a medium-pore acid zeolite catalyst.[58] The catalyst may be H-ZSM-5. Its shape selectivity will ensure that the higher hydrocarbons produced after OLI are not excessively branched hydrocarbons. The diesel boiling range hydrocarbons thus produced are predominantly methyl branched with a small amount ethyl branching. The reactor used for the OLI process operates between 5 and 8 MPa and between 200 and 340 °C. It incorporates continuous catalyst regeneration to overcome the gradual coking of the catalyst during operation, which is the main differentiating benefit of this invention.

A modification of an SPA-catalyzed OLI process has been suggested for reduction in refinery benzene.[50] This invention has been demonstrated on an industrial scale with FT feed.[59] The benzene-containing refinery material is co-fed with light FT alkenes at a high alkene-to-aromatics ratio. By doing so, the benzene is alkylated without disrupting the OLI process.

Metathesis of butene-containing feed materials from FTS in order to produce propene has been proposed as an upgrading strategy.[60] In this proposed process, metathesis is conducted over transition metal oxide catalysts, such as WO_3/SiO_2.

10.3.2 Naphtha Conversion

There is overlap between the conversion processes for light alkenes and those for naphtha-range alkenes, with a broad range of alkenes generally being claimed in the patent literature. Conversions of alkenes specifically to produce distillates and lubricating oils are topics of many inventions by Chevron.

A process was described for making a lube base stock from a light and a medium alkene fraction.[61] In this process, the light alkene fraction is brought into contact with the first OLI catalyst in an OLI zone to produce the first alkene product. The medium alkene fraction and the first alkene product are then contacted with a second OLI catalyst to produce a second alkene product. This second alkene product is separated into a light by-product fraction and a heavy product fraction. The latter fraction includes hydrocarbons in the lubricant base oil range. The first OLI catalyst can be the same as or different from the second OLI catalyst. The OLI catalysts can be nickel on ZSM-5. Alternatively, the OLI catalysts can include an acidic ionic liquid. In practice, it is likely that two different catalysts should be used. The first OLI step provides a pathway for converting the light alkenes to heavier alkenes and will benefit from a catalyst with pore-constraining geometry to reduce branching. The second OLI step involves typical 'polyalphaolefin' (PAO)-type OLI, which generally requires a very different catalyst with good accessibility to allow the formation of heavy products.

Alkenes can also be prepared by dehydrogenation of paraffinic feeds. The alkenes thus produced can subsequently be used as feed for an OLI process. For example, they can be oligomerised to produce lubricating base oil.[62] If necessary, the oligomerised product is hydrogenated to eliminate any remaining alkenes. In another related invention disclosed by O'Rear *et al.*,[63] the olefinic feedstock prepared by dehydrogenation of FTS products is oligomerised to obtain a lube base oil fraction. The OLI catalyst includes a zeolitic support and a Group VIII metal, for example, ZSM-5 and Ni.

According to a process disclosed by Moore and van Gelder,[64] the alkenes and oxygenates present in material from FTS can be hydrotreated to form alkanes before the alkanes are subjected to HIS to form branched alkanes. Hydrocarbons with chain lengths above a desired value, for example C_{24}, are hydrocracked. The hydrogenolysis that would otherwise form undesired C_1–C_4 compounds is minimised by the judicious selection of noble metal catalysts.

A patent from Chevron describes a process for converting Fischer–Tropsch products comprising oxygenates and C_{6+} alkenes to valuable light alkenes, such as propene, butenes and some pentenes, while leaving the alkanes largely unconverted.[65] The light alkenes thus formed can easily be separated and used for a variety of purposes. The acidic alkene cracking catalyst is a zeolite having 10-membered ring pores, such as ZSM-5 or ZSM-11, and containing a binder.

Conversion of the alcohols and alkenes in FT naphtha over an acid catalyst to produce ethers has been proposed by O'Rear *et al.* of Chevron.[66] This conversion is conducted in the presence of alkanes and leaves the alkanes unconverted. The ethers thus formed are higher boiling than the feed and can easily be separated by distillation. The ethers can then be hydrolysed with water over an acid catalyst to regenerate the alcohols, and the alcohols can be used as lubricity enhancers for distillate range fuels. Preferred acid catalysts for alcohol-rich feeds are zeolites, whereas those for alkene-rich feeds and for hydrolysis of ethers to alcohols are acidic resins.

A combined process for hydrotreating and isomerising a C_4–C_7 feedstock was disclosed by Schmidt and Haizman.[67] In this process, the feed is contacted in a hydrotreater with a catalyst comprising a Group VIB metal and a Group VIII metal on an alumina support to remove sulfur (not present in material from FTS) and oxygen. The effluent from the hydrotreater passes to a first separator that separates the effluent into a gas stream comprising hydrogen, hydrogen sulfide and water and a treated stream comprising C_4–C_7 hydrocarbons. The treated stream is mixed with a second hydrogen-containing stream and becomes the isomerisation feed. The isomerisation feed is contacted with a HIS catalyst comprising a crystalline aluminosilicate and a Group VIII metal under typical HIS conditions. The effluent from the reaction zone enters a stabiliser where it is separated into a product stream of C_4–C_7 hydrocarbons and a second gas stream which is removed from the process. This is essentially a Fischer–Tropsch HIS process with feed pretreatment.

Another conversion process with Fischer–Tropsch naphtha that has been patented is aromatic alkylation to produce alkylbenzenes. A process is described in which a combined alkene and alkane mixture from FTS can be used for benzene alkylation.[68] The alkenes are directly alkylated and the alkanes are recycled, chlorinated and then used for alkylation as chloroalkanes over $AlCl_3$. In this way, the alkenes and alkanes are alkylated in the same reactor. This patent demonstrates how the benefit of having alkenes in the feed can be exploited in combination with alkanes.

Rangarajan *et al.* suggested a number of upgrading pathways that may be considered for FT naphtha refining.[69] A portion of the FT naphtha stream may be aromatised to produce an aromatic hydrocarbon stream with improved octane number. A portion of the aromatic hydrocarbon stream may also be isomerised to produce a hydrocarbon stream with an even better octane rating. Alternatively, the method can be individually applied to at least one of three naphtha cuts: C_4–C_5, C_6–C_8 and C_9–C_{11}. Furthermore, the C_6–C_8 stream can be either aromatised to form an aromatic hydrocarbon stream with a higher octane number or it can be subjected to steam cracking to produce alkenes. Similarly, the C_9–C_{11} stream may be cracked to produce alkenes. Alternatively, a portion of the C_9–C_{11} stream can be sold as solvents.

A way to overcome the poor performance of LTFT naphtha in conventional catalytic reforming over $PtRe/Cl^-/Al_2O_3$ catalysts (Section 8.1.1) has been suggested by Baird of Exxon.[70] It was recognised that the liquid yield during catalytic reforming can be improved if the lean naphtha from FTS is blended with materials having more cyclic compounds before being reformed. Essentially it provides a way to improve the $N+2A$ by blending.

10.3.3 Middle Distillate Conversion

Middle distillate fractions separated from FTS usually require upgrading to attain desirable diesel fuel performance characteristics, such as good cold flow properties and storage stability. Much of the patent literature is focused on HIS and HCR when the feed includes heavier fractions with the distillate. Benefits are claimed for a variety of HIS and HCR catalyst systems. Some blending solutions are also suggested in the patent literature.

A process has been proposed for the conversion of hydrotreated 175–455 °C FT material over a metal-promoted H-ZSM-5 catalyst to produce mainly jet fuel.[71,72] In this invention, it is a prerequisite that the feed must be hydrogenated to saturate alkenes and oxygenates in order to avoid deactivation of the HIS catalyst. The preferred HIS catalyst contains a 0.5–5% loading of Ni on H-ZSM-5, but the patent was not restricted to Ni/H-ZSM-5 catalysts. Typical operating conditions are 260–425 °C, 0.7–5.5 MPa and LHSV 0.5–5 h^{-1}, and some HCR is implied. No examples were given that provided properties of the jet fuel thus produced.

A process for converting a Fischer–Tropsch light oil stream into jet fuel was disclosed by Wittenbrink *et al.*[73,74] In this process, the oil stream flows counter-current to a hydrogen-containing gas while contacting a HIS catalyst. The HIS catalysts may comprise a metal that is active for HYD and an acidic support. The active metals are selected from Groups IB, VIB and VIII, for example Cu, Mo and Pd. A modification of the process may include a HIS reactor upstream of a dewaxing reactor, both operating in counter-current flow mode.

A process for producing a winter diesel fuel consisting of two reaction zones was disclosed by Berlowitz *et al.*[75] In this process, the effluent from the first zone, containing a HIS catalyst, enters the second reaction zone with a catalytic dewaxing catalyst. The catalytic dewaxing catalyst is a molecular sieve with one-dimensional channels containing a 10-membered ring structure. The dewaxing catalyst is selected from the group consisting of SAPO-11, SAPO-41, ZSM-22, ZSM-23, ZSM-48, ZSM-57, SSZ-31, SSZ-32, SSZ-41 and SSZ-43.

Miller *et al.* patented a process for producing a diesel fuel having a branched to linear alkane mole ratio of 5:1 or higher.[76–78] This highly isomerised distillate is produced from feed containing at least 40% C_{10} and heavier *n*-alkanes and at least 20% C_{26} and heavier *n*-alkanes. It is produced during the HIS/HCR of the feed at 340–420 °C and around 2 MPa H_2 pressure over a catalyst comprising a molecular sieve and a noble metal. Preferred molecular sieves include SAPO-11, SAPO-31, SAPO-41 and/or their mixtures in combination with Pt.

The process described by Wittenbrink and co-workers[79–82] involves the separation of FT syncrude into a light and heavy fraction. The heavy fraction (>370 °C) is subjected to HIS/HCR and then recombined with the untreated light fraction (distillate). By doing so, the product has excellent lubricity, good oxidative stability, high cetane number and good cold flow properties. This is essentially a process that exploits some of the benefits of retaining oxygenates as discussed in Section 7.3. Any bifunctional catalyst consisting of a metal HYD component and an acidic component and that is useful in HIS or HCR may be satisfactory for the conversion of the heavy fraction. For example, supported Pt and Pd catalysts or catalysts containing Ni or Co may be suitable. Preferred supports include alumina, silica–alumina, silicoaluminophosphates and ultrastable Y-zeolites.

A process for producing diesel oil by blending the FT distillate with a similar fraction of a petroleum origin has been proposed to achieve a diesel fuel with acceptable density.[83] The blending is also effective in reducing the sulfur content of the petroleum-derived fraction.

Rosenbaum *et al.* patented a process for treating nitrogen-containing alkane-rich products derived from FTS.[84,85] Oxygen and other impurities are removed in combination with nitrogen. The nitrogen content of the purified product is monitored and the conditions of the purification step are adjusted to increase nitrogen removal if the nitrogen content of the purified product exceeds a preselected value. Different HYD catalysts can be used for the purification. For example, a noble metal from Group VIIIA, such as Pt or Pd, on an alumina or siliceous support or unsulfided Group VIIIA and Group VIB metals, such as Ni and Mo, on an alumina or siliceous support are all suitable catalysts.

A process patented by Chevron describes co-processing of products from FTS with petroleum-derived liquids.[86] According to this process, one or more fractions from FTS are blended with one or more petroleum-derived fractions. If necessary, the crude oil fractions can be pretreated to lower the sulfur content so that the blend has an acceptable sulfur level. The fraction from FTS may include different fractions, for example, C_5–C_{20} hydrocarbons, C_{20} and heavier hydrocarbons or C_5 and heavier hydrocarbons. In this process, the hydroprocessing catalysts contain noble metals. Based on a similar concept, Moore and van Gelder disclosed a process for processing C_4 and lighter and C_5 and heavier fractions isolated from natural gas.[87] The C_4 and lighter fraction is converted into syngas for FTS. The C_5 and heavier fraction is blended with a similar fraction from FTS to obtain a blend containing less than 200 $\mu g\,g^{-1}$ of sulfur. If necessary, the blend can be additionally processed to attain an acceptable sulfur level. Further hydroprocessing employs a noble metal containing catalyst that is stable with feed containing less than 200 $\mu g\,g^{-1}$ of sulfur.

10.3.4 Residue and Wax Conversion

A portion of the boiling range of waxy hydrocarbon liquids and wax from FTS overlaps with that of lubricating oil base stock. Catalytic dewaxing emerged as an important field in the patent literature. Ways to upgrade FT feed to lube base oil have been described and the production of the middle distillates and gasoline from FT feeds also received attention.

A refinery configuration employing FCC of LTFT wax (Figure 10.3) has already been discussed. More recently, Shell described an analogous FCC-based process for the production of motor gasoline.[88]

The advantage of employing low-pressure HCR of FT wax has been claimed in patents by Chevron[89] and UOP.[90] Mechanistically this makes a lot of sense, since it increases the alkene concentration on the catalytic surface and thereby the reaction rate, while exploiting the low coking tendency of FT wax to limit catalyst deactivation in the more alkene-rich operating environment.

An invention by de Haan *et al.* relates to an HCR process for producing middle distillates having good cold flow properties and a high cetane number.[91] The middle distillate produced by the process contains predominantly methyl, ethyl and/or propyl branched alkanes. Catalysts for the HCR step are of the bifunctional type and contain sites active for cracking and for HYD. Catalytic metals active for HYD include Group VIII noble metals, such as Pt and Pd, or sulfided Group VIII base metals, such as Ni and Co, which may or may not include a sulfided Group VI metal, such as Mo. The support for the metals can be any refractory oxide, such as silica, alumina, titania, zirconia, vanadia and other Group III, IV, VA and VI oxides, alone or in combination with other refractory oxides. Alternatively, the support can consist partly or totally of zeolite. However, for this invention the preferred support is ASA. Essentially the patent states than any typical HCR catalyst can be employed to convert heavy FT feed materials into middle distillates.

Tsao *et al.* disclosed a process that is suitable for selectively producing distillate with increased cetane number from a hydrocarbon feedstock.[92] The process involves contacting the feedstock with a catalyst consisting of a large-pore crystalline molecular sieve having a faujasite structure and an α-acidity of about 0.3 or less. The catalyst also contains a dispersed Group VIII noble metal such as Pt, which catalyses the HYD/HCR of the aromatic and naphthenic species in the feedstock. This type of conversion is not specific to FT material, but is relevant to the conversion of HTFT residue, which is rich in cyclic compounds.

Moore disclosed an integrated method for producing liquid fuels from primary FT products.[93] According to this method, the primary products are separated into a light fraction and a heavy fraction. The latter is subjected to

HCR through multiple catalyst beds, to reduce the chain length. The HCR products are directed to the last bed comprised of a HIS catalyst. After the last bed, the products are combined with the light fraction. The combined fractions are subjected to hydroprocessing to remove double bonds, reduce oxygenates to alkanes and, if necessary, final HDS and HDN.[94] The preferred HCR and HIS catalyst systems include one or more of zeolite Y, zeolite ultrastable Y, SAPO-11, SAPO-31, SAPO-37, SAPO-41, ZSM-5, ZSM-11, ZSM-48 and SSZ-32. The deHYD/HYD component may comprise Mo, Ni, Pt, Pd, Co and/or their mixtures. Conceptually the proposal by Moore is very similar to that by Wittenbrink and co-workers that was discussed earlier.[79–82]

A process for HIS and dewaxing a hydrocarbon feed was described that employs a large pore size, small crystal size molecular sieve and an intermediate pore size, small crystal size molecular sieve to produce a dewaxed product with a reduced pour point and a reduced cloud point.[95] In this process, the feed is contacted with the molecular sieves sequentially, first with the large-pore sieve followed by the intermediate-pore sieve. Preferably, the intermediate-pore crystalline molecular sieve is selected from the group consisting of ZSM-23, ZSM-48 and SAPO-11, whereas the large-pore crystalline molecular sieve is Beta-zeolite.

Dewaxing processes for hydrocarbon feedstocks were also disclosed using catalysts comprised of non-zeolitic molecular sieves, amongst some other SAPO-type materials.[96,97] The products of the dewaxing processes are characterised by lower pour points than the hydrocarbon feedstock.

A process for dewaxing a liquid hydrocarbon from FTS using a particulate solid dewaxing catalyst dispersed in the feed was disclosed.[98] The preferred dewaxing catalyst includes a shape-selective crystalline zeolite, such as a metal-exchanged ZSM-5, although other similar zeolites may also be suitably employed as a catalyst material. The pour point and wax content of waxy feed can also be reduced under standard catalytic dewaxing conditions using an aluminosilicate catalyst with a very low crystallinity.[99] Such materials are derived from crystalline aluminosilicate zeolites exchanged with cations.

It has been observed that dewaxing catalysts can be selectively activated by treatment with oxygenates.[100–102] The HIS activity of the catalyst was enhanced by contacting it with a stream containing oxygenates at a level of least $100\,\mu g\,g^{-1}$ as oxygen. Oxygenates such as alcohols, carboxylic acids, esters, aldehydes and ketones can be used. In related disclosures, the selectively activated catalysts that were pretreated with oxygenates, when used to dewax waxy hydrocarbons, improved the yield of isomerate at equivalent pour point over a dewaxing catalyst that has not been oxygenate treated.[101–103] Such treated catalysts were used in the process disclosed by Grove *et al.* for catalytic dewaxing and catalytic HCR of hydrocarbon streams containing waxy components and having an end boiling point above 340 °C.[99] The feed was contacted at super-atmospheric H_2 partial pressure, with a HIS/dewaxing catalyst that included ZSM-48 and with a HCR catalyst to produce an upgraded product with a reduced wax content. In an analogous process disclosed by Bishop *et al.*,[104] the waxy hydrocarbons are hydrodewaxed with a reduced conversion to lower boiling hydrocarbons in the presence of H_2 using an unsulfided 10-membered ring, one-dimensional zeolite catalyst. The catalyst support can be selected from one of ZSM-22, ZSM-23, ZSM-35 ZSM-48, ZSM-57, SSZ-32 or rare earth-exchanged ferrierite and a Group VIII metal component. In this case, the catalyst was reduced and then contacted with the synthesised hydrocarbons containing one or more oxygenates, including indigenous oxygenates.

Baker and Dougherty disclosed a two-stage process for catalytically dewaxing products from FTS with minimal ageing of the dewaxing catalyst.[105] In this process, the feed is treated with a catalyst system comprising of a hydrotreating stage upstream of the dewaxing stage. The hydrotreating catalyst is loaded with noble metals. The highly shape-selective dewaxing catalyst is comprised of a constrained intermediate-pore crystalline material, which is loaded with a noble metal.

A process for reducing the wax content of the wax-containing hydrocarbon feedstocks to produce middle distillates, including a low freezing point jet fuel and/or low pour point and low cloud point diesel fuel and heating oil, was disclosed by Sonnemans *et al.*[106] The process involves contacting the feedstock with an HCR catalyst containing Groups VIB and VIII metals and a large-pore zeolite such as a Y-type zeolite. Subsequently, the effluent enters a dewaxing zone comprising of a fixed bed of the catalyst containing a crystalline, intermediate pore size molecular sieve selected from metallosilicates and silicoaluminophosphates.

A two-stage process for producing high-octane naphtha range branched alkanes from waxy distillates was disclosed by Girgis and Tsao.[107] In the first stage, linear and branched alkanes having two or fewer alkyl substituents is hydroisomerised to give multi-branched alkanes. The multi-branched alkanes from the first stage are then selectively cracked in a second stage to naphtha range multi-branched alkanes. The resulting branched alkanes are more branched than those obtained by HCR alone, resulting in a naphtha with a higher octane number. Suggested catalysts for HCR are NiW-, Pd- or Pt-promoted USY zeolites. Hydroisomerisation of the feed is conducted over a sulfided HIS catalyst.

According to the process patented by Berlowitz *et al.*,[108] a clean diesel fuel or diesel blending stock is produced from FT wax by separating the wax into heavier and lighter fractions, followed by a HIS step. Suitable catalysts are catalysts containing a supported noble metal, such as Pt and Pd, and catalysts containing one or more Group VIII base metals (Ni, Co), which may or may not include a Group VI metal (Mo). The support for the metals can be any refractory oxide or zeolite or their mixtures, such as silica, alumina, silica–alumina. silicoaluminophosphates, titania, zirconia, vanadia and other Group III, IV or VA or VI oxides, and also Y sieves, such as ultrastable Y sieves. (The general description of possible catalysts is very similar to that in the patent by de Haan *et al.*[91]) Preferred supports include alumina and silica–alumina where the silica concentration of the bulk support is less than about 50 mass%. If a winter diesel fuel is the final product, the HIS step is followed by a dewaxing step to attain better cold flow properties.[109]

An integrated process for producing a liquid hydrocarbon stream from FTS wax without removing particulate contaminants, such as catalyst fines (Section 10.1.2), was disclosed by O'Rear *et al.*[110] The wax is subjected to HIS in an upflow reactor under typical HIS conditions. The design of the catalyst bed is such that it permits passage of the particulate contaminants. The particulates are then removed from the upgraded liquid product by filtration, distillation and/or centrifugation. Removal of the particulate contaminants from the upgraded liquid hydrocarbon products is significantly easier than removing the particulate matter from the unprocessed heavy waxy products.

A petroleum wax-containing feed can be converted to a high-grade middle distillate by employing a homogeneous pretreatment before dewaxing and HCR.[111] The feed is pretreated by contacting it with a homogeneous solution of an acid diluted in an alcohol–water mixture. The pretreated feed is then contacted in the presence of hydrogen with a hydrodewaxing (HIS) catalyst followed by a HCR catalyst in sequence and with no intermediate separation. The hydrodewaxing catalyst is typically an intermediate-pore molecular sieve such as metallosilicates and silicoaluminophosphates, having a pore diameter in the range 0.5–0.7 nm. The HCR catalyst is typically a large-pore zeolite with a pore diameter in the range 0.7–1.5 nm. This invention indicates the similarity between the upgrading of FT wax and that of a waxy fraction of petroleum origin.

Heavy paraffinic feeds (petroleum wax, FT wax and deoiled waxes) with an end boiling point exceeding 650 °C could be converted into a good-quality base oil by HIS.[112] If necessary, HIS may be preceded by hydroprocessing for heteroatom removal. Among the HIS catalysts, silica–alumina-based zeolites and aluminophosphates (SAPO and MAPO) were identified as exhibiting good activity for such applications. With a suitable catalyst, lube base oil may be an attractive outlet for FT waxes. Indeed, a patent by Miller describes the IS of the FT wax to obtain a blending component with a petroleum-derived base oil.[113] The resulting blend had a lower pour point and cloud point, and also higher viscosity index, compared with the individual blending components.

O'Rear and Biscardi disclosed a process for the preparation of lube base stocks from heavy FT fractions.[114] The process involves feed material from FTS that has a T95 boiling point below 630 °C. The feed is catalytically dewaxed. One or more of the fractions can also be obtained from other sources, for example, *via* distillation of crude oil. Catalysts that are useful for dewaxing are typically 12- and 10-membered ring zeolites. Zeolites of these classes include ZSM-5, ZSM-11, ZSM-22, ZSM-23, ZSM-35 and MOR.

According to a process disclosed by Degnan and Mazzone,[115] FT waxes can be converted into high viscosity index lubricants by HIS over a low-acidity molecular sieve containing a noble metal. The HIS stage is operated at high pressure, at least 7 MPa of H_2, and around 340 °C. If desirable, a final dewaxing step to obtain a better pour point may be used.

A process for preparing a lubricating oil base stock having good cold flow properties was described by Leta *et al.*[116] The process includes an ASA-based HIS catalyst having a pore volume less than 0.99 ml g^{-1}, an alumina content in the range 30–50 mass% and an isoelectric point in the range 4.5–6.5. The silica–alumina may be modified with a rare earth oxide or yttrium, boron and magnesia. Partially isomerised feed is subjected to a catalytic dewaxing step using an intermediate-pore crystalline molecular sieve such as a metallosilicate or metallophosphate.

Miller and Rosenbaum disclosed a method for producing lubricant base oils by separating light and heavy base oil fractions from the primary FTS products and HIS of the fractions over a medium-pore molecular sieve catalyst to produce an isomerised light lubricant base oil fraction and a heavy fraction, both having desirable pour points and cloud points.[117,118] The medium-pore molecular sieve catalyst comprises a molecular sieve selected from SAPO-11, ZSM-3, ZSM-22, ZSM-23 and SSZ–32.

In an invention by Berlowitz *et al.*,[119] a waxy feed such as obtained during FTS and/or derived from paraffinic crudes does not require a hydroprocessing step before HIS. After distilling off the products boiling below 350 °C, the hydroisomerate is subjected to catalytic dewaxing to obtain lubricating base oil.

A process for HCR of heavy hydrocarbon feeds using a catalyst containing a HYD/deHYD component such as a noble metal and an acidic solid component including a Group IVB metal oxide modified with an oxyanion of a Group VIB metal was described.[120] The HCR product had high branched-to-linear alkane ratios. Moreover, at high conversions, ethane and methane formation was minimal. The HCR step is useful in processes for producing high-quality lubricating oil base stocks, along with naphtha and distillate products.

A process for preparing hydrocarbons in the lube base oil range from a fraction with an average molecular weight above a target molecular weight and a fraction with an average molecular weight below a target molecular weight *via* molecular averaging has been described.[121] The fractions can be obtained from FTS and/or the distillation of crude oil. Molecular averaging converts the fractions into a product with a desirable molecular weight distribution, for use in preparing a lube oil composition. If necessary, the product can be isomerised to attain desirable cold flow properties.

A number of patents involving petroleum wax are included here to indicate a similarity in upgrading conditions compared with FTS wax.[122–124] According to invention of Marler and Mazzone,[124] petroleum wax feeds can be converted into high viscosity index lubricants by a two-step HCR–HIS process. During this process, the wax feed is initially subjected to HCR under mild conditions with a conversion to non-lube range products of no more than about 40 mass% of the feed using an amorphous or mesoporous crystalline catalyst. This catalyst preferentially removes the aromatic components present in the initial feed. The hydrocracked effluent is then subjected to HIS in a second step using a low-acidity Beta-zeolite-based HIS catalyst, which results in preferential HIS of the *n*-alkanes to produce less waxy, high viscosity index branched alkanes. The second-stage conversion is carried out in the presence of a catalyst which contains a HYD component, preferably a noble metal such as platinum, on a mesoporous support material. The mesoporous support material (for the first and second steps) is comprised of a non-layered, porous, crystalline aluminosilicate material having a uniform, hexagonal arrangement of pores with diameters of at least about 1.3 nm.

The desirability of cycloalkanes for the production of on-specification EN590:2004 diesel fuel from FTS has been pointed out.[125] The patent application of ChevronTexaco,[126] which discloses a process for converting FTS wax to produce a lubricating oil with a mono-cycloalkane content of more than 10%, is consequently also relevant to diesel fuel production. This type of conversion can be achieved with catalysts such as Pt/SAPO-11 and Pt/SSZ-32.

Hard wax has been finding numerous industrial applications. In this regard, several patents indicate that a purification step is necessary before specifications of the final product can be attained. In some cases, the objective is a decrease in the melting point of wax to desirable levels. A high-purity wax from an FT slurry process can be prepared according to the invention of Wittenbrink and Ryan.[127] As part of this process, the synthesis slurry comprising liquid product and catalyst particles is purified in a treatment zone by contacting it with hydrogen and/or a hydrogen-containing gas to removes impurities. Purified wax is separated and removed *in situ*. This minimises the need for further treating of the wax product.

Inventions by Wittenbrink and co-workers describe a mild hydrotreating process which removes the oxygenates, alkenes and any aromatic species that may be present in a raw FT wax.[128,129] At the same time, the hardness of the wax is reduced. The process involves passing the raw wax over a HIS catalyst under mild conditions such that chemical conversions (HYD and mild HIS) take place, while less than 10% boiling point conversion (HCR) occurs, thus preserving the overall yield of the wax product. The HIS catalyst comprises a non-noble Group VIII metal in conjunction with a Group VI metal such as Mo, supported on an acidic support such as silica–alumina.

The preparation of microcrystalline waxes by HIS of FT wax has been described in a Schümann–Sasol patent.[130] It was claimed that HIS of FT wax over a metal-promoted zeolite catalyst with pore size in the range 0.5–0.8 nm would produce a microcrystalline wax. Preferred operating conditions are 230–270 °C, 3–8 MPa and LHSV 0.2–0.8 h^{-1}.

A different method for the production of microcrystalline wax has been described by Shell.[131] The FT wax is converted over a noble metal-promoted porous silica–alumina carrier material. The catalyst preferably has 5–50% macroporosity (pores $>$ 10 nm), with Pt, Pd or a combination of both as noble metal promoter. Typical operating conditions are 250–350 °C, 3–6 MPa and WHSV 0.5–5 h^{-1}.

10.3.5 Aqueous Product Conversion

An aqueous product refinery flow scheme (Figure 10.4) was proposed by Holland and Tabak of Mobil.[132] The nonacidic oxygenates are separated from the bulk of the water and carboxylic acids by distillation. The remainder of the oxygenates are dehydrated, preferably over γ-alumina, to produce an alkene-rich product. This alkene-rich product is then distilled, with the pentene and

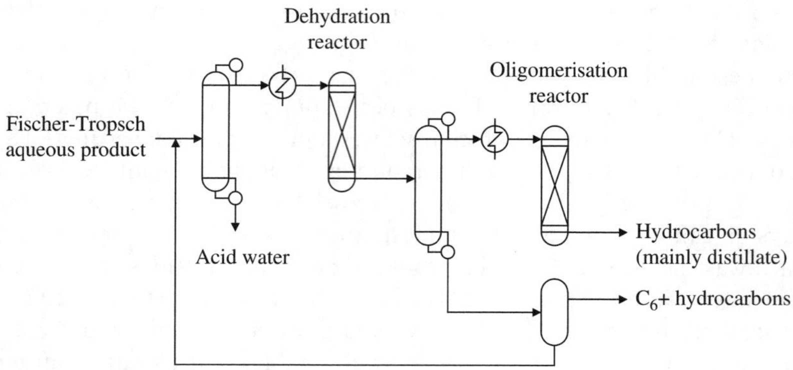

Figure 10.4 Fischer–Tropsch aqueous product refinery configuration for the recovery and conversion of oxygenates into alkenes.

lighter boiling material becoming a feed for OLI over a zeolite catalyst, which is preferably H-ZSM-5. The main product from such an OLI process is middle distillates. The heavier than pentene boiling material is phase separated to recover the hydrocarbons and to recycle the oxygenates.

The alcohols separated from the FT aqueous product may be etherified over an acidic resin catalyst with alkenes to produce a fuel additive.[133] Such a fuel additive has better water tolerance than just the alcohols.

Processes for the purification of water from the aqueous product (reaction water) of FTS were disclosed by Sasol.[134–138] The processes involve an equilibrium staged separation for removing nonacidic oxygenates and a secondary treatment stage comprising at least one membrane separation process for removing some suspended solids and acidic oxygenates. The tertiary treatment stage is used to remove dissolve salts from water. In another patent, the tertiary treatment stage involves a biological treatment for removing acidic oxygenates and a quartic treatment stage comprising solid–liquid separation for removing solids from at least a portion of the tertiary water-enriched stream. Another process for the production of highly purified water from FT reaction water includes distillation as a primary treatment stage, evaporation as a secondary treatment stage, aerobic treatment as a tertiary treatment stage, solid–liquid separation as a quartic treatment stage and membrane separation as a final treatment stage. Another modification of the processes includes a biological treatment using anaerobic and aerobic digestion as a secondary treatment stage following distillation before solid–liquid separation and the removal of dissolved salts and organics as the final stage. Alternatively, the acid water may be beneficially employed as feed for the microbial production of γ-linolenic acid.[139] A specific distillation design for the primary separation stage has also been suggested.[140]

An approach that involves thermal oxidation of the products dissolved in the FTS aqueous product has been proposed by Chevron.[141] The aqueous product that is obtained by condensation from FTS is vaporised by indirect heat exchange with the hot FT product. The vaporised aqueous product is then converted in a thermal oxidiser to produce flue gas.

The acid-containing aqueous product from FTS may also be beneficially employed for gasification[142] or steam cracking,[143] thereby beneficially using the oxygenates contained therein.

References

1. C. Tugendhat and A. Hamilton, *Oil. The Biggest Business*, Eyre Methuen, London, 1975.
2. G. A. Olah, A. Goeppert and G. K. S. Prakash, *Beyond Oil and Gas: the Methanol Economy*, Wiley-VCH, Weinheim, 2006.
3. G. P. Hamner, *US Pat.*, 4832819, 1989.
4. G. H. Dieckmann and D. J. O'Rear, *US Pat.*, 7501019, 2009.
5. G. H. Dieckmann, M. R. Buetzow and D. J. O'Rear, *US Pat.*, 7488411, 2009.
6. G. H. Dieckmann, M. R. Buetzow and D. J. O'Rear, *US Pat.*, 7479216, 2009.
7. B. Smith, R. Aviani and D. J. O'Rear, *US Pat.*, 6881760, 2005.
8. J. A. Brennan, A. W. Chester and F. Y. Chu, *US Pat.*, 4605678, 1986.
9. C. M. White, M. S. Quiring, K. L. Jensen, R. F. Hickey and L. D. Gillham, *US Pat.*, 5827903, 1998.
10. M. D. Spena, D. S. Jack and D. Fraenkel, *US Pat. Appl.*, 2006/0111232, 2006.
11. H. R. Khakdaman, B. A. Horri, H. M. Varkiani, S. S. Noori and K. J. Jozani, *US Pat. Appl.*, 2007/0039852, 2007.
12. Anon., *Pet. Econ.*, 2008, **75** (6), 36.
13. A. P. Darling, J. P. Coetzee and D. K. Yoell, *Int. Pat. Appl.*, WO 2008/116233, 2008.
14. A. de Klerk, *Catal. Today*, 2008, **130**, 439.
15. J. F. Mayer, A. Rainis, R. O. Moore Jr., *US Pat.*, 7150823, 2006.
16. D. R. Johnson, *US Pat.*, 7332073, 2008.
17. P. Z. Havlik, N. Jannasch, P. Ahner and H. L. Tomlinson, *US Pat. Appl.*, 2007/0215521, 2007.
18. A. Kuperman, L. M. Bul, D. J. O'Rear and D. Kuehne, *US Pat.*, 7416656, 2008.
19. A. P. Vogel and H. G. Nel, *Int. Pat. Appl.*, WO 2006/053350, 2006.
20. J. M. Botha, *Int. Pat. Appl.*, WO 2006/005084, 2006.
21. J. M. Botha, D. O. Leckel, J. L. Visagie, H. Preston and D. Smook, *Int. Pat. Appl.*, WO 2006/005085, 2006.
22. R. O. Moore Jr. and P. D. Cambern, *US Pat.*, 6709569, 2004.
23. C. J. Helmers, A. Clark and R. C Alden, *Oil Gas J.*, 1948, **47** (26), 86.
24. Standard Oil Development Company, *GB Pat.*, 641121, 1950.
25. J. V. C. Kuo, D. Prater and J. J. Wise, *US Pat.*, 4041094, 1977.
26. P. D. Caesar, W. E. Garwood, W. F. Krudewig and J. J. Wise, *US Pat.*, 4111792, 1978.

27. J. S. Buchanan, J. G. Santiesteban and W. O. Haag, *J. Catal.*, 1996, **158**, 279.
28. H. R. Ireland, A. W. Peters and T. R. Stein, *US Pat.*, 4052477, 1977.
29. H. R. Ireland, A. W. Peters and T. R. Stein, *US Pat.*, 4045505, 1977.
30. A. de Klerk, in *Catalysis in Application*, ed. S. D. Jackson, J. S. J. Hargreaves and D. Lennon, Royal Society of Chemistry, Cambridge, 2003, p. 24.
31. A. D. Pienaar and A. de Klerk, *Ind. Eng. Chem. Res.*, 2008, **47**, 4962.
32. W. R. Derr, J. R. McClernon, S. J. McGovern and F. A. Smith, *US Pat.*, 4059648, 1977.
33. W. R. Derr, J. R., McClernon, S. J. McGovern and F. A. Smith, *US Pat.*, 4080397, 1978.
34. A. de Klerk, *Prepr. Pap. Am. Chem. Soc. Div. Pet. Chem.*, 2008, **53**(2), 105.
35. D. O. Leckel, *Energy Fuels*, 2009, **23**, 2342.
36. H. R. Ireland and T. R. Stein, *US Pat.*, 4044063, 1977.
37. J. C. Kuo, *US Pat.*, 4041096, 1977.
38. J. C. Kuo, D. Prater and J. J. Wise, *US Pat.*, 4049741, 1977.
39. M. A. M. Boersma and S. T. Sie, *GB Pat.*, 2021145, 1979.
40. J. C. Kuo, *US Pat.*, 4046830, 1977.
41. J. C. Kuo, *US Pat.*, 4046831, 1977.
42. W. O. Haag and T. J. Huang, *US Pat.*, 4159995, 1979.
43. J. A. Brennan, P. D. Caesar, J. Ciric and W. E. Garwood, *US Pat.*, 4304871, 1981.
44. Y. F. Chu, T. S. Chou and A. W. Chester, *US Pat.*, 4471145, 1984.
45. F. G. Botes and W. Böhringer, *Appl. Catal. A*, 2004, **267**, 217.
46. F. G. Botes, *Appl. Catal. A*, 2005, **284**, 21.
47. W. R. Derr, W. E. Garwood, J. C. Kuo, T. M. Leib, D. M. Nace and S. A. Tabak, *US Pat.*, 4684756, 1987.
48. B. Jager and R. Espinoza, *Catal. Today*, 1995, **23**, 17.
49. A. de Klerk, *Int. Pat. Appl.*, WO 2008/144782, 2008.
50. A. de Klerk, R. J. J. Nel and T. M. Sakuneka, *Int. Pat. Appl.*, WO 2007/068008, 2007.
51. A. de Klerk, *Int. Pat. Appl.*, WO 2008/124852, 2008.
52. J. F. Mayer and D. J. O'Rear, *US Pat. Appl.*, 2008/0319094, 2008.
53. A. Abazajian, *US Pat. Appl.*, 2004/0176654, 2004.
54. D. J. O'Rear and W. L. Schinsky, *US Pat.*, 6392109, 2002.
55. A. de Klerk, *Energy Fuels*, 2006, **20**, 439.
56. N. M. Prinsloo, *Fuel Process. Technol.*, 2006, **87**, 437.
57. D. J. Ward, *US Pat.*, 4013737, 1977.
58. F. B. du Toit, *US Pat.*, 7271304, 2007.
59. T. M. Sakuneka, R. J. J. Nel and A. de Klerk, *Ind. Eng. Chem. Res.*, 2008, **47**, 7178.
60. J. M. Botha, M. M. J. Mbatha, B. S. Nkosi, A. Spamer and J. Swart, *US Pat.*, 6586649, 2003.
61. S. J. Miller, D. J. O'Rear, T. V. Harris and R. R. Krug, *US Pat.*, 6686511, 2004.

62. D. J. O'Rear and T. V. Harris, *US Pat.*, 6398846, 2002.
63. D. J. O'Rear, T. V. Harris, S. J. Miller, R. R. Krug and B. K. Lok, *US Pat.*, 6706936, 2004.
64. R. O. Moore Jr. and R. D. van Gelder, *US Pat.*, 6515032, 2003.
65. D. J. O'Rear and S. J. Miller, *US Pat.*, 6768037, 2004.
66. D. J. O'Rear, T. V. Harris and C. Y. Chen, *US Pat.*, 7402187, 2008.
67. R. J. Schmidt and R. S. Haizmann, *US Pat.*, 5026950, 1991.
68. W. K. Griesinger, H. A. Sorgenti and G. W. Schell, *US Pat.*, 3674885, 1972.
69. P. Rangarajan, S. R. McDonald, J. D. Allison, K. H. Lawson, O. A. Odueyungbo, D. S. Jack and R. L. Espinoza, *US Pat.*, 7541504, 2009.
70. W. C. Baird Jr, *US Pat.*, 4579648, 1986.
71. D. Milstein and T. R. Stein, *US Pat.*, 4044064, 1977.
72. D. Milstein and T. R. Stein, *US Pat.*, 4071574, 1978.
73. R. J. Wittenbrink, S. M. Davis and L. L. Iaccino, *US Pat.*, 5888376, 1999.
74. R. J. Wittenbrink, S. M. Davis and L. L. Iaccino, *US Pat.*, 5882505, 1999.
75. P. J. Berlowitz, D. F. Ryan, R. J. Wittenbrink and J. W. Johnson, *US Pat.*, 6787022, 2004.
76. S. J. Miller, A. J. Dahlberg, K. R. Krishna and R. R. Krug, *US Pat.*, 6723889, 2004.
77. S. J. Miller, A. J. Dahlberg, K. R. Krishna and R. R. Krug, *US Pat.*, 6458265, 2002.
78. S. J. Miller, A. J. Dahlberg, K. R. Krishna and R. R. Krug, *US Pat.*, 6204426, 2001.
79. R. J. Wittenbrink, R. F. Bauman, P. J. Berlowitz and B. R. Cook, *US Pat.*, 6296757, 2001.
80. R. J. Wittenbrink, R. F. Bauman, P. J. Berlowitz and B. R. Cook, *US Pat.*, 6274029, 2001.
81. R. J. Wittenbrink, P. J. Berlowitz and B. R. Cook, *US Pat.*, 57667274, 1998.
82. P. J. Berlowitz, B. R. Cook and R. J. Wittenbrink, *US Pat.*, 5689031, 1997.
83. A. R. Maund, D. L. Smith and K. H. Lawson, *US Pat.*, 7345210, 2008.
84. J. M. Rosenbaum, C. A. Simmons and D. J. O'Rear, *US Pat.*, 6635171, 2003.
85. J. M. Rosenbaum, C. A. Simmons and D. J. O'Rear, *US Pat.*, 6900366, 2005.
86. R. O. Moore Jr. and R. D. van Gelder, *US Pat.*, 6515034, 2003.
87. R. O. Moore Jr. and R. D. van Gelder, *US Pat.*, 6515934, 2003.
88. X. Dupain, R. A. Krul, M. Makkee and J. A. Moulijn, *US Pat. Appl.*, 2006/0283778, 2006.
89. C. J. Egan, *US Pat.*, 4097364, 1978.
90. J. A. Petri, *US Pat. Appl.*, 2009/0065394, 2009.
91. A. R. de Haan, L. P. Dancaurt, M. J. Prins and E. W. de Wet, *US Pat.*, 7294253, 2007.
92. Y.-Y. P. Tsao, T. J. Huang and P. J. Angevine, *US Pat.*, 6210563, 2001.

93. R. O. Moore Jr., *US Pat.*, 6583186, 2003.
94. D. Farshid and R. O. Moore Jr., *US Pat.*, 7507326, 2009.
95. T. F. Degnan and P. J. Angevine, *US Pat.*, 6652735, 2003.
96. R. J. Pellet, J. A. Rabo, G. N. Long, F. P. Gortsema and A. R. Springer, *US Pat.*, 4906351, 1990.
97. R. J. Pellet, J. A. Rabo, G. N. Long, F. P. Gortsema and A. R. Springer, *US Pat.*, 4880760, 1989.
98. A. Y. Kam, J. C. Kuo, Q. N. Le and J. Smith, *US Pat.*, 4647369, 1987.
99. M. T. Grove, R. D. Partridge, T. E. Helton, D. A. Pappal, P. J. Angevine and D. N. Mazzone, *US Pat.*, 7261805, 2007.
100. I. A. Cody, *US Pat.*, 4515681, 1985.
101. I. A. Cody, W. J. Murphy and S. Hantzer, *US Pat.*, 7125818, 2006.
102. I. A. Cody, W. J. Murphy and S. Hantzer, *US Pat.*, 7087153, 2006.
103. I. A. Cody, W. J. Murphy and S. Hantzer, *US Pat.*, 7220350, 2007.
104. A. R. Bishop, W. B. Genetti, J. W. Johnson, L. L. Ansell and N. M. Page, *US Pat.*, 7201838, 2007.
105. C. L. Baker Jr. and R. C. Dougherty, *US Pat.*, 5951848, 1999.
106. J. W. M. Sonnemans, F. M. Nooij and J. F. Jacques, *US Pat.*, 5935414, 1999.
107. M. J. Girgis and Y.-Y.P. Tsao, *US Pat.*, 5284985, 1994.
108. P. J. Berlowitz, R. J. Wittenbrink and B. R. Cook, *US Pat.*, 6822131, 2004.
109. P. J. Berlowitz, D. F. Ryan, R. J. Wittenbrink, W. B. Genneti and J. W. Johnson, *US Pat.*, 6787022, 2004.
110. D. J. O'Rear, K. Parimi and R O Moore Jr., *US Pat.*, 6359018, 2002.
111. C. Olivier and J. Grootjans, *US Pat.*, 5730858, 1998.
112. S. J. Stephen and J. M. Rosenbaum, *US Pat.*, 6962651, 2005.
113. S. J. Miller, *US Pat.*, 7053254, 2006.
114. D. J. O'Rear and J. A. Biscardi, *US Pat.*, 6773578, 2004.
115. T. F. Degnan and D. N. Mazzone, *US Pat.*, 6190532, 2001.
116. D. P. Leta, S. L. Soled, G. B. McVicker and S. Hantzer, *US Pat.*, 6457374, 2002.
117. S. J. Miller and J. M. Rosenbaum, *US Pat.*, 7198710, 2007.
118. J. S. Miller and J. M. Rosenbaum, *US Pat.*, 6962651, 2005.
119. P. J. Berlowitz, J. J. Habeeb and R. J. Wittenbrink, *US Pat.*, 6475960, 2002.
120. C. D. Chang, S. Han, D. J. Martenak, J. G. Santiesteban and D. E. Walsh, *US Pat.*, 6217747, 2001.
121. D. J. O'Rear and R. R. Krug, *US Pat.*, 6562230, 2003.
122. S. W. Borghard, T. F. Degnan Jr. and D. N. Mazzone, *US Pat.*, 5643440, 1997.
123. M. R. Apelian, T. F. Degnan Jr., D. O. Marler and D. N. Mazzone, *US Pat.*, 5264118, 1993.
124. D. O. Marler and D. N. Mazzone, *US Pat.*, 5288395, 1994.
125. A. de Klerk, *Energy Fuels*, 2009, **23**, 4593.
126. J. M. Rosenbaum, N. J. Bertrand, S. Deskin, K. Krishna, S. J. Miller and S. M. Abernathy, *US Pat. Appl.*, 2006/0289337, 2006.

127. R. J. Wittenbrink and D. F. Ryan, *US Pat.*, 6695965, 2004.
128. R. J. Wittenbrink and D. F. Ryan, *US Pat.*, 6776898, 2004.
129. R. J. Wittenbrink, L. L. Ansell, D. F. Ryan and L. F. Burns, *US Pat.*, 6846848, 2005.
130. M. Matthäi, G. Hildeband, H. Schulze-Trautmann and T. Butz, *Int. Pat. Appl.*, WO 02/096842, 2002.
131. A. Hoek and H. Schadenberg, *Int. Pat. Appl.*, WO 02/102941, 2002.
132. R. E. Holland and S. A. Tabak, *US Pat.*, 4547601, 1985.
133. W. M. Sweeney and S. Herbstman, *US Pat.*, 4356001, 1982.
134. L. P. F. Dancuart Kohler, G. H. du Plessis, F. J. du Toit, E. L. Koper, T. D. Phillips and J. van der Walt, *US Pat.*, 7147775, 2006.
135. L. P. F. Dancuart Kohler, G. H. du Plessis, F. J. du Toit, E. L. Koper, T. D. Phillips and J. van der Walt, *US Pat.*, 7150831, 2006.
136. L. P. F. Dancuart Kohler, G. H. du Plessis, F. J. du Toit, E. L. Koper, T. D. Phillips and J. van der Walt, *US Pat.*, 7153393, 2006.
137. L. P. F. Dancuart Kohler, G. H. du Plessis, F. J. du Toit, E. L. Koper, T. D. Phillips and J. van der Walt, *US Pat.*, 7153432, 2006.
138. L. P. F. Dancuart Kohler, G. H. du Plessis, F. J. du Toit, E. L. Koper, T. D. Phillips and J. van der Walt, *US Pat.*, 7166219, 2007.
139. J. L. F. Kock and A. Botha, *US Pat.*, 5429942, 1995.
140. A. J. Roelofse and R. A. Russell, *Int. Pat. Appl.*, WO 03/048272, 2003.
141. R. B. Pruet, *US Pat.*, 6887908, 2005.
142. L. S. Shah, *US Pat.*, 6533945, 2003.
143. K. H. Lawson and S. A. Scholten, *US Pat. Appl.*, 2005/0187415, 2005.

CHAPTER 11

Future Perspectives

In order to make predictions about the future of catalysis for the refining of Fischer–Tropsch syncrude, it is instructive to look at the past. There are three aspects to consider, namely developments in catalysis, refining and Fischer–Tropsch technology. These areas evolved in parallel, but not without some interdependence.

In the present context, it is difficult to separate the developments in catalysis from the needs and progress made in refining. Most of this effort was directed at improving the conversion of conventional crude oil into transportation fuels. The change drivers were both economical and legislative.

The need for high-octane aviation gasoline in the 1930s and 1940s stimulated many improvements in both refining technology and catalysis. However, this period of rapid advancement came to an end in the 1950s, with readily available and cheap crude oil. This is not to say that there were no further developments, but there was little incentive to drive innovation. This all changed in the 1970s with the 'Oil Crisis' (the period at the end of 1973 when there was a six-fold increase in the price of crude oil). Further impetus for improvement was provided by a growing awareness of the environmental impact that human behaviour has had and specifically the deterioration of air quality in densely populated areas. In turn this spurred changes in transportation fuel specifications. The legislative demands placed on the composition of transportation fuels out of necessity led to developments in catalysis and refining technology. However, most of these changes focused exclusively on crude oil refining.

The developments in Fischer–Tropsch technology were more localised and often motivated by strategic needs.

Transportation fuel forms an integral part of how present-day society is structured. Access to transportation fuels is therefore of strategic and economic importance to all countries. Countries that are not self-sufficient in terms of crude oil supply do not have energy security. This vulnerability of energy security with respect to transportation fuels can be addressed in two ways: fundamentally altering the energy carrier or employing an alternative carbon source that is locally available.

RSC Catalysis Series No. 4
Catalysis in the Refining of Fischer–Tropsch Syncrude
By Arno de Klerk and Edward Furimsky

Published by the Royal Society of Chemistry, www.rsc.org

The notion of a hydrogen economy and the use of battery-operated electric vehicles are attempts to alter the energy carrier fundamentally. However, the infrastructure and global vehicle ownership base is too large to change without significant incentive. As a consequence, hydrocarbon-based motor gasoline, jet fuel, diesel fuel and fuel oil are likely to remain the main transportation fuels in the foreseeable future.

In the short to medium term, energy security will more likely be addressed by substituting alternative carbon sources for crude oil in the production of transportation fuels. Historically, this led to the development of coal-to-liquids (CTL) and gas-to-liquids (GTL) technologies. In this respect, little has changed, except that the feedstock base will be expanded to include other carbon sources too, such as biomass (renewable carbon based energy sources) and carbon-rich waste. Among the processing pathways devised for these feed-to-liquid (XTL) conversions, Fischer–Tropsch synthesis is industrially the most widely applied.

11.1 Future Interest in Fischer–Tropsch Synthesis

Throughout history, Fischer–Tropsch facilities were mostly justified by strategic reasons that were related to energy security.

At the end of the Second World War, Germany had eleven CTL plants located at nine different sites that employed FTS. In addition to the Fischer–Tropsch plants, there were also seven direct coal liquefaction plants. Together these facilities produced 100 000 barrels per day of synthetic fuels, over one-third of Germany's transportation fuel requirements. A further five Fischer–Tropsch plants based on German Fischer–Tropsch technology were constructed under licence in France, Japan and Manchuria during the War years. All of these facilities were constructed to provide energy security.

After the Second World War, energy security was no longer an issue and the last German Fischer–Tropsch plant, that of Schering AG at Bergkamen, was closed in 1962 for economic reasons. A similar fate befell the American Hydrocol Fischer–Tropsch facility at Brownsville in Texas. Crude oil was too cheap and too readily available. The only country where a series of Fischer–Tropsch facilities were constructed in the second half of the 20th century was South Africa. Due to its political dispensation, South Africa did not have ready access to crude oil and these facilities were justified by strategic reasons.

It is possible to economically justify an investment in FTS, but in order to do so there must be a considerable difference in the price of the carbon source used as raw material and the price of crude oil. For example, the GTL Fischer–Tropsch facilities in Qatar and Nigeria could be economically justified due to the availability and the low price of natural gas in those regions. Affordable crude oil generally makes it difficult to justify investment in a Fischer–Tropsch facility economically.

Future interest in FTS will likely be governed by either of these two drivers: energy security, to produce transportation fuels from alternative carbon sources, or economics, which is very dependent on the price differential between crude oil and alternative carbon-based energy sources.

11.2 Future Interest in Fischer–Tropsch Refining

In order to gauge future interest in Fischer–Tropsch refining, let us look at the current interest in it, and also the change drivers that will promote interest in the industrial application of FTS. The latter provides justification for developments in both FTS and the refining of products from FTS.

11.2.1 Energy Security

Fischer–Tropsch synthesis produces synthetic crude oil and not transportation fuels. It is therefore necessary to refine Fischer–Tropsch syncrude to transportation fuels, just as it is necessary to refine conventional crude oil to transportation fuels. Consequently, strategic interest in Fischer–Tropsch technology is dependent not only on the ability to convert alternative carbon sources into liquid products, but also on the ability to refine the Fischer–Tropsch liquids to on-specification transportation fuels.

The catalysts and processes to refine Fischer–Tropsch syncrude are enabling developments, without which FTS cannot provide energy security. Any interest in FTS based on strategic considerations should by definition also promote interest in refining. This partly served as justification for the present work dealing with catalysis for the refining of Fischer–Tropsch syncrude.

11.2.2 Economic Justification

If one considers a Fischer–Tropsch-based facility, excluding the Fischer–Tropsch refinery, it essentially produces a synthetic crude oil. This synthetic crude oil potentially has a similar market value to a good-quality conventional crude oil. Based on energy value alone, Fischer–Tropsch syncrude is less valuable than crude oil, since it has a lower volumetric energy density.

Depending on the cost of the raw material used for FTS, whether it is natural gas, coal, biomass or a carbon-rich waste, it may be possible to justify economically investment in an FTS facility without an associated refinery. In these instances, the value addition is based purely on the difference in feed cost and crude price. This approach is seriously considered by some, as can be seen from the patents dealing with the conversion of syncrude to a pumpable product (Section 10.1.1). A pumpable synthetic crude oil can be sold just like crude oil for refining elsewhere, rather than refining at its origin.

Whether such an investment is competitive with an investment in conventional crude oil exploration and production is altogether a different matter. The complexity and capital cost associated with a Fischer–Tropsch facility, including its associated raw material logistics, far exceed those associated with crude oil production.

Much of the economic justification for investment in a Fischer–Tropsch facility comes from the products that are marketed after refining of the syncrude. Although the refinery associated with FTS represents only 10–20% of the total capital cost of such a facility, it is the refinery that provides most of the value addition. It is only in the refinery that the real benefits of syncrude compared with conventional crude oil are realised. An analogous situation is found when comparing direct coal liquefaction with FTS. Direct coal liquefaction is more carbon efficient when transforming coal into liquid products, but coal liquids are more difficult to refine to final products than Fischer–Tropsch syncrude. As a result, many of the benefits of direct coal liquefaction compared with FTS are lost during refining.

The ability to refine Fischer–Tropsch syncrude efficiently to valuable products, whether they are transportation fuels or chemicals, depends on a good understanding of Fischer–Tropsch refining catalysis. Advances made in the development of catalysts and processes for the refining of Fischer–Tropsch syncrude directly influences the profitability of Fischer–Tropsch-based facilities. Even practitioners of Fischer–Tropsch technology do not always appreciate this fact.

11.2.3 Status of Fischer–Tropsch Refining

Few processes have been developed specifically for the refining of Fischer–Tropsch syncrude, as was very clear from the previous chapters. One may find this surprising considering the key role of refining. Without a proper Fischer–Tropsch refinery, FTS itself does not provide energy security and the economics of FTS must be directly compared with that of crude oil production. It is necessary to put this situation into perspective.

At the time of writing, the global crude oil refining capacity is around 85 million barrels per day and the total installed production capacity of syncrude by Fischer–Tropsch synthesis is around a 250 000 barrels per day. The market for Fischer–Tropsch refining technology (catalysts and processes) is very small compared with that for crude oil and it is consequently seen as a niche application. Unless interest in Fischer–Tropsch technology increases meaningfully, there will be limited incentive for technology suppliers to develop catalysts and processes specifically for the refining of Fischer–Tropsch syncrude.

Companies that practise Fischer–Tropsch synthesis industrially (for example, Shell, Sasol and PetroSA) have a vested interest in Fischer–Tropsch refining technology and may see sufficient commercial benefit to engage in such development. However, this is not necessarily the case. For example, Sasol relies on external companies such as Axens, UOP (Universal Oil Products), KBR (Kellogg, Brown and Root) and Chevron to supply it with crude oil refining technologies for syncrude refining. It conducts only limited research and development in the field of Fischer–Tropsch refining catalysis, despite a history of operating industrial Fischer–Tropsch based facilities since the 1950s.

Some companies that are interested in entering the field of Fischer–Tropsch-based XTL conversion (for example, Eni, ExxonMobil, Statoil, Rentech/UOP and Velocys) may have sufficient commercial interest to co-develop Fischer–Tropsch refining technology. By doing so, these companies may be able to differentiate their technology offering and compete successfully with companies already practising FTS commercially.

In general, the recent developments in Fischer–Tropsch refining catalysis and refining technology have focused mainly on the HCR and HIS of waxes to produce distillates and lubricant base oils. If this is seen as indicative of the vision of companies that are active in the field of Fischer–Tropsch-based XTL, then Fischer–Tropsch syncrude is at best considered to be an incremental crude oil supplement in the future. Such a view undermines the tremendous potential of Fischer–Tropsch syncrude to be refined to on-specification transportation fuels and chemicals.

The lack of developments to permit the efficient refining of Fischer–Tropsch syncrude to on-specification transportation fuels has partly eroded the justification for investing in FTS in the first place, namely to provide energy security. The trend to produce distillate blending stock by HCR of wax, rather than the production of on-specification fuels, has not encouraged investment in FTS. It created the impression that Fischer–Tropsch syncrude is less valuable than crude oil, since crude oil can at least be refined to marketable transportation fuels. However, limiting the refining investment associated with FTS has not always been the design approach. In the past, at one stage, all of the Fischer–Tropsch facilities that were constructed in South Africa produced on-specification transportation fuels for the local market. This is no longer the case.

11.2.4 Advantages Offered by Fischer–Tropsch Refining

It was pointed out earlier (Section 4.6) that Fischer–Tropsch syncrude has some inherent advantages over conventional crude oil for refining to fuels and chemicals.

A Fischer–Tropsch refinery, unlike a crude oil refinery, is not burdened by sulfur and nitrogen compounds. Fischer–Tropsch syncrude consists only of hydrocarbons and oxygenates, since other heteroatoms are removed during synthesis gas purification before FTS. The benefit for fuels production is clear, since the composition of transportation fuels is generally restricted to only hydrocarbons and oxygenates.

Fischer–Tropsch syncrude is more reactive than crude oil on account of its alkene and oxygenate content. This allows refinery conversion at lower temperatures and by technologies that are less energy intensive. The environmental footprint of a Fischer–Tropsch refinery is therefore smaller than that of a conventional crude oil refinery. It also permits refining pathways that one would not normally associate with crude oil refining, for example, aromatic alkylation, which is very useful for meeting stringent benzene specifications for motor gasoline without a loss in fuel octane number.

The alkenes and oxygenates in Fischer–Tropsch syncrude present opportunities for the direct recovery of chemicals, and also for the synthesis of chemicals. It thereby allows the inclusion of refinery units that can produce compounds that may be employed either as fuels or as chemicals, for example, cumene and ethanol. When properly designed, a Fischer–Tropsch refinery can offer tremendous refining flexibility.

One may therefore gain advantages in a Fischer–Tropsch refinery that can partially offset some of the detractors noted previously. However, in order to realise these benefits, one has to develop catalysts and processes that exploit the feed advantages offered by Fischer–Tropsch syncrude. Many of these feed advantages are forfeited when employing standard crude oil refining technology.

11.3 Future Interest in Catalysis to Refine Fischer–Tropsch Syncrude

Throughout the discussion of catalysis for the refining of Fischer–Tropsch syncrude, the differentiating qualities have been highlighted. These all affect catalysis and create specific opportunities for catalyst and process development:

1. *Alkene content.* The availability of alkenes permits the use of catalytic processes that would not normally be considered in crude oil refining due to the limited availability of alkenes. Alkenes confer a synthetic ability on the syncrude. Alkenes are also reactive, which implies that conversion can be conducted at lower temperatures. Temperature-sensitive catalysts and catalysts that are less active, but more selective, may consequently be employed.
2. *Oxygenate content.* Oxygenates are also reactive compounds with synthetic ability. The strong competitive adsorption of oxygenates in a hydrocarbon mixture (as in syncrude) may allow the preferential conversion of oxygenates over appropriate catalysts. Oxygenates can also be employed to improve selectivity, by strongly adsorbing on catalytic sites that may catalyse unwanted hydrocarbon side-reactions. The type of catalysts that can be used with oxygenates is restricted to water-tolerant materials, however. Many oxygenate conversions produce water and a meaningful fraction of the oxygenates from FTS are present in the Fischer–Tropsch aqueous product. The influence of oxygenates on catalysis has therefore been pointed out repeatedly (see, for example, Sections 5.1.7.1, 5.2.4.1 and 5.3.4.1 and Chapter 7). In this respect, there is strong commonality between Fischer–Tropsch and biomass refining. Efficient refining of these unconventional oxygenate-rich materials requires catalysis that can exploit the reactivity of oxygenates and catalysts that are water tolerant.
3. *Linear skeletal structure.* Fischer–Tropsch syncrude contains little cyclic aliphatic and aromatic material, the atmospheric residue fraction

from HTFT synthesis being the exception. Linear molecules and specifically linear hydrocarbons are more resistant to coking. This resistance to coking holds benefits for catalysis that may be exploited. For example, catalysts may be operated at higher alkene partial pressure or lower hydrogen partial pressure without risk of catalyst deactivation by coking. Some catalysts also gain a performance boost with linear material, for example nonacidic Pt/L-zeolites. Depending on the application, linear materials may be desirable and the linear 1-alkenes, linear alcohols and waxes present in Fischer–Tropsch syncrude all have value as chemical commodities or chemical intermediates. There is consequently a rich field of catalysis opportunities to exploit.

4. *Absence of sulfur and nitrogen.* The absence of sulfur compounds allows sulfur-sensitive catalysts to be used, such as reduced base metal- and noble metal-containing catalysts. These catalysts in turn are potentially more selective or more active. The absence of nitrogen compounds, in particular basic nitrogen compounds, allows acid catalysis to play an important role in Fischer–Tropsch refining.

Despite these opportunities, the potential value addition may still be insufficient to encourage catalyst development for the refining of Fischer–Tropsch syncrude. Familiarity with conventional crude oil and the availability of affordable crude oil provide disincentives to embark on catalysis research in a field that is clearly a niche application based on market size. However, there are drivers that may result in research that would also stimulate interest in catalysis to refine Fischer–Tropsch syncrude. Foremost of these change drivers are biomass conversion, regulations concerning carbon dioxide emissions and the chemicals market. Each of these will be considered separately.

11.3.1 Biomass Conversion

The ambitious targets set by politicians to substitute crude oil by renewable energy have stimulated research interest in the upgrading and refining of biomass. Biomass refining is being actively investigated at present and as a result attention has been focused on the catalysis of oxygenate conversion.

Biomass-derived liquids are rich in carbon, hydrogen and oxygen, the same elements as found in syncrude from FTS. Many of the catalysis challenges found in biomass refining consequently have parallels in Fischer–Tropsch refining.

It is anticipated that history will repeat itself. Just as the catalysis know-how related to hydrodeoxygenation received a boost from interest in coal liquefaction (especially the flurry of activity after the 'Oil Crisis'), interest in biomass conversion will result in advances being made in catalysis related to oxygenate conversion. These will have similar benefits for the understanding of Fischer–Tropsch refining catalysis.

Other aspects that are expected to attract attention due to interest in biomass conversion are process intensification of refining processes and the catalysis for small-scale applications of FTS. The feed logistics associated with the transportation of biomass as feed material for biomass-to-liquids (BTL) conversion are a significant contributor to the overall cost of a BTL process. Biomass has a low energy density and it places a limitation on the feed supply radius that can be economically considered for a BTL facility. One way to overcome this limitation is to shrink the size of the facility. Small-scale BTL facilities require a different approach to FTS and refining. The mindset associated with 'economy of scale' does not apply, because the economics are dominated by feed supply cost.

Catalysis can be employed in non-traditional ways to shrink the refinery effectively; the term 'one-pot synthesis' comes to mind. However, small-scale BTL facilities may not be economically viable when producing intermediate products, which opens the door for creative catalysis solutions that would allow the production of marketable final products. Such facilities can likewise not indulge in a plethora of products.

Investment cost is another hurdle in the way of FTS and one that may be overcome by small-scale BTL facilities. The high capital cost associated with large scale Fischer–Tropsch syncrude production increases the financial risk associated with such projects. Smaller companies often do not have access to the capital required to finance such facilities, which creates a high barrier to entry and keeps global interest in FTS limited. Reducing the absolute capital cost may stimulate investment in FTS, which in turn may stimulate interest in the development of catalysts for the refining of Fischer–Tropsch syncrude.

11.3.2 Regulation of Carbon Dioxide Emissions

Global climate change is attracting much attention from both politicians and scientists. One of the factors that has been singled out as a driver for global warming is the correlation between increasing carbon dioxide (CO_2) concentration in the atmosphere over time and the increase in the average global temperature over time. One can scientifically question the prudence of using time as a correlating variable, but politically the link between CO_2 emissions and global warming is an established reality.

Depending on how the legislative framework around CO_2 emissions evolves, the low carbon efficiency of Fischer–Tropsch facilities may dampen future interest in FTS as a way to exploit alternative carbon sources. During feed-to-syncrude conversion, about half of the carbon in the feed material is converted into CO_2. Although this sounds unacceptably high, it is not very different from other carbon-based energy conversion technologies, such as coal- or gas-fired power plants. However, FTS is a very obvious contributor to CO_2 and politically this may saddle Fischer–Tropsch technology with negative political and public opinion. In fact, even in the technical literature FTS is seen as a low carbon efficiency technology for XTL conversion. Such analyses ignore the quality of the syncrude and the contribution of refining to maintain or degrade the carbon efficiency calculated on a feed to final product basis, the caveat being that the syncrude is indeed refined to final products.

If carbon dioxide emissions are taxed, the impact on the economics of Fischer–Tropsch-based XTL will be similar to that caused by a decrease in the feed–product price differential. In such a legislative environment, the cost of carbon becomes more important than the cost of energy *per se*. Carbon efficiency, rather than thermal efficiency, will then have to guide design decisions. Of necessity, the importance of the refinery and the catalysis in the refinery to make efficient use of the carbon in the syncrude will increase.

The situation regarding CO_2 can also be exploited to the benefit of FTS. The design of Fischer–Tropsch gas loops with CO_2 and H_2 as pseudo-syngas instead of CO and H_2 may be politically advantageous. Iron-based Fischer–Tropsch catalysts are water gas shift active and such catalysts are able to convert a CO_2 and H_2 pseudo-syngas into syncrude. The same is not true for cobalt-based catalysts. Although such an approach may not change the overall CO_2 emissions much, it may meet with more positive political and public opinion. Analogous approaches may be considered to boost interest in refining catalysis.

11.3.3 Chemicals Production

Significant differentiation between the competitiveness of Fischer–Tropsch syncrude and crude oil as feedstocks is possible in a refinery context. This is especially apparent for the production of chemicals.

Many industrial facilities based on FTS include co-production of chemicals. The chemicals are often directly recovered from the syncrude and the refinery processes to achieve this are mainly based on separation technology. However, syncrude presents many opportunities for the application of catalysis to produce chemicals in ways that are more efficient than its production from crude oil. The reactive nature of the Fischer–Tropsch syncrude, and also the absence of sulfur and nitrogen compounds noted above, provide a meaningful competitive refining advantage.

At present, FTS is not promoted for its value as a petrochemical platform, but in the future this may change. Such a change will stimulate interest in both FTS and catalysts for the refining of Fischer–Tropsch products. By focusing on chemicals, justification for investment in FTS will be strengthened in terms of economics and energy security. Chemicals are generally higher value products than fuels, thereby increasing the feed–product price differential and improving the process economics. Energy security is provided indirectly, by freeing up crude oil that would otherwise have been needed for the production of chemicals. Fischer–Tropsch refineries can also provide supply security for some strategically important chemicals (the history of synthetic rubber development being a case in point).

11.4 Concluding Remarks

In the preceding chapters, the current state of catalysis for the refining of Fischer–Tropsch syncrude was reviewed. It is clear that many areas of catalysis pertinent to the refining of Fischer–Tropsch syncrude were neglected, with little or no Fischer–Tropsch specific research being published in either the patent or journal literature.

Historically, interest in FTS rose and fell with the availability and price of conventional crude oil. This situation has not changed, nor is it expected to change in the future. The waxing and waning of interest in FTS is a natural consequence of political expedience and a profit-driven economy.

Interest in Fischer–Tropsch refining catalysis is unfortunately intrinsically linked to the fortunes of FTS. The development of catalysts for refining Fischer–Tropsch syncrude suffers from the further disadvantage that there is limited commercial access to the raw materials, namely Fischer–Tropsch syncrudes. As a consequence, industrial and academic programmes dealing with the refining of Fischer–Tropsch products generally lag behind developments in FTS, rather than occurring in parallel. This is ironic, since Fischer–Tropsch refining catalysis has the potential to dramatically improve the prospects for investment and interest in FTS. Furthermore, what is perceived to be good FTS performance may not translate into good overall performance when taking the refinery into consideration. A case in point is the deactivation behaviour of Fe-LTFT catalysts, which is negatively perceived in isolation, but is actually beneficial for refining.

Variability in economic incentives and political support for research dealing with the conversion of alternative carbon sources into fuels and chemicals causes a dilemma. It requires sustained interest (and funding) to make progress in catalysis, and sustaining interest in catalysis to refine the products from FTS is difficult with the stop–start–stop–start interest in Fischer–Tropsch technology. Lack of commercially available Fischer–Tropsch refining technology has led some industrial practitioners of FTS to adopt crude oil technology for the refining of syncrude, despite the inefficiency of such an approach. This does not help the situation at all – in fact it deflects attention and interest away from relevant catalysis research.

What is required is for Fischer–Tropsch refining catalysis to be developed before it is needed in Fischer–Tropsch facilities. In this way, it can help to maintain interest in FTS by allowing the design of more efficient Fischer–Tropsch-based XTL facilities. It is our hope that this book will stimulate some interest in the topic and that research in Fischer–Tropsch refining catalysis will receive continued attention and support in anticipation for when it will be needed in the future.

Subject Index